THE FRONTIERS COLLECTION

THE FRONTIERS COLLECTION

Series Editors:
A.C. Elitzur   L. Mersini-Houghton   M. Schlosshauer   M.P. Silverman   J. Tuszynski
R. Vaas   H.D. Zeh

The books in this collection are devoted to challenging and open problems at the forefront of modern science, including related philosophical debates. In contrast to typical research monographs, however, they strive to present their topics in a manner accessible also to scientifically literate non-specialists wishing to gain insight into the deeper implications and fascinating questions involved. Taken as a whole, the series reflects the need for a fundamental and interdisciplinary approach to modern science. Furthermore, it is intended to encourage active scientists in all areas to ponder over important and perhaps controversial issues beyond their own speciality. Extending from quantum physics and relativity to entropy, consciousness and complex systems – the Frontiers Collection will inspire readers to push back the frontiers of their own knowledge.

For a full list of published titles, please see back of book or springer.com/series 5342

Series home page – springer.com

Yemima Ben-Menahem · Meir Hemmo
Editors

Probability in Physics

 Springer

Editors
Yemima Ben-Menahem
Department of Philosophy
The Hebrew University of Jerusalem
Jerusalem
Israel
msbenhy@mscc.huji.ac.il

Meir Hemmo
Department of Philosophy
University of Haifa
Haifa
Israel
meir@research.haifa.ac

Series Editors:
Avshalom C. Elitzur
Bar-Ilan University, Unit of Interdisciplinary Studies, 52900 Ramat-Gan, Israel
email: avshalom.elitzur@weizmann.ac.il

Laura Mersini-Houghton
Dept. Physics, University of North Carolina, Chapel Hill, NC 27599-3255, USA
email: mersini@physics.unc.edu

Maximilian A. Schlosshauer
Institute for Quantum Optics and Quantum Information, Austrian Academy of Sciences,
Boltzmanngasse 3, 1090 Vienna, Austria
email: schlosshauer@nbi.dk

Mark P. Silverman
Trinity College, Dept. Physics, Hartford CT 06106, USA
email: mark.silverman@trincoll.edu

Jack A. Tuszynski
University of Alberta, Dept. Physics, Edmonton AB T6G 1Z2, Canada
email: jtus@phys.ualberta.ca

Rüdiger Vaas
University of Giessen, Center for Philosophy and Foundations of Science, 35394 Giessen,
Germany
email: ruediger.vaas@t-online.deH.

Dieter Zeh
Gaiberger Straße 38, 69151 Waldhilsbach, Germany
email: zeh@uni-heidelberg.de

ISSN 1612-3018
ISBN 978-3-642-43692-5   ISBN 978-3-642-21329-8 (eBook)
DOI 10.1007/978-3-642-21329-8
Springer Heidelberg Dordrecht London New York

Printed on acid-free paper

Springer is part of Springer Science+Business Media (www.springer.com)

Dedicated to the memory of Itamar Pitowsky

Preface

The notion of probability has become fundamental to modern physics, playing a crucial role in both quantum mechanics and statistical mechanics. Understanding probability is thus an essential part of research into the foundations of contemporary physics. Notably, the meaning of the notion of probability has been a matter of controversy ever since its conception, and is still being debated quite independently of the role of probability in physics. At the center of this debate is the question of whether probabilities are 'objective,' that is, do they characterize events and states of affairs 'in the world,' or are they 'subjective,' representing only degrees of our knowledge or belief about such states? With the integration of probabilistic laws into physics, this traditional question has become all the more pertinent. Moreover, new possibilities come to mind: On the one hand, a specific interpretation of the notion of probability might change our conception of the physical theory in which it figures. For example, a subjective interpretation of probability could turn a theory previously considered a description of reality into an account of what one can know, or is entitled to believe, about reality. On the other hand, the integration of probability into physics may transform the very notion of probability, turning it from a concept constrained solely by *a priori* considerations, into an empirical concept. Many of the papers in this collection engage these intriguing possibilities and their implications.

This volume is dedicated to the memory of Itamar Pitowsky, a beloved friend and colleague of many of the present contributors. Itamar Pitowsky was born in Jerusalem in 1950. He was awarded the B.Sc. in physics and mathematics, and M.A. in mathematics from the Hebrew University of Jerusalem, and received his Ph.D. from the University of Western Ontario. His dissertation entitled *The Logic of Fundamental Processes* was written under the direction of Jeffrey Bub. Itamar Pitowsky was the Eleanor Roosevelt Professor of the History and Philosophy of Science at the Hebrew University of Jerusalem. His untimely death in 2010 is a great loss to the international community of philosophers of physics. He is survived by his wife Liora Lurie Pitowsky and his daughters Noga and Michal Pitowsky. Itamar Pitowsky devoted much of his intellectual life to thinking about probability and its role in physics. He worked on the foundations of probability theory, statistical mechanics, computation theory and, above all, the foundations of quantum mechanics. His interpretation of quantum mechanics is described in the introduction to this collection. Some of the papers published here originate in a conference held in honor of Pitowsky at the van Leer Jerusalem Institute in December 2008. All the papers are published here for the first time. We wish to thank Avshalom Elitzur for establishing contact with Springer, and Angela Lahee, our editor, for her professional advice and unfailing support in bringing this project to completion.

July 2011   Yemima Ben-Menahem
                                                                                Meir Hemmo

Contents

Contributors

David Albert Department of Philosophy, Columbia University, New York, NY, USA

Yemima Ben-Menahem Department of Philosophy, The Hebrew University of Jerusalem, Jerusalem, Israel

Joseph Berkovitz Institute of History and Philosophy of Science and Technology, University of Toronto, Toronto, ON, Canada

Jeffrey Bub Philosophy Department and Institute for Physical Science and Technology, University of Maryland, College Park, MD, USA

William Demopoulos Department of Philosophy, University of Western Ontario, London, ON, Canada

Alon Drory Department of Basic Sciences, Afeka College of Engineering, Tel-Aviv, Israel

John Earman Department of History and Philosophy of Science, University of Pittsburgh, Pittsburgh, PA, USA

Avshalom C. Elitzur UJF-Grenoble 1, CNRS, Laboratoire TIMC-IMAG UMR 5525, F-38041 Grenoble, France

Eric Fanchon UJF-Grenoble 1, CNRS, Laboratoire TIMC-IMAG UMR 5525, F-38041 Grenoble, France

Roman Frigg Department of Philosophy, Logic and Scientific Method and Centre for Philosophy of Natural and Social Science, London School of Economics, London, UK

Christopher A. Fuchs Perimeter Institute for Theoretical Physics, Waterloo, ON, Canada

Sheldon Goldstein Departments of Mathematics, Physics, and Philosophy – Hill Center Rutgers, The State University of New Jersey, Piscataway, NJ, USA

Meir Hemmo Department of Philosophy, University of Haifa, Haifa, Israel

Wayne C. Myrvold Department of Philosophy, The University of Western Ontario, London, ON, Canada

Klil Ha-Horesh Neori Department of Physics, University at Albany (SUNY), 1400 Washington Ave, Albany, NY 12222, USA

Itamar Pitowsky Edelstein Center, The Hebrew University, Jerusalem, Israel

Daniel Rohrlich Physics Department, Ben Gurion University of the Negev, Beersheba 84105, Israel

Laura Ruetsche Department of Philosophy, University of Michigan, Ann Arbor, MI, USA

Rüdiger Schack Department of Mathematics, Royal Holloway, University of London, Egham, Surrey, UK

Orly Shenker Department of Philosophy, The Hebrew University of Jerusalem, Jerusalem, Israel

Lev Vaidman School of Physics and Astronomy, Tel Aviv University, Tel Aviv, Israel

Charlotte Werndl Department of Philosophy, Logic and Scientific Method, London School of Economics, London, UK

Alexander Wilce Department of Mathematical Sciences, Susquehanna University, Selinsgrove, PA USA

Chapter 1
Introduction

Yemima Ben-Menahem and Meir Hemmo

Questions concerning the meaning of probability and its applications in physics are notoriously subtle. In the philosophy of the exact sciences, the conceptual analysis of the foundations of a theory often lags behind the discovery of the mathematical results that form its basis. The theory of probability is no exception. Although Kolmogorov's axiomatization of the theory [1] is generally considered definitive, the meaning of the notion of probability remains a matter of controversy.[1] Questions pertain both to gaps between the formalism and the intuitive notions of probability and to the inter-relationships between the intuitive notions. Further, although each of the interpretations of the notion of probability is usually intended to be adequate throughout, independently of context, the various applications of the theory of probability pull in different interpretative directions: some applications, say in decision theory, are amenable to a subjective interpretation of probability as representing an agent's degree of belief, while others, say in genetics, call upon an objective notion of probability that characterizes certain biological phenomena. In this volume we focus on the role of probability in physics. We have the dual goal and challenge of bringing the analysis of the notion of probability to bear on the meaning of the physical theories that employ it, and of using the prism of physics to study the notion of probability.

We thank Orly Shenker for comments on an earlier draft.

[1] As far as interpretation is concerned, the drawback of the axiom system is that, in formalizing measure theory in general, it captures more than the intuitive notion of probability, including such notions as length and volume.

Y. Ben-Menahem (✉)
Department of Philosophy, The Hebrew University of Jerusalem, Jerusalem, Israel
e-mail: msbenhy@mscc.huji.ac.il

M. Hemmo
Department of Philosophy, University of Haifa, Haifa, Israel
e-mail: meir@research.haifa.ac.il

Y. Ben-Menahem and M. Hemmo (eds.), *Probability in Physics*, The Frontiers Collection,
DOI 10.1007/978-3-642-21329-8_1, © Springer-Verlag Berlin Heidelberg 2012

The concept of probability is indispensable in contemporary physics. On the micro-level, quantum mechanics, at least in its standard interpretation, describes the behavior of elementary particles such as the decay of radioactive atoms, the interaction of light with matter and of electrons with magnetic fields, by employing probabilistic laws as its first principles. On the macro-level, statistical mechanics appeals to probabilities in its account of thermodynamic behavior, in particular the approach to equilibrium and the second law of thermodynamics. These two entries of probability into modern physics are quite distinct. In standard quantum mechanics, probabilistic laws are taken to *replace* classical mechanics and classical electrodynamics. Quantum probabilities are here understood as reflecting genuine stochastic behavior, ungoverned by deterministic laws.[2] By contrast, in classical statistical mechanics, probabilistic laws are *supplementary* to the underlying deterministic mechanics, or perhaps even reducible to it. These probabilities may therefore be the understood as reflecting our ignorance about the details of the microstates of the world or as a byproduct of our coarse-grained descriptions of these states.

Despite this radical difference, the probabilistic laws in both theories pertain to the behavior of real physical systems. When quantum mechanics ascribes a certain probability to the decay of a radium atom, it must be saying something about the atom, not only about our beliefs, expectations or knowledge regarding the atom. Likewise, when classical statistical mechanics ascribes a high probability to the spreading of a gas throughout the volume accessible to it, it is purportedly saying something about the gas, not only about our subjective beliefs about the gas. In this sense the probabilistic laws in both quantum and statistical mechanics are supposed to have some genuine objective content. What this objective content is, however, and how it is related to epistemic notions such as ignorance, rational belief and the accuracy of our descriptions are open issues, hotly debated in the literature, and reverberating through this volume. Before getting into these issues in the physical context, let us briefly review some of the general problems confronting the interpretation of probability.[3]

1.1 The Notion of Probability

Consider the paradigmatic example of a game of chance, a flip of a coin in which in each flip there is a fixed probability of 1/2 for getting tails and 1/2 for getting heads. How are we to understand the term 'probability' in this context? At least three

---

[2] This does not apply to Bohmian quantum mechanics, which is deterministic. See also Earman [2] for an analysis of determinism and for an unorthodox view about both classical and quantum mechanics regarding their accordance with determinism.

[3] See Fine [3] for the various approaches to probability, and Hajek [4] for an overview and references.

distinct answers can be found in the literature. First, we could be referring to the objective *chance* for getting heads or tails in each flip. This objective chance is supposed to pertain to the coin, or the flip, or both, or, more generally, to the set up of the coin flip. On this understanding, the chance is a kind of property, a tendency (propensity) of the set up, analogous to other tendencies of physical systems. The analogy suggests that, just as fragility is a tendency to break, chance is the tendency to... The problem in completing this sentence is that fragile objects sometimes in fact break whereas a chance 1/2 coin always lands on one of its faces. In other words, a single outcome never instantiates the chance in the way a broken glass instantiates fragility.

The response to this problem leads to the second notion of probability according to which when we say that the probability of tails (heads) in each game is 1/2 we refer to the *relative frequency* of tails (heads) in a (finite or infinite) series of repeated games. The intuition here is that it is the relative frequency that links the probability of an event to the actual occurrences. While the link cannot be demonstrated in any single event, it is manifest in the long run by a series of events. How does this response affect the notion of chance? We may either conclude that chance has been eliminated in favor of relative frequency, or hold on to the notion of chance along with the caveat that it is instantiated (and thus verified) only by relative frequency. An example of the first kind is the relative frequency analysis by von Mises [5], where probability is identified with the infinite limit (when it exists) of the relative frequency along what von Mises called *random* sequences. Of course, the notion of a random sequence is itself in need of a precise and non-circular characterization. Moreover, the relation between a single case chance and relative frequency is not yet as close as we would like it to be, for any objective chance, say 1/2 for tails, is compatible with *any* finite sequence of heads and tails, no matter how long it is.

The third notion of probability is known as *subjective* or *epistemic* probability. On this interpretation, championed by Ramsey, de Finetti and Savage, probabilities represent degrees of belief of *rational* agents, where rationality is defined as acting so as to maximize profit. The rationality requirement places normative constraints on subjective degrees of belief: the beliefs of rational agents, it is claimed, must obey the axioms of probability theory in the following sense. Call a series of bets each of which is acceptable to the agent a *Dutch-book* if they collectively yield a sure loss to the agent, regardless of the outcomes of the games. It has been proved independently by de Finetti and Ramsey that if the degrees of belief (the subjective probabilities) of the agent violate the axioms of probability theory, then a Dutch-book can be tailored against her. And conversely, if the subjective probabilities of the agent conform to the axioms of probability, then no Dutch book can be made against her. That is, obeying the axioms of probability is necessary and sufficient to guarantee Dutch-book coherence, and in this sense, rationality. Thus construed, the theory of probability – as a theory of partial belief – is actually an extension of logic!

One of the problems that the subjective interpretation faces is that without further assumptions, the consistency of our beliefs (in the above sense of avoiding a Dutch book, or complying with the axioms of probability theory) does not guarantee their reasonableness. In order to move from consistency to reasonableness, it would seem, we need some guidance from 'reality', namely, we need a procedure for adjusting our subjective probabilities to objective evidence. Here Dutch book considerations are less effective – they do not even compel rational agents to update their beliefs by conditioning on the evidence (see van Fraassen [6]). To do that, one needs a so-called *diachronic* (rather than synchronic) Dutch book argument. More generally, the question arises of whether the evidence can be processed without recourse to objectivist notions of probability.[4] If it cannot, then subjective probability is not, after all, the entirely self-sufficient concept that extreme subjectivists have in mind. Suppose we accept this conclusion and seek to link the subjective notion of probability to objective matters of fact. Ideally, it seems that the subjective probability assigned by a rational agent to a certain outcome should converge on the objective probability of that outcome, where the objective probability is construed either in terms of chance or in terms of relative frequency. Can we appeal to the laws of large numbers of probability theory in order to justify this idea and close the gaps between the three interpretations of probability? Not quite, for the following reasons.

According to Bernoulli's weak law of large numbers, for any ε there is a number n such that in a sequence of (independent) n flips, the relative frequency of tails will, with some fixed *probability*, be in the interval $1/2 \pm \varepsilon$. And if we increase the number of flips without bound then according to the weak law, the relative frequency of tails and heads will approach their objective probability with probability that approaches one. That is, the relative frequency would *probably* be close to the objective probability. But this means that the convergence of relative frequency to objective chance is itself *probabilistic*. Threatening to lead to a regress, these second-order probabilities cast doubt on the identification of the concept of objective chance with that of relative frequency. A similar problem arises for the subjectivist: are the second order probabilities to be understood subjectively or objectively? The former reply is what we would expect from a confirmed subjectivist, but it leaves the theorem rather mysterious.

Moreover, consider a situation in which the objective chance is *unknown* and one wishes to form an opinion about the chance relying on the observed evidence: Can the law of large numbers support such an inference? Unfortunately, the answer is negative, essentially because any finite relative frequency is compatible with infinitely many objective chances. Suppose that we want to find out whether the objective chance in our coin flip is 1/3 for tails and 2/3 for heads, or whether it is 1/2 for both. Even if we assume that the flips are independent of each other (so that the

---

[4] It has been shown that the subjective probabilities of an agent who updates her beliefs in accordance with Bayes' theorem, converge on the observed relative frequencies no matter what her prior subjective probabilities are. But this is different from converging on the chances or the relative frequencies in the infinite limit. Similar considerations apply also to the so-called 'logical' approach to probability on which probabilities are quantitative expressions of the degree of support of a statement conditional on the evidence.

law of large numbers applies), we cannot infer the chance from long run relative frequencies unless we put *weights* on all possible sequences of tails and heads. Placing such weights, however, amounts to presupposing the objective chances we are looking for.[5] We will encounter several variations on this question in the context of quantum mechanics and classical statistical mechanics. In both cases, the choice of the probability measure over the relevant set of sequences is designed to allow for the derivation of the chances from the relative frequencies. Let us now turn to these theories.

1.2 Statistical Mechanics

Statistical mechanics, developed in the late nineteenth century by Maxwell, Boltzmann, Gibbs and others, is an attempt to understand thermodynamics in terms of classical mechanics. In particular, it aims to explain the irreversibility typical of thermodynamic phenomena on the basis of the laws of classical mechanics. Thus, while in classical thermodynamics, the irreversibility characteristic of the approach to equilibrium and the second law of thermodynamics is put forward in addition to the Newtonian laws of classical mechanics, in statistical mechanics the laws of thermodynamics are expected to be reducible to the Newtonian laws. This, at least, was the aspiration underlying the theory. One of the major difficulties in this respect is that the laws of classical mechanics (as well as those of other fundamental theories) are time-symmetric, whereas the second law of thermodynamics is time-asymmetric. How can we derive an asymmetry in time from time-symmetric laws?

In order to appreciate the full force of this question, it is instructive to consider Boltzmann's approach in the early stages of the kinetic theory of gases.[6] In his famous H-theorem, Boltzmann attempted to prove a mechanical version of the second law of thermodynamics on the basis of the mechanical equations of motion (describing the evolution of a low density gas in terms of the distribution of the velocities of the particles that make up the gas). Roughly, according to the H-theorem, the dynamical evolution of the gas as described by Boltzmann's equation is bound to approach the distribution that maximizes the entropy of the gas, that is, the Maxwell-Boltzmann equilibrium distribution. Essentially, what this startling result was meant to show was that an isolated low-density gas in a non-equilibrium state evolves *deterministically* towards equilibrium, and therefore its entropy increases with time. However, Boltzmann's H-theorem turned out to be

---

[5] See van Fraassen [7], p. 83. Note again that the condition of Dutch-book coherence is of no help here since the objective probabilities in the situation we consider are *unknown*.

[6] Our account below is not meant to be historically rigorous. We essentially follow the Ehrenfest and Ehrenfest [8] reconstruction of Boltzmann's ideas in a very schematic way. See Uffink [9] for a detailed historical account of statistical mechanics and references.

inconsistent with the fundamental time-symmetric principles of mechanics. This was the thrust of the reversibility objection raised by Loschmidt: given the time-symmetry of the classical equations of motion, for any trajectory passing through a sequence of thermodynamic states along which entropy increases with time, there is a corresponding trajectory which travels through the same sequence of states in the *reversed* direction (i.e., with reversed velocity), along which entropy therefore *decreases* in the course of time.[7]

It is at this juncture that probability came to play an essential role in physics. In the face of the reversibility objections, Boltzmann concluded that his H-theorem must be interpreted probabilistically. The initial hope was that an analysis of the behavior of many-particles systems in probabilistic terms would reveal a straight-forward linkage between probability and entropy. If one then makes the seemingly natural assumption that a system tends to move from less probable to more probable states, a direction in the evolution of thermodynamic systems towards high entropy states would emerge. In other words, what Boltzmann now took the H-theorem to prove was that although it is possible for a thermodynamic system to evolve away from equilibrium, such an anti-thermodynamic evolution is highly unlikely or improbable.

To see how this idea can be made to work, consider the following. First, thermodynamic magnitudes such as volume, pressure and temperature, are associated with regions in the phase space called *macrostates*, where a macrostate is conceived as an equivalence class of all the *microstates* that realize it. Second, the thermodynamic entropy of a macrostate is identified with the number of microstates that realize that macrostate, as measured by the Lebesgue measure (or volume) of the phase space region associated with the macrostate. Third, a dynamical hypothesis is put forward to the effect that the trajectory of a thermodynamic system in the phase space is *dense* in the sense that the trajectory passes arbitrarily close to every microstate in the energy hypersurface. This idea, which goes back to Boltzmann's so-called ergodic hypothesis, was rigorously proved around 1932 by Birkhoff and von Neumann. According to their ergodic theorem, a system is ergodic if and only if the relative time its trajectory spends in a measurable region of the phase space is equal to the relative Lebesgue measure of that region (i.e., its volume) in the limit of infinite time. This feature holds for all initial conditions except for a set of Lebesgue measure zero. Now, since the *probability* of a macrostate (or rather the relative frequency of that macrostate) along a typical infinite trajectory can be thought of as the relative time the trajectory spends in the macrostate, it seems to follow that in the long run the probability of a macrostate is equal to its entropy. And this in turn seems to imply that an ergodic system will most *probably* follow trajectories that in the course of time pass from low-entropy macrostates to high-entropy macrostates.

---

[7] In fact, there are other arguments, which show that Boltzmann's deterministic approach in deriving the H-theorem could not be consistent with the classical dynamics, e.g. the historically famous objection by Zermelo based on the Poincaré recurrence theorem. It was later discovered that one of the premises in Boltzmann's proof was indeed time-asymmetric.

And so, if thermodynamic systems are in fact ergodic, we seem to have a probabilistic version of the second law of thermodynamics.

This admittedly schematic outline gives the essential idea of how the thermodynamic time-asymmetry was thought to follow from the time-symmetric laws of classical mechanics. Despite its ingenuity, the probabilistic construal of the second law still faces intriguing questions. Current research in the foundations of statistical mechanics is particularly beleaguered by the problems generated by the (alleged?) reduction of statistical mechanics to the fundamental laws of physics. Here are some of the questions that are addressed in this volume.

1. The first challenge is to give a mathematically *rigorous* proof of a probabilistic version of the second law of thermodynamics. Some attempts are based on specific assumptions about the initial conditions characterizing thermodynamic systems (as in Boltzmann's H-theorem), while others appeal to general features characterizing the dynamics of thermodynamic systems (as in the ergodic approach). The extent to which these attempts are successful is still an open question. In this volume, Roman Frigg and Charlotte Werndl defend a number of variations on the ergodic approach.

   A related issue is that of Maxwell's Demon. Maxwell introduced his thought experiment – portraying a Demon who violates the second law – to argue that it is impossible to derive a universal proof of this law from the principles of mechanics. He concluded that, while the law is generally (probabilistically) valid, its violation under specific circumstances, such as those described in his thought experiment, is in fact possible. In the literature, however, the standard view is that Maxwell was wrong. That is, it is argued that the operation of the Demon in bringing about a local decrease of entropy is inevitably counterbalanced by an appropriate increase of entropy in the environment (including the Demon's own entropy).[8] The article by Eric Fanchon, Klil Ha-Horesh Neori and Avshalom Elitzur analyses some new aspects of the Demon's operation in defense of this view.

2. Another question concerns the status of the probabilities in statistical mechanics. Given that the classical equations of motion are completely deterministic, what exactly do probabilities denote in a classical theory? Of course, classical deterministic dynamics applies to the *micro*structure of physical systems whereas Boltzmann's probabilities are assigned to *macro*states, which can be realized in numerous ways by various microstates. Nonetheless, the determinism of classical mechanics implies that anything that happens in the world is fixed by the world's actual trajectory. And so, at first sight, the probabilities in statistical mechanics can only represent the ignorance of observers with respect to the microstructure. But if so, it is not clear what could be *objective* about statistical

---

[8] See Leff and Rex [10] and Maroney [11] for reviews and the recent literature on the Demon question. For a rigorous recent account of Maxwell's Demon supporting Maxwell in the context of Boltzmann's approach, see Albert [12] and Hemmo and Shenker [13, 14].

mechanical probabilities and how they could be assigned to physical states and processes.[9] Since we expect these probabilities to account for thermodynamic behavior– the approach to thermodynamic equilibrium and the second law of thermodynamics – which are as physically objective as anything we can get, the epistemic construal of probability is deeply puzzling. Responding to this challenge, David Albert provides an over-arching account of the structure of probability in physics in terms of single case chances. Wayne Myrvold, in turn, proposes a synthesis of objective and subjective elements, construing probabilities in statistical mechanics as objective, albeit epistemic, chances.

3. We noted that the appeal to probability was meant to counter the reversibility objection. But does it? What we would like to get from statistical mechanics is not only a high probability for the increase of entropy towards the future but also an asymmetry in time – an 'arrow of time' (or rather an arrow of entropy in time) – so that the same probabilistic laws would also indicate a *decrease* of entropy towards the past. And once again, the time-reversal symmetry of classical mechanics stands in our way. Whatever probability implies with respect to evolutions directed forward in time must be equally true with respect to evolutions directed backward in time. In particular, whenever entropy increases towards the future, it also increases towards the past. If this is correct, it implies that the present entropy of the universe, for any present moment, must always be the minimal one. That is to say, one would be justified to infer that the cup of coffee in front of me was at room temperature a few minutes ago and has spontaneously warmed up as to infer that it will cool off in a few minutes. To fend off this absurd implication, Richard Feynman ([25], p. 116) said, "I think it necessary to add to the physical laws the hypothesis that in the past the universe was more ordered, in the technical sense, than it is today." Here Feynman is introducing the so-called *past-hypothesis* in statistical mechanics,[10] which in classical statistical mechanics is the only way to get a temporal arrow of entropy. Note that the past-hypothesis amounts to adding a distinction between past and future, as it were, by hand. It is an open question as to whether quantum mechanics may be more successful in this respect.[11] The past-hypothesis and its role in statistical mechanics are subject to a thorough examination and critique in Alon Drory's paper.

4. In the assertion that thermodynamic behavior is highly probable, the high probability pertains to subsystems of the universe. It is generally assumed that the trajectory of the entire universe, giving rise to this high probability, is itself fixed by laws and initial conditions. Questions now arise about these initial conditions and their probabilities. Must they be highly probable in order to confer high probability on the initial conditions of subsystems? Would

---

[9] See Maudlin [15] for a more detailed discussion of this problem.

[10] See Albert [12], Chap. 4 for some variations on this idea.

[11] The situation, however, is somewhat disappointing, since in quantum mechanics the question of how to account for the past is notoriously hard due to the problematic nature of retrodiction in standard quantum mechanics.

improbable initial conditions of the universe make its present state inexplicable? But even before we answer these questions, it is not clear that it even makes sense to consider a probability distribution over the initial conditions of the universe. A probability distribution suggests some sort of a random sampling, but with respect to the initial conditions of the universe, this random sampling seems an empirically meaningless fairy tale. An attempt to answer these questions goes today under the heading of the *typicality* approach. The idea is to justify the set of conditions that give rise to thermodynamic behavior in statistical mechanics (or to quantum mechanical behavior in Bohmian mechanics) by appealing to the high measure of this set in the space of all possible initial conditions, where the high measure is *not* understood as high probability. Two articles in this volume contribute to this topic: Sheldon Goldstein defends the non-probabilistic notion of typicality in both statistical and Bohmian mechanics; Meir Hemmo and Orly Shenker criticize the typicality approach.

5. Finally, a probability distribution over a continuous set of points (e.g. the phase space) requires a choice of measure. The standard choice in statistical mechanics (relative to which the probability distribution is uniform) is the Lebesgue measure. But in continuous spaces infinitely many other measures (including some that do not even agree with the Lebesgue measure on the measure zero and one sets) are mathematically possible. Such measures could lead to predictions that differ significantly from the standard predictions of statistical mechanics. Are there mathematical or physical grounds that justify the choice of the Lebesgue measure? This problem is intensified when combined with the previous question about the probability of the initial conditions of the universe. And here too there are attempts to use the notion of typicality to justify the choice of a particular measure. In his contribution to this volume, Itamar Pitowsky justifies the choice of the Lebesgue measure, claiming it to be the *mathematically natural extension* of the counting (combinatorial) measure in discrete cases to the continuous phase space of statistical mechanics.

1.3 Quantum Mechanics

The probabilistic interpretation of the Schrödinger wave equation, put forward by Max Born in 1926, has become the cornerstone of the standard interpretation of quantum mechanics. Two (possibly interconnected) features distinguish quantum mechanical probabilities from their classical counterparts. First, on the standard interpretation of quantum mechanics, quantum probabilities are irreducible, that is, the probabilistic laws in which they appear are not superimposed on an underlying deterministic theory. Second, the structure of the quantum probability space differs from that of the classical space. Let us look at each of these differences more closely.

According to the von Neumann-Dirac formulation of quantum mechanics, the ordinary evolution of the wave function is governed by the deterministic

Schrödinger equation. Upon measurement, however, the wave function undergoes a genuinely stochastic, instantaneous and nonlocal *'collapse,'* yielding a definite value – one out of a (finite or infinite) series of possible outcomes. The probabilities of these outcomes are given by the quantum mechanical algorithm (e.g. the modulus square of the amplitude of Born's rule). Formally, this process is construed as a projection of the system into an eigenstate of the operator representing the measured observable, where the measured value represents the corresponding eigenvalue of this state. Thus, the collapse of the wave function, unlike its ordinary evolution, is said to be governed by a 'projection postulate'. Among other implications, this 'projection' means that in contrast with the classical picture, where a measurement yields a result predetermined by the dynamics and the initial conditions of the system at hand, on the standard interpretation of quantum mechanics, measurement results are not predetermined. Moreover, they come about through the measurement process (although how precisely this happens is an open question) and in no way reflect the state of the system prior to measurement.

The collapse of the wave function is the Achilles heel of the standard interpretation. Is it a physical process or just a change in the state of our knowledge? Why does it occur specifically during measurement? Can it be made Lorentz invariant in a way that ensures compatibility with special relativity? Above all, can quantum mechanics be formulated in a more uniform way, one that does not single out measurement as a distinct process and makes no recourse to 'projection'? These questions are the departure point of alternative, non-standard, interpretations of quantum mechanics that seek to solve the above problems either by altogether getting rid of the collapse or by providing a dynamical account that explains it. As it turns out, rival interpretations also provide divergent accounts of the meaning of quantum probabilities. While the standard interpretation has been combined with practically all the probabilistic notions, relative frequency, single case chance and epistemic probability, the non-standard alternatives are generally more genial to a specific interpretation.

Bohm's [16] theory is a deterministic theory, empirically equivalent to quantum mechanics. Here there is no collapse. Rather, the probability distribution $||\psi(x)\rangle|^2$, introduced in order to reproduce the empirical predictions of standard quantum mechanics, reflects ignorance about the exact positions of particles (and ultimately about the exact initial position of the entire universe). In this sense the probabilities in Bohm's theory play a role very similar to that played by probability in statistical mechanics, and can thus be construed along similar lines. By contrast, stochastic theories, such as that proposed by Ghirardi et al. (GRW, [17]), prima facie construe quantum probabilities as single case chances. In these theories, probability-as-chance enters quantum mechanics not only through the measurement process but through a purely stochastic dynamics of spontaneous 'jumps' of the wave function under general dynamical conditions. Many worlds approaches, based on Everett's relative-state theory [18], take quantum mechanics to be a deterministic theory involving neither collapse nor ignorance over additional variables. Rather, the unitary dynamics of the wave function is associated with a peculiarly quantum mechanical process of branching (fission and fusion) of worlds. However, since the branching is *un*related to the quantum probabilities, the role of probability in this theory remains obscure and is widely debated in the literature. Addressing this issue, Lev Vaidman's article secures a place for probabilities (obeying Born's rule) in the many worlds theory.

In addition, there are subjectivist approaches, according to which quantum probabilities are constrained only by Dutch-book coherence. Such an epistemic approach was defended by Itamar Pitowsky [19] and is advocated in this volume by Christopher Fuchs and Rüdiger Schack. A subjectivist approach inspired by de Finetti is the subject of Joseph Berkovitz's paper. He discusses the implications of de Finetti's verificationism for the understanding of the quantum mechanical probabilities in general and Bell's nonlocality theorem in particular.

The second characteristic that distinguishes quantum mechanical probabilities from classical probabilities is the logical structure of the quantum probability space. This feature, first identified by Schrödinger in 1935, was the basis of Itamar Pitowsky's interpretation of quantum mechanics. In a series of papers culminating in 2006, he urged that quantum mechanics is to be understood primarily as a non-classical theory of probability. On this approach, it is the non-classical nature of the probability space of quantum mechanics that is at the basis of other characteristic features of quantum mechanics, such as the non-locality exemplified by the violation of the Bell inequalities. Since these probabilities obey non-classical axioms, the expectation values they lead to deviate from expectations derived from classical ignorance interpretations.

Pitowsky believed that the difference between quantum and classical mechanics is already manifested at the phenomenological level, that is, the level of events and their correlations. He therefore insisted that before *saving* the quantum phenomena by means of theoretical terms such as superposition, interference, nonlocality, probabilities over initial conditions, collapse of the wavefunction, fission of worlds, etc. (let alone the vaguer notions of duality or complementarity), we should first have the phenomena themselves in clear view. The best handle on the phenomena, he maintained, is provided by the non-classical nature of the probability space of quantum events. Note that what is at issue here is not the familiar claim that the laws of quantum mechanics, unlike those of classical mechanics, are irreducibly probabilistic, but the more radical claim that quantum probabilities deviate in significant ways from classical probabilities. Hence the project of providing an axiom system for quantum probability. The very notion of non-classical probability, that is, the idea that there are different notions of probability captured by different axiom systems, has far-reaching implications not only for the interpretation of quantum mechanics, but also for the theory of probability and the understanding of rational belief.

Pitowsky's probabilistic interpretation brings together suggestions made by a number of earlier theorists, among them Schrödinger and Feynman, who saw quantum mechanics as a new theory of probability, and Birkhoff and von Neumann, who sought to ground quantum mechanics in a non-classical logic. However, Pitowsky goes beyond these earlier works in spelling out in detail the non-classical nature of quantum probability. In addition, his work differs from that of his predecessors in three important respects. First, Pitowsky clarified the relation between quantum probability and George Boole's work on the foundations of classical probability. Second, Pitowsky explored in great detail the geometrical structure of the non-classical probability space of quantum mechanics, an exploration that sheds new light on central features of quantum mechanics, and non-locality in particular. Finally, unlike any of the above mentioned theorists, Pitowsky endorsed an epistemic interpretation of probability [19, 20]. It is this epistemic component of his interpretation, arguably its most controversial component, that he took to be necessary in order to challenge rival interpretations of quantum mechanics, such as the many worlds and Bohmian interpretations. Once quantum mechanics is seen as a non-classical theory of probability, he argued, and once probability is construed subjectively, as a measure of credence, the puzzles that gave rise to these rival interpretations lose much of their bite. Specifically, Pitowsky thought we may no longer need to worry about the notorious measurement problem. For, as Pitowsky and Bub put it in their joint paper [24], we may now safely reject "the two dogmas of quantum mechanics," namely, the concept of the quantum state as a *physical* state on a par with the classical state, and the concept of measurement as a physical process that must receive a *dynamical* account. Instead, we should view the formalism of quantum mechanics as a 'book-keeping' algorithm that places constraints on (rational) degrees of belief regarding the possible results of measurement.

To motivate this point of view, Pitowsky and Bub draw an analogy between quantum mechanics and the special theory of relativity: According to special relativity, effects such as Lorenz contraction and time dilation, previously thought to require *dynamical* explanations, are now construed as inherent to the relativistic *kinematics* (and therefore the very concept of motion). Theories that do provide dynamical accounts of the said phenomena may serve to establish the inner consistency of special relativity, but do not constitute an essential part of this theory. Similarly, on the probabilistic approach to quantum mechanics, the puzzling effects of quantum mechanics need no deeper 'physical' explanation over and above the fact that they are entailed by the non-classical probability structure. And again, theories that do provide additional structure, e.g. collapse dynamics of the measurement process, should be regarded as consistency proofs of the quantum formalism rather than an essential part thereof. As it happens, the debate over the purely kinematic understanding of special relativity has recently been renewed (see Brown [21]). Following Pitowsky's work, a similar debate may be conducted in the context of quantum mechanics. A number of papers in this volume contribute to such a debate. While Laura Ruetsche and John Earman examine the applicability of Pitowsky's interpretation to quantum field theories, William Demopoulos and Yemima Ben-Menahem address issues related to its philosophical underpinnings.

To illustrate the difference between classical and quantum probability, it is useful to follow Pitowsky and revisit George Boole's pioneering work on the "conditions of possible experience." Boole raised the following question: given a series of numbers $p_1, p_2, \ldots p_n$, what are the conditions for the existence of a probability space and a series of events $E_1, E_2, \ldots E_n$, such that the given numbers represent the probabilities of these events. (Probabilities are here understood as frequencies in finite samples represented by numbers between 0 and 1). The answer given by Boole was that the envisaged necessary and sufficient conditions could be expressed by a set of linear equations in the given numbers. All, and only, numbers that satisfy these equations, he argued, can be the obtained from experience, i.e. from actual frequencies of events. In the simple case of two events E_1 and E_2, with probabilities p_1 and p_2 respectively, an intersection $E_1 \cap E_2$ whose probability is p_{12}, and a union $E_1 \cup E_2$, Booles conditions are:

$$p_1 \geq p_{12}$$
$$p_2 \geq p_{12}$$
$$p(E_1 \cup E_2) = p_1 + p_2 - p_{12} \leq 1$$

Prima facie, these conditions seem self-evident or *a priori*. Evidently, if we have an urn of balls containing, among others, red, wooden and red-and-wooden balls, the probability of drawing a ball that is either red or wooden (or both) cannot exceed the sum of the probabilities for drawing a red ball and a wooden ball. And yet, violations of these predictions are predicted by quantum mechanics and demonstrated by experiments. In the famous two-slit experiment, for instance, there are areas on the screen that receive more hits than allowed by the condition that the probability of the union event cannot exceed the sum of the probabilities of the individual events. We are thus on the verge of a logical contradiction. The only reason we do not actually face an outright logical contradiction is that the probabilities in question are obtained from different samples, or different experiments (in this case, one that has both slits open at once, the other that has them open separately). It is customary to account for the results of the two-slit experiment by means of the notion of superposition and the wave ontology underlying it. But as noted, Pitowsky's point was that acknowledging the deviant phenomena should take precedence over their explanation, especially when such an explanation involves a dated ontology.

Pitowsky [22] showed further that the famous Bell-inequalities (and other members of the Bell inequalities family such as the Clauser, Horn, Shimony and Holt inequality) can be derived from Boole's conditions. Thus, violations of Bell-inequalities actually amount to violations of Boole's supposedly *a priori* "conditions of possible experience"! This derivation provides the major motivation for Pitowsky's interpretation of quantum mechanics, for, if Bell's inequalities characterize classical probability, their violation (as predicted by quantum mechanics) indicates a shift to an alternative, non-classical, theory of probability. Note that the classical assumption of 'real' properties, existing prior to measurement and discovered by it, underlies both Pitowsky's derivation of the Bell inequalities via Boole's conditions and the standard derivations of the inequalities. Consequently, in both cases, the violation of the inequalities suggests renunciation of a realist understanding of states and properties taken for granted in classical physics. In this respect, Pitowsky remained closer to the Copenhagen tradition than to its more realist rivals – Bohmian, Everett and GRW mechanical theories.

Examining the geometrical meaning of Boole's classical conditions, Pitowsky showed that the probabilities satisfying Boole's linear equations lie within n-dimensional polytopes whose dimensions are determined by the number of the events and whose facets are determined by the equations. The geometrical expression of the fact that the classical conditions can be violated is the existence of quantum probabilities lying *outside* of the corresponding classical polytope. The violation of Bell-type inequalities implies that in some experimental set-ups we get higher correlations than those permitted by classical considerations, that is, we get nonlocality. If, as Pitowsky argued, the violation of the inequalities is just a manifestation of the non-classical structure of the quantum probability space, nonlocality is likewise such a manifestation, integral to this structure and requiring no further explanation.

The next step was to provide an axiom system for quantum probability. The core of such a system had been worked out early on by Birkhoff and von Neumann [23], who considered it a characterization of the non-Boolean logic of quantum mechanics. Pitowsky endorsed the basic elements of the Birkhoff-von Neumann axioms, in particular the Hilbert space structure with its lattice of its subspaces. He also followed Birkhoff and von Neumann in identifying the failure of the classical axiom of reducibility as the distinct feature of quantum mechanics. By employing a number of later results, however, Pitowsky managed to strengthen the connection between the Birkhoff-von Neumann axioms and the theory of probability. To begin with, Gleason's celebrated theorem ensures that the only non-contextual probability measure definable on an n-dimensional Hilbert Space (n $\geq$ 3) yields the Born rule. On the basis of this theorem, Pitowsky was able to motivate the non-contextuality of probability, namely the claim that the probability of an event is independent of that of the context in which it is measured. Further, Pitowsky linked this non-contextuality to what is usually referred to as 'no signaling,' the relativistic limit on the transmission of information between entangled systems even though they may exhibit nonlocal correlations. In his interpretation, then, both nonlocality and no signaling are derived from formal properties of quantum probability rather than from any physical assumptions. It is the possibility of such a formal derivation that inspired the analogy between the probabilistic interpretation of quantum mechanics and the kinematic understanding of special relativity.

Several articles in this volume further explore the probabilistic approach and the formal connections between the different characteristics of quantum mechanics. Alexander Wilce offers a derivation of the logical structure of quantum mechanical probabilities (in a quantum world of finite dimensions) from four (and a half) probabilistically motivated axioms. Daniel Rohrlich attempts to invert the logical order by taking nonlocality as an axiom and *derive* standard quantum mechanics (and its probabilities) from nonlocality and no signaling. Jeffrey Bub takes up Wheeler's famous question 'why the quantum?' and uses the no signaling principle to explain why, despite the fact that stronger violations of locality are logically possible, the world still obeys the quantum mechanical bound on correlations.

From a philosophical point of view, Pitowsky's approach to quantum mechanics constitutes a landmark not only in the interpretation of quantum mechanics, but also in our understanding of the notion of probability. Traditionally, the theory of probability is conceived as an extension of logic, in the sense that both logic and the theory of probability lay down rules for rational inference and belief. Like logic, the theory of probability is therefore viewed as *a priori*. But if Pitowsky's conception of quantum mechanics as a non-classical theory of probability is correct, then the question of which is the right theory of probability is an empirical question, contingent on the way the world is. The shift from an *a priori* to an empirical construal has its precedents in the history of science: Geometry as well as logic have been claimed to be empirical rather than *a priori*, the former in the context of the theory of relativity, the latter in the context of quantum mechanics. Pitowsky's approach to quantum mechanics suggests a similar reevaluation of the status of probability and rational belief.

References

1. Kolmogorov, A.N.: Foundations of Probability. Chelsea, New York (1933). Chelsea 1950, Eng. Trans
2. Earman, J.: A Primer on Determinism. University of Western Ontario Series in Philosophy of Science, vol. 32. Reidel, Dordrecht (1986)
3. Fine, T.: Theories of Probability: An Examination of Foundations. Academic, New York (1973)
4. Hajek, A.: Interpretations of probability, Stanford Encyclopedia of Philosophy. http://plato.stanford.edu/entries/probability-interpret/ (2009)
5. von Mises, R.: Probability, Statistics and Truth. Dover, New York (1957), second revised English edition. Originally published in German by J. Springer in 1928
6. van Fraassen, B.: Belief and the will. J. Philos. **81**, 235–256 (1984)
7. van Fraassen, B.: Laws and Symmetry. Clarendon, Oxford (1989)
8. Ehrenfest, P., Ehrenfest, T.: The Conceptual Foundations of the Statistical Approach in Mechanics. Cornell University Press, New York (1912), 1959 (Eng. Trans.)
9. Uffink, J.: Compendium to the foundations of classical statistical physics. In: Butterfield, J., Earman, J. (eds.) Handbook for the Philosophy of Physics, pp. 923–1074. Elsevier, Amsterdam (2007). Part B
10. Leff, H., Rex, A.: Maxwell's Demon 2: Entropy, Classical and Quantum Information, Computing. Institute of Physics Publishing, Bristol (2003)
11. Maroney, O.: Information Processing and Thermodynamic Entropy. http://plato.stanford.edu/entries/information-entropy/ (2009)
12. Albert, D.: Time and Chance. Harvard University Press, Cambridge (2000)
13. Hemmo, M., Shenker, O.: Maxwell's Demon. J. Philos. The Journal of Philosophy **107**, 389–411 (2010)
14. Hemmo, M., Shenker, O.: Szilard's perpetuum mobile, Philosophy of Science **78**, 264–283 (2011)
15. Maudlin, T.: What could be objective about probabilities? Stud. Hist. Philos. Mod. Phys. **38**, 275–291 (2007)
16. Bohm, D.: A suggested interpretation of the quantum theory in terms of 'Hidden' variables. Phys. Rev. **85**, 166–179 (1952) (Part I), 180–193 (Part II)

17. Ghirardi, G.C., Rimini, A., Weber, T.: Unified dynamics for microscopic and macroscopic systems. Phys Rev D **34**, 470–479 (1986)
18. Everett, H.: 'Relative state' formulation of quantum mechanics. Rev. Mod. Phys. **29**, 454–462 (1957)
19. Pitowsky, I.: Betting on the outcomes of measurements: a Bayesian theory of quantum probability. Stud. Hist. Philos. Mod. Phys. **34**, 395–414 (2003)
20. Pitowsky, I.: Quantum mechanics as a theory of probability. In: Demopoulos, W., Pitowsky, I. (eds.) Physical Theory and Its Interpretation: Essays in Honor of Jeffrey Bub. The Western Ontario Series in Philosophy of Science, pp. 213–239. Springer, Dordrecht (2006)
21. Brown, H.: Physical Relativity. Oxford University Press, Oxford (2005)
22. Pitowsky, I.: George Boole's 'Conditions of possible experience' and the quantum puzzle. Br. J. Philos. Sci. **45**, 95–125 (1994)
23. Birkhoff, G., von Neumann, J.: The logic of quantum mechanics. Ann. Math. **37**(4), 823–843 (1936)
24. Bub, J., Pitowsky, I.: Two dogmas about quantum mechanics. In: Saunders, S., Barrett, J., Kent, A., Wallace, D. (eds.) Many Worlds? Everett, Quantum Theory and Reality. Oxford University Press, Oxford (2010), pp. 433–459
25. Feynman, R.: The Character of Physical Law. MIT Press, Cambridge (1965)

Chapter 2
Physics and Chance

David Albert

Abstract I discuss the role of chance in the fundamental physical picture of the world, and in the connections between that fundamental picture and the various other pictures of the world that we have from the special sciences, and I make a few remarks about the sort of thing that chance would need to be in order to be able to play that role.

2.1 Chance

Suppose that the world consisted entirely of point masses, moving in perfect accord with the Newtonian law of motion, under the influence of some particular collection of inter-particle forces. And imagine that that particular law, in combination with those particular forces, allowed for the existence of relatively stable, extended, rigid, macroscopic arrangements of those point masses – chairs (say) and tables and rocks and trees and all of the rest of the furniture of our everyday macroscopic experience.[1] And consider a rock, traveling at constant velocity, through an otherwise empty infinite space, in a world like that. And note that nothing whatsoever in the Newtonian law of motion, together with the laws of the interparticle forces, together with a stipulation to the effect that those interparticle forces are all the forces there are, is going to stand in the way of that rock's suddenly ejecting one of its trillions of elementary particulate constituents at enormous speed and careening off in an altogether different direction, or (for that matter) spontaneously disassembling itself into statuettes of the British royal family, or (come to think of it) reciting the Gettysburg address.

---

[1] And this, of course, is not true. And it is precisely because Newtonian Mechanics appears not to allow for the existence of these sorts of things or, even for the stability of the very atoms that make them up, that it is no longer entertained as a candidate for the fundamental theory of the world. But put all that aside for the moment.

D. Albert (✉)
Department of Philosophy, Columbia University, New York, NY, USA
e-mail: da5@columbia.edu

Y. Ben-Menahem and M. Hemmo (eds.), *Probability in Physics*, The Frontiers Collection,   17
DOI 10.1007/978-3-642-21329-8_2, © Springer-Verlag Berlin Heidelberg 2012

It goes without saying that none of these is in fact a serious possibility. And so the business of producing a scientific account of anything at all of what we actually know of the behaviors of rocks, or (for that matter) of planets or pendula or tops or levers or any of the traditional staples of Newtonian mechanics, is going to call for something over and above the deterministic law of motion, and the laws of the inter-particle forces, and a stipulation to the effect that those inter-particle forces are all the forces there are – something along the lines of a probability-distribution over microconditions, something that will entail, in conjunction with the law of motion and the laws of the inter-particle forces and a stipulation to the effect that those forces are all the forces there are, that the preposterous scenarios mentioned above – although they are not impossible – are nonetheless immensely unlikely.

And there is a much more general point here, a point which has nothing much to do with the ontological commitments or dynamical peculiarities or empirical inadequacies of the mechanics of Newtonian point masses, which goes more or less like this: Take any fundamental physical account of the world on which a rock is to be understood as an arrangement, or as an excitation, or as some more general collective upshot of the behaviors of an enormous number of elementary microscopic physical degrees of freedom. And suppose that there is some convex and continuously infinite set of distinct exact possible microconditions of the world – call that set {R} – each of which is compatible with the macrodescription "a rock of such and such a mass and such and such a shape is traveling at such and such a velocity through an otherwise empty infinite space". And suppose that the fundamental law of the evolutions of those exact microconditions in time is completely deterministic. And suppose that the fundamental law of the evolutions of those exact microconditions in time entails that for any two times $t_1 < t_2$, the values of all of the fundamental physical degrees of freedom at t_2 are invariably some continuous function of the values of those degrees of freedom at t_1. If all that is the case, then it gets hard to imagine how {R} could possibly fail to include a continuous infinity of distinct conditions in which the values of the elementary microscopic degrees of freedom happen to be lined up with one another in precisely such a way as to produce more or less any preposterous behavior you like – so long as the behavior in question is in accord with the basic ontology of the world, and with the conservation laws, and with the continuity of the finial conditions as a function of the initial ones, and so on. And so the business of discounting such behaviors as implausible – the business (that is) of underwriting the most basic and general and indispensable convictions with which you and I make our way about in the world – is again going to call for something over and above the fundamental deterministic law of motion, something along the lines, again, of a probability-distribution over microconditions.

If the fundamental microscopic dynamical laws themselves have chances in them, then (of course) all bets are off. But there are going to be chances – or that (at any rate) is what the above considerations suggest – at one point or another. Chances are apparently not to be avoided. An empirically adequate account of a world even remotely like ours in which nothing along the lines of a fundamental probability ever makes an appearance is apparently out of the question. And questions of precisely where and precisely how and in precisely what form such probabilities enter into nature are apparently going to need to be reckoned with in any serviceable account of the fundamental structure of the world.

2.2   The Case of Thermodynamics

Let's see what there is to work with.

The one relatively clear and concrete and systematic example we have of a fundamental probability-distribution over microconditions being put to useful scientific work is the one that comes up in the statistical-mechanical account of the laws of thermodynamics.

One of the monumental achievements of the physics of the nineteenth century was the discovery of a simple and beautiful and breathtakingly concise summary of the behaviors of the temperatures and pressures and volumes and densities of macroscopic material systems. The name of that summary is thermodynamics – and thermodynamics consists, in its entirety, of two simple laws. The first of those laws is a relatively straightforward translation into thermodynamic language of the conservation of energy. And the second one, the famous one, is a stipulation to the effect that a certain definite function of the temperatures and pressures and volumes and densities of macroscopic material systems – something called the entropy – can never decrease as time goes forwards. And it turns out that this second law in and of itself amounts to a complete account of the inexhaustible infinity of superficially distinct time-asymmetries of what you might call ordinary macroscopic physical processes. It turns out – and this is something genuinely astonishing – that this second law in and of itself entails that smoke spontaneously spreads out from and never spontaneously collects into cigarettes, and that ice spontaneously melts and never spontaneously freezes in warm rooms, and that soup spontaneously cools and never spontaneously heats up in a cool room, and that chairs spontaneously slow down but never spontaneously speed up when they are sliding along floors, and that eggs can hit a rock and break but never jump off the rock and re-assemble themselves, and so on, without end.

In the latter part of the nineteenth century, physicists like Ludwig Boltzmann in Vienna and John Willard Gibbs in New Haven began to think about the relationship between thermodynamics and the underlying complete microscopic science of elementary constituents of the entirety of the world – which was presumed (at the time) to be Newtonian Mechanics. And the upshot of those investigations is a beautiful new science called statistical mechanics.

Statistical mechanics begins with a postulate to the effect that a certain very natural-looking measure on the set of possible exact microconditions of any classical-mechanical system is to be treated or regarded or understood or put to work – of this hesitation more later – as a probability-distribution over those microconditions. The measure in question here is (as a matter of fact) the simplest imaginable measure on the set of possible exact microconditions of whatever system it is one happens to be dealing with, the standard Lebesgue measure on the phase-space of the possible exact positions and momenta of the Newtonian particles that make that system up. And the thrust of all of the beautiful and ingenious arguments of Boltzmann and Gibbs, and of their various followers and collaborators, was to make it plausible that the following is true:

Consider a true thermodynamical law, any true thermodynamical law, to the effect that macrocondition A evolves – under such-and-such external circumstances and over such-and-such a temporal interval – into macrocondition B. Whenever such a law holds, the overwhelming majority of the volume of the region of phase-space associated with macrocondition A – on the above measure, the simple measure, the standard measure, of volume in phase-space – is taken up by microconditions which are sitting on deterministic Newtonian trajectories which pass, under the allotted circumstances, at the end of the allotted interval, through the region of the phase space associated with the macrocondition B.

And if these arguments succeed, and if Newtonian mechanics is true, then the above-mentioned probability-distribution over microconditions will underwrite great swaths of our empirical experience of the world: It will entail (for example) that a half-melted block of ice alone in the middle of a sealed average terrestrial room is overwhelmingly likely to be still more melted towards the future, and that a half-dispersed puff of smoke alone in a sealed average terrestrial room is overwhelmingly likely to be still more dispersed towards the future, and that a tepid bowl of soup alone in a sealed average terrestrial room is overwhelmingly likely to get still cooler towards the future, and that a slightly yellowed newspaper alone in a sealed average terrestrial room is overwhelmingly likely to get still more yellow towards the future, and uncountably infinite extensions and variations of these, and incomprehensibly more besides.

But there is a famous trouble with all this, which is that all of the above-mentioned arguments work just as well in reverse, that all of the above-mentioned arguments work just as well (that is) at making it plausible that (for example) the half-melted block of ice I just mentioned was more melted towards the past as well. And we are as sure as we are of anything that that's not right.

And the canonical method of patching that trouble up is to supplement the dynamical equations of motion and the statistical postulate with a new and explicitly non-time-reversal-symmetric fundamental law of nature, a (so-called) past-hypothesis, to the effect that the universe had some particular, simple, compact, symmetric, cosmologically sensible, very low entropy initial macrocondition. The patched-up picture, then, consists of the complete deterministic microdynamical laws and a postulate to the effect that the distribution of probabilities over all of the possible exact initial microconditions of the world is uniform, with respect to the Lebaguse measure, over those possible microconditions of the universe which are compatible with the initial macrocondition specified in the past-hypothesis, and zero elsewhere. And with that amended picture in place, the arguments of Boltzmann and Gibbs will make it plausible not only that paper will be yellower and ice cubes more melted and people more aged and smoke more dispersed in the future, but that they were all less so (just as our experience tells us) in the past. With that additional stipulation in place (to put it another way) the arguments of Boltzmann and Gibbs will make it plausible that the second law of thermodynamics remains in force all the way form the end of the world back to its beginning.

*

What we have from Boltzmann and Gibbs, then, is a probability-distribution over possible initial microconditions of the world which – when combined with the exact deterministic microscopic equations of motion – apparently makes good empirical predictions about the values of the thermodynamic parameters of macroscopic systems. And there is a question about what to make of that success: We might take that success merely as evidence of the utility of that probability-distribution as an instrument for the particular purpose of predicting the values of those particular parameters, or we might take that success as evidence that the probability-distribution in question is literally true.

And note (and this is something to pause over) that if the probability-distribution in question were literally true, and if the exact deterministic microscopic equations of motion were literally true, then that probability-distribution, combined with those equations of motion, would necessarily amount not merely to an account of the behaviors of the thermodynamic parameters macroscopic systems, but to the complete scientific theory of the universe – because the two of them together assign a unique and determinate probability-value to every formulable proposition about the exact microscopic physical condition of whatever physical things there may happen to be. If the probability-distribution and the equations of motion in question here are regarded not merely as instruments or inference-tickets but as claims about the world, then there turns out not to be any physical question whatever on which they are jointly agnostic. If the probability-distribution and the equations of motion in question here are regarded not merely as instruments or inference-tickets but as claims about the world, then they are either false or they are in some sense (of which more in a minute) all the science there can ever be.

And precisely the same thing will manifestly apply to any probability-distribution over the possible exact microscopic initial conditions of the world, combined with any complete set of laws of the time-evolutions of those macroconditions.[2]

---

[2] Shelly Goldstein and Detlef Durr and Nino Zhangi and Tim Maudlin have worried, with formidable eloquence and incisiveness, that probability-distributions over the initial conditions of the world might amount to vastly more information than we could ever imaginably have a legitimate epistemic right to. Once we have a dynamics (once again) a probability-distribution over the possible exact initial conditions of the world will assign a perfectly definite probability to the proposition that I am sitting precisely here writing precisely this precisely now, and to the proposition that I am doing so not now but (instead) 78.2 s from now, and to the proposition that the Yankees will win the world series in 2097, and to the proposition that the zodiac killer was Mary Tyler Moore, and to every well-formed proposition whatever about the physical history of the world. And it will do so as a matter of fundamental physical law. And the worry is that it may be mad to think that there could be a fundamental physical law as specific as that, or that we could ever have good reason to believe anything as specific as that, or that we could ever have good reason to believe anything that logically implies anything as specific as that, even if the calculations involved in spelling such an implication out are prohibitively difficult.

Moreover, there are almost certainly an enormous number of very different probability-distributions over the possible initial conditions of the world which are capable of underwriting the laws of thermodynamics more or less as well as the standard, uniform, Boltzmann-Gibbs distribution does. And the reasons for that will be worth rehearsing in some detail.

Call the initial macrocondition of the world M. And let R_M be that region of the exact microscopic phase-space of the world which corresponds to M. And let ^aR_M be the sub-region of R_M which is taken up with "abnormal" microconditions – microconditions (that is) that lead to anomalously widespread violations of the laws of thermodynamics. Now, what the arguments of

Boltzmann and Gibbs suggest is as a matter of fact not only that the familiarly calculated volume of $^a R_M$ is overwhelmingly <u>small</u> compared with the familiarly calculated volume of R_M – which is what I have been at pains to emphasize so far – but also that $^a R_M$ is <u>scattered</u>, in unimaginably tiny clusters, more or less at random, all over R_M. And so the percentage of the familiarly calculated volume of any regularly shaped and <u>not</u> unimaginably tiny sub-region of which is taken up with abnormal microconditions will be (to an extremely good approximation) the same as the percentage of the familiarly calculated volume of R_M as a whole which is taken up by $^a R_M$. And so any reasonably smooth probability distribution over the microconditions in R_M – any probability-distribution over the microconditions in R_M (that is) that varies slowly over distances two or three orders of magnitude larger than the diameters of the unimaginably tiny clusters of which $^a R_M$ is composed – will yield (to an extremely good approximation) the same overall statistical propensity to thermodynamic behavior as does the standard uniform Boltzmann-Gibbs distribution over R_M as a whole. And exactly the same thing, or much the same thing, or something in the neighborhood of the same thing, is plausibly true of the behaviors of pin-balls and adrenal glands and economic systems and everything else as well.

   The suggestion (then) is that we proceed as follows: Consider the <u>complete set</u> of those probability-distributions over the possible exact initial conditions of the world – call it $\{P_f\}$ – which can be obtained from the uniform Boltzmann-Gibs distribution over R_M by multiplication by any relatively smooth and well-behaved and appropriately normalized function f of position in phase space. And formulate your fundamental physical theory of the world in such a way as to commit it to the truth of all those propositions on which every single one of the probability-distributions in $\{P_f\}$, combined with the dynamical laws, <u>agree</u> – and to leave it resolutely agnostic on everything else.

   If everything works as planned, and if everything in the paragraph before last is true – a theory like that will entail that the probability of smoke spreading out in a room, at the usual rate, is very high, and it will entail that the probability of a fair and well flipped coin's landing on heads is very nearly 1/2, and it will entail (more generally) that all of the stipulations of the special sciences are very nearly true. And yet (and this is what's different, and this is what's cool) it will almost entirely abstain from the assignment of probabilities to universal initial conditions. It <u>will</u> entail – and it had <u>better</u> entail – that the probability that the initial condition of the universe was one of those that lead to anomalously widespread violations of the laws of thermodynamics, that the probability that the initial condition of the universe lies (that is) in $^a R_M$, is overwhelmingly small. But it is going to assign <u>no probabilities whatsoever</u> to any of the <u>smoothly-bounded</u> or <u>regularly-shaped</u> or <u>easily-describable</u> proper subsets of the microconditions compatible with M.

   Whether or not a theory like that is ever going to look as simple and as serviceable and as perspicuous as the picture we have from Boltzmann and Gibbs (on the other hand) is harder to say. And (anyway) I suspect that at the end of the day it is <u>not</u> going to spare us the awkwardness of assigning of a definite probability, as a matter of fundamental physical law, to the proposition that the Zodiac Killer was Mary Tyler Moore. I suspect (that is) that <u>every single one</u> of the probability-distributions over R_M that suffice to underwrite the special sciences are going end up assigning very much the <u>same</u> definite probability to the proposition that the Zodiac Killer was Mary Tyler Moore as the standard, uniform, Boltzmann-Gibbs distribution does. And if that's <u>true</u>, then a move like the one being contemplated here may end up buying us very little.

   And beyond that, I'm not sure what to say. In so far as I can tell, our present business is going proceed in very much the same way, and arrive at very much the same conclusions, whether it starts out with the standard, uniform, Boltzmann-Gibbs probability-distribution over the microconditions in R_M, or with any other particular one of the probability-distributions in $\{P_f\}$, or with $\{P_f\}$ as a whole. And the first of those seems by far the easiest and the most familiar and the most intuitive and the most explanatory and (I guess) the most advisable. Or it does at first glance. It does for the time being. It does unless, or until, we find it gets us into trouble.

*

I want to look into the possibility that the probability-distribution we have from Boltzmann and Gibbs, or something like it, something more up-to date, something adjusted to the ontology of quantum field theory or quantum string theory or quantum brane theory, is true.

And this is a large undertaking.

Let's start slow.

Here are three prosaic observations.

The laws of thermodynamics are not quite true. If you look closely enough, you will find that the temperatures and pressures and volumes of macroscopic physical systems occasionally fluctuate away from their thermodynamically predicted values. And it turns out that precisely the same probability-distribution over the possible microconditions of such a system that accounts so well for the overwhelming reliability of the laws of thermodynamics accounts for the relative frequencies of the various different possible transgressions <u>against</u> those laws <u>as well</u>. And it turns out that the particular <u>features</u> of that distribution that play a pivotal role in accounting for the overwhelming reliability of the laws of thermodynamics are largely <u>distinct</u> from the particular features of that distribution that play a pivotal role in accounting for the relative frequencies of the various possible transgressions <u>against</u> those laws. It turns out (that is) that the relative frequencies of the transgressions give us information about a different <u>aspect</u> of the underlying microscopic probability-distribution (if there is one) than the overwhelming reliability of the laws of thermodynamics does, and it turns out that both of them are separately <u>confirmatory</u> of the empirical rightness of the distribution as a whole.

And consider a speck of ordinary dust, large enough to be visible with the aid of a powerful magnifying glass. If you suspend a speck like that in the atmosphere, and you watch it closely, you can see it jerking very slightly, very erratically, from side to side, under the impact of collisions with individual molecules of air. And if you carefully keep tabs on a large number of such specks, you can put together a comprehensive statistical picture of the sorts of jerks they undergo – as a function (say) of the temperatures and pressures of the gasses in which they are suspended. And it turns out (again) that precisely the same probability-distribution over the possible microconditions of such a system that accounts so well for the overwhelming reliability of the laws of thermodynamics accounts for the statistics of those jerks too. And it turns out (again) that the particular features of that distribution that play a pivotal role in accounting for the reliability of the laws of thermodynamics are largely distinct from the particular features of that distribution that play a pivotal role in accounting for the statistics of the jerks. And so the statistics of the jerks give us information about yet <u>another</u> aspect of the underlying microscopic probability-distribution (if there is one), and that new information turns out to be confirmatory, yet again, of the empirical rightness of the distribution as a whole.

And very much the same is true of isolated pin-balls balanced atop pins, or isolated pencils balanced on their points. The statistics of the directions in which such things eventually <u>fall</u> turn out to be very well described by precisely the same probability-distribution over possible microconditions, and it turns out (once more) that the particular features of that distribution that play a pivotal role in accounting for the reliability of the laws of thermodynamics are distinct from the particular features of that distribution that play a pivotal role in accounting for the statistics of those fallings.

And so the standard statistical posit of Boltzmann and Gibbs – when combined with the microscopic equations of motion – apparently has in it not only the thermodynamical science of melting, but also the quasi-thermodynamical science of chance fluctuations away from normal thermodynamic behavior, and (on top of that) the quasi-mechanical science of unbalancing, of breaking the deadlock, of pulling infinitesimally harder this way or that. And these sorts of things are manifestly going to have tens of thousands of other immediate applications. And it can now begin to seem plausible that this standard statistical posit might in fact have in it the entirety of what we mean when we speak of anything's happening at random or just by coincidence or for no particular reason.

2.3   The Special Sciences in General

The thermodynamic parameters all have straightforward and explicit and complete translations into the fundamental physical languages of Newtonian point mechanics, or of non-relativistic quantum mechanics, or of relativistic quantum field theory, or what have you. But that's not the case, and perhaps it will never be the case, and perhaps it can never be the case, for (say) economics, or epidemiology, or reader reception theory.

Let's take that for granted, then. Let's suppose that there can be no explicit translations from the languages of the various special sciences into the language of fundamental physics. But let's suppose as well that there is some fundamental physical language in which the world can be described completely, in at least the minimal sense that no two physically possible worlds can have different descriptions in the languages of any of the special sciences unless they have different descriptions in that fundamental physical language too. And let's imagine that we have in hand a complete microscopic dynamics, a complete theory (that is) of the time-evolutions of the elementary particles and fields or the elementary strings and branes or the elementary quantum-mechanical wave-functions or whatever the elementary physical constituents of the world turn out to be. And let's imagine (just for the sake of keeping things simple, and just for the moment) that that theory is fully deterministic.

And now take a computer, or (rather) super-computer, or (rather) a super-duper-computer, or whatever sort of a computer it is that the operations about to be described might turn out to require. And enter the dynamics into the computer. And enter the exact microscopic physical conditions of (say) the entirety of the solar system, at precisely 8:00 P.M., on the evening of a certain particular formal dinner party, into the computer.[3] And instruct the computer to perform a calculation (which nothing on the level of principle will now stand in the way of it's successfully carrying out, with the benefit of no further input whatever, and to whatever accuracy we might like) of the fundamental physical conditions obtaining near the surface of the earth throughout the period between (say) 10:00 P.M. and 10:15 P.M. later that same evening. And instruct the computer to output that information in such a way as to make it possible for its human operators to walk at will, in real time, about a virtual reconstruction of the barometric and electromagnetic conditions in the room where the dinner-party is taking place, throughout the interval between 10:00 P.M. and 10:15 P.M. later that same evening, and look here, and listen there, and just sort of take the whole thing in.

A sufficiently powerful computer (then) equipped with nothing over and above the fundamental physical laws, and provided (on the particular occasion in question) with nothing over and above an appropriate set of fundamental physical initial conditions, can show us, can display, for us, how any particular party (or war, or election, or painting, or investigation, or marriage, or whatever) comes out.

Suppose we want something fancier. Suppose (that is) that we want a computer that can do more than merely show us or display for us how the dinner party in question (or any other one) is going to come out. Suppose that what we want is a computer that can evaluate for us, and report to us, and predict for us, in English – equipped (mind you) with nothing over and above the fundamental physical theory, and provided with nothing over and above an appropriate set of fundamental physical initial conditions (of which more in a minute) – whether the party succeeds or fails.

Here's how to do that: Take a computer of the sort we were talking about before. And enter the dynamics into the computer. And enter the exact microscopic physical conditions of the entirety of the solar system, at precisely 8:00 P.M., on the evening of the party, into the computer. And instruct the computer (as before) to perform a calculation of the fundamental physical conditions obtaining near the surface of the earth throughout the period between 10:00 P.M. and 10:15 P.M. later that same evening. And instruct the computer to output that information in precisely the same sort of virtual-reality format as we described above. And instruct the computer's human operator to survey whatever portions of that output he needs to in order to evaluate whether the party turns out to be a success or a failure, and to report his findings in writing, in English, on a piece of paper, and to place that piece of paper in a certain particular (otherwise empty) box. And now enter the dynamics into a second computer. And enter into that second computer the exact microscopic

---

[3] Or (at any rate) enter in the conditions of as much of the world, at 8:00 P.M. of the evening in question, as would need to be taken into account in the process of calculating the physical conditions on the surface of the earth some hours later – enter in (that is) the conditions throughout the cross-section, at 8:00 P.M. of the evening in question, of the backward light cone of the surface of the earth some hours later.

fundamental physical conditions of the entirety of (say) the sealed interstellar spaceship containing the first computer and its output mechanisms and its human operator and the piece of paper and the empty box and whatever else happens to be in there at (say) the moment just after all of the instructions described in the earlier part of this paragraph have been delivered. And instruct the second computer to calculate and reproduce for us the physical contents of the box 20 min or so hence.

The output of this second computer, then, will consist of a piece of paper on which either "the party is a success" or "the party is a failure" appears. And the report on that piece of paper will (with extremely rare exceptions, of which more in a minute) prove accurate – once the actual historical facts are in. And the input (once again) consists of nothing over and above the fundamental physical laws and the fundamental physical initial conditions of various parts of the world back at the time the party in question first got underway.

Suppose we want something fancier still. Suppose we want a computer that can deduce for us, given nothing over and above the fundamental physical laws, the laws (if there are any) of the special science of the success or failure of formal dinner parties. Suppose (for example) that we want a computer that can deduce for us, given nothing over and above the fundamental physical laws, whether inviting an odd number of guests, or inviting an even number, is more likely to produce a better party.

Here's how to work that: Take a computer. And enter the fundamental physical dynamical laws into it. And set things up in such a way as to allow the operator of the computer to take a virtual tour – of just the sort we talked about above – of whichever of the possible exact physical microconditions of the solar system he chooses.[4] And instruct the operator to survey the space of those exact physical microconditions by means of this technique, and to identify the regions of that space which correspond to circumstances in which odd numbers of guests are being invited to a formal dinner party, and to identify the regions of that space which correspond to circumstances in which even numbers of guests are being invited to a formal dinner party, and to order the computer to calculate, and to display for him, how all of those dinner-parties come out, and to prepare a written report, and to put it into the box we were talking about above. And now (just as before) enter the dynamics into a second computer. And enter into that second computer the exact microscopic fundamental physical conditions of the entirety of the sealed interstellar spaceship containing the first computer and its output mechanisms and its human operator and the piece of paper and the empty box and whatever else happens to be in there at the moment just after all of the instructions described in the earlier part of this paragraph have been delivered. And instruct the second computer to calculate and reproduce for us the physical contents of the box 20 min (or 20 years, or however long it might imaginably take) hence.

Now a few remarks are in order.

---

[4] Things might be set up in such a way as to allow the operator to point and click on a map of the space of states.

Note that the roles of the various computers in the above three scenarios are merely to function as concrete realizations of the implicative structures of the fundamental physical theories – the roles of those computers are merely to make vivid what sorts of information about the world those theories have in them, what sorts of things they can predict, what sorts of things they can account for. Nothing whatever hinges on the possibility (which is presumably immensely remote – both for practical reasons and for reasons of principle as well) of such computers ever actually being constructed!

And note that the role of the actual living human operator of the computer in the second and third of the scenarios is merely to serve as a catalogue of fundamental physical initial conditions. Neither of those two scenarios involves any living human being's ever actually evaluating the success or failure of a virtual dinner party, or surveying sets of possible initial conditions, or preparing a written report, or anything of the sort – all of that gets done by a simulator, by a subroutine – call it the 'human operator subroutine' – which exists only in cyberspace, and of which the living human being in question is merely the plan, merely the template, merely the set of instructions. The living human being in question can perfectly well be far away, or asleep, or dead, long before the simulated evaluations or surveyings or reportings ever get underway. It is nothing whatever over and above the computer, nothing (that is) over and above the implicative structure of the fundamental physical laws, that's doing all the work.

These subroutines (by the way) – precisely because and precisely in so far as they can faithfully reproduce the behavioral dispositions of the living human beings on which they are modeled, beings which (after all) can lie and make mistakes and get moody and have heart attacks – will necessarily be imperfect as speakers of the language of dinner parties. But they can be very good. They can be exactly as good as the best of us, or as whole committees of us, or as whole societies of us. And the very idea of doing any better, the very idea of coming up with a program which somehow instantiates a formal and mechanical and algorithmic scheme for translating from the language of fundamental physical theory to the language of dinner parties is (by hypothesis) out of the question.

And there's one more thing – and this is precisely the thing that the previous two sections of this chapter were about. Recall the third of the scenarios we talked about above – the one where the computer derives the general laws (if – once again – there are any) of the success and failure of dinner parties, and the full apparatus for explaining why this or that particular dinner party succeeded or failed, from nothing over and above the fundamental laws of physics. The computer first surveys the space of possible exact physical microconditions of the solar system, and then it identifies the regions of that space which correspond to circumstances in which odd numbers of guests are being invited to a formal dinner party and the regions of that space which correspond to circumstances in which even numbers of guests are being invited to a formal dinner party, and then it calculates how all those dinner parties come out, and finally it prepares a written report on the respective probabilities that even or odd numbers of invitations will (as a matter of fundamental physical law) result in success. And what I want to draw attention to at the moment is just that the fundamental physical laws in question here are going to have to include stipulations of a kind that we have so far neglected to bring up in this section, that the fundamental physical laws in question here are going to have to amount (again) to something over and above the complete theory of the time-evolutions of the fundamental physical systems that we talked about some pages back. And the reason (of course) is this: A complete theory of the time-evolutions of the fundamental physical systems – although it will settle all questions as to which initial microconditions lead to success and which to failure, although it will settle all questions as to how many of the microconditions which correspond to (say) an even number of invitations being sent out lead to success and how many lead to failure – will have nothing whatever to say about the probability of any particular one of those microconditions actually obtaining, given (say) that an even number of invitations are sent out. And without that we have nothing of any macroscopic use at all. And so (again) we are going to need an other and altogether different sort of fundamental physical law, a law which will apparently need to take the form of a probability-distribution – or something like a probability-distribution – over initial conditions.

   And the alluring possibility is (again) that the law we need here is in fact the one we already have, the one suggested by altogether different sorts of considerations, the one that seemed to show some promise of making concrete and explicit and quantitative sense of what we mean when we speak of anything's happening at random or just by coincidence or for no particular reason, the one (that is) due to Boltzmann and Gibbs.

2.4   Explanation

Here is a line of argument – one of many – aimed directly against the sort of universality and completeness of physics that I was trying to imagine in the previous section. It comes from Science, Truth, and Democracy, by my friend and teacher Philip Kitcher. The worry here is not about the capacities of fundamental physical theories to predict – which Philip is willing to grant – but about their capacities to explain. Philip directs our attention to

> .... the regularity discovered by John Arbuthnot in the early eighteenth century. Scrutinizing the record of births in London during the previous 82 years, Arbuthnot found that in each year a preponderance of the children born had been boys: in his terms, each year was a "male year". Why does this regularity hold? Proponents of the Unity-of Science view can offer a recipe for the explanation, although they can't give the details. Start with the first year (1623); elaborate the physicochemical details of the first copulation-followed-by-pregnancy showing how it resulted in a child of a particular sex; continue in the same fashion for each pertinent pregnancy; add up the totals for male births and female births and compute the difference. It has now been shown why the first year was "male"; continue for all subsequent years.

Even if we had this "explanation" to hand, and could assimilate all the details, it would still not advance our understanding. For it would not show that Arbuthnot's regularity was anything more than a gigantic coincidence. By contrast, we can already give a satisfying explanation by appealing to an insight of R. A. Fisher. Fischer recognized that, in a population in which sex ratios depart from 1:1 at sexual maturity, there will be a selective advantage to a tendency to produce the underrepresented sex. It will be easy to show from this that there should be a stable evolutionary equilibrium at which the sex ratio at sexual maturity is 1:1. In any species in which one sex is more vulnerable to early mortality than the other, this equilibrium will correspond to a state in which the sex ratio at birth is skewed in favor of the more vulnerable sex. Applying this analysis to our own species, in which boys are more likely than girls to die before reaching puberty, we find that the birth sex ratio ought to be 1.104:1 in favor of males – which is what Arbuthnot and his successors have observed. We now understand *why* [my italics], for a large population, all years are overwhelmingly likely to be male.

The key word here, the word that carries the whole burden of Philip's argument, is 'coincidence'.

And that will be worth pausing over, and thinking about.

The moral of the first section of this chapter (remember) was that the fundamental physical laws of the world, merely in order to get the narrowest imaginable construal of their 'work' done, merely in order to get things right (that is) about projectiles and levers and pulleys and tops, will need to include a probability-distribution over possible microscopic initial conditions. And once a distribution like that is in place, all questions of what is and isn't <u>likely</u>, all questions of what was and wasn't to be <u>expected</u>, all questions of whether or not this or that particular collection of events happened merely 'at random' or 'for no particular reason' or 'as a matter of coincidence', are (in principle) <u>settled</u>. And (indeed) it is only <u>by reference</u> to a distribution like that that talk of coincidence can make any precise sort of sense in the <u>first</u> place – it is only <u>against the background</u> of a distribution like that that questions of what is or is not coincidental can even be <u>brought up</u>.

It goes without saying that we do not (typically, consciously, explicitly) <u>consult</u> that sort of a distribution when we are engaged in the practical business of making judgments about what is and is not coincidental. But that is no evidence at all against the hypothesis that such a distribution <u>exists</u>, and it is no evidence at all against the hypothesis that such a distribution is the sole ultimate arbiter of what is and is not <u>coincidental</u>, and it is no evidence at all against the hypothesis that such a distribution informs <u>every single one</u> of our billions of everyday deliberations. If anything along the lines of the complete fundamental theory we have been trying to imagine here is true (after all) some crude, foggy, reflexive, largely unconscious but perfectly serviceable acquaintance with that distribution will have been hard-wired into us as far back as when we were fish, as far back (indeed) as when we were <u>slime</u>, by natural selection – and lies buried at the very heart of the deep instinctive primordial unarticulated feel of the world. If anything along the lines of the complete fundamental theory we have been trying to imagine here is true (after all) the penalty for expecting anything <u>else</u>, the penalty for expecting anything to the <u>contrary</u>, is extinction.

And if one keeps all this in the foreground of one's attention, it gets hard to see what Philip can possibly have in mind in supposing that something can amount to a 'gigantic coincidence' from the standpoint of the true and complete and universal fundamental physical theory of the world and yet (somehow or other) not be.

*

If anything along the lines of the picture we are trying to imagine here should turn out to be true, then any correct special-scientific explanation whatsoever can in principle be uncovered, can in principle be descried, in the fundamental physical theory of the world, by the following procedure:

Start out with a distribution of probabilities which is uniform, on the standard statistical-mechanical measure, over all of the possible exact initial microconditions of the world which are compatible with the past-hypothesis, and zero elsewhere. And conditionalize that distribution on whatever particular features of the world play a role in the special-scientific explanation in question – conditionalize that distribution (that is) on whatever particular features of the world appear either explicitly or implicitly among the explanans of the special-scientific explanation in question.[5] And check to see whether or not the resultant distribution – the conditionalized distribution, makes the explanandum likely. If it does, then we have recovered the special-scientific explanation form the fundamental physical theory – and if it doesn't, then either the fundamental theory, or the special-scientific explanation, or both, are wrong.

Consider (for example) the evolution of the total entropy of the universe over the past 10 min. That entropy (we are confident) is unlikely to have gone down over those 10 min. The intuition is that the entropy's having gone down over those 10 min would have amounted to a gigantic coincidence. The intuition is that the entropy's having gone down over those 10 min would have required detailed and precise and inexplicable correlations among the positions and velocities of all of the particles that make the universe up. And questions of whether or not correlations like that are to be expected, questions of whether or not correlations like that amount to a coincidence, are matters (remember) on which the sort of fundamental physical theory we are thinking about here can by no means be agnostic. And it is part and parcel of what it is for that sort of a theory to succeed that it answers those questions correctly. It is part and parcel of what it is for that sort of a theory to succeed (that is) that it transparently captures, and makes simple, and makes elegant, and makes precise, the testimony our intuition, and our empirical experience of the world, to the effect that correlations like that are in fact fantastically unlikely, that they are not at all to be expected, that they do indeed amount to a

---

[5] Those explanans, of course, are initially going to be given to us in the language of one or another of the special sciences. And so, in order to carry out the sort of conditionalization we have in mind here, we are going to need to know which of those special-scientific explanans correspond to which regions of the space of possible exact physical microconditions of the world. And those correspondences can be worked out – not perfectly, mind you, but to any degree of accuracy and reliability we like – by means of the super-duper computational techniques described in Sect. 2.3.

gigantic coincidence. And there is every reason in the world to believe that there is a fundamental physical theory that can <u>do</u> that. It was precisely the achievement of Boltzmann and Gibbs (after all) to make it plausible that the Newtonian laws of motion, together with the statistical-postulate, together with the past-hypothesis, all of it conditionalized on a proposition to the effect that the world was not swarming, 10 min ago, with malevolent Maxwellian Demons, can do, precisely, that.

And now consider the descent of man. The first humans (we are confident) are unlikely to have condensed out of swamp gas, or to have grown on trees, or to have been born to an animal incapable of fear. The first (after all) would require detailed and precise and inexplicable correlations among the positions and velocities of all of the molecules of swamp gas, and of the surrounding air, and the ground, and god knows what else. And the second would require a vast, simultaneous, delicately co-ordinated unimaginably fortuitous set of mutations on a single genome. And the third would require that every last one of a great horde of mortal dangers all somehow conspire to avoid the animal in question – with no help whatever from the animal herself – until she is of age to deliver her human child. And it is <u>precisely</u> because the account of the descent of man by <u>random mutation</u> and <u>natural selection</u> involves vastly <u>fewer</u> and more <u>minor</u> and less <u>improbable</u> such coincidences than any of the imaginable others that it strikes us as the best and most plausible <u>explanation</u> of that descent we have. And (indeed) it is precisely the relative <u>paucity</u> of such coincidences, and it is precisely the relative <u>smallness</u> of whatever such coincidences there <u>are</u>, to which words like 'random' and 'natural' are meant to direct our attention! And questions about what does and what does not amount to a coincidence are matters (once again) on which the sort of fundamental physical theory we are imagining here can by no means be agnostic. And it is part and parcel of what it is for that sort of a theory to <u>succeed</u> (once again) that it answers every last one of those questions <u>correctly</u>.

Now, compelling arguments to the effect that this or that particular fundamental physical theory of the world is actually going to be able to <u>do</u> all that are plainly going to be harder to come by here than they were in the much more straightforward case of the entropy of the universe. All we have to go on are small intimations – the ones mentioned above, the ones you can make out in the behaviors of pin-balls and pencils and specks of dust – that perhaps the exact microscopic laws of motion together with the statistical postulate together with the past-hypothesis has in it the entirety of what we mean when we speak of anything's happening <u>at random</u> or <u>for no particular reason</u> or <u>just by coincidence</u>.

But if all that should somehow happen to <u>pan out</u>, if there <u>is</u> a true and complete and fundamental physical theory of the sort that we have been trying to imagine here, then it is indeed going to follow directly from the fundamental laws of motion, together with the statistical-postulate, together with the past-hypothesis, all of it conditionalized on the existence of our galaxy, and of our solar system, and of the earth, and of life, and of whatever else is implicitly being taken for granted in scientific discussions of the descent of man, that the first humans are indeed extraordinarily unlikely to have condensed out of swamp gas, or to have grown on trees, or to have been born to an animal incapable of fear.

And very much the same sort of thing is going to be true of the regularity discovered by Arubthot.

What Fisher has given us (after all) is an argument to the effect that it would amount to a gigantic coincidence, that it would represent an enormously improbable insensitivity to pressures of natural selection, that it would be something very much akin to a gas spontaneously contracting into one particular corner of its container, for sex ratios to do anything other than settle into precisely the stable evolutionary equilibrium that he identifies. And questions about what does and what does not amount to a coincidence are (for the last time) matters on which the sort of fundamental physical theory we are imagining here can by no means be agnostic. And it is part and parcel of what it is for that sort of a theory to <u>succeed</u> that it answers every last one of those questions <u>correctly</u>.

And once again, compelling arguments to the effect that this or that particular fundamental physical theory of the world is actually going to be able to <u>do</u> all that are plainly going to be hard to come by – and all we are going to have to go on are the small promising intimations from pin-balls and pencils and specks of dust.

But consider how things would stand if all that should somehow happen to pan out. Consider how things would stand if there <u>is</u> a true and complete and fundamental physical theory of the sort that we have been trying to imagine here.

Start out – as the fundamental theory instructs us to do – with a distribution of probabilities which is uniform, on the standard statistical-mechanical measure, over all of the possible exact initial microconditions of the world which are compatible with the past-hypothesis, and zero elsewhere. And evolve that distribution – using the exact microscopic deterministic equations of motion – up to the stroke of midnight on December 31st of 1623. And conditionalize that evolved distribution on the existence of our galaxy, and of our solar system, and of the earth, and life, and of the human species, and of cities, and of whatever else is implicitly being taken for granted in any scientific discussion of the relative birth rates of boys and girls in London in the years following 1623. And call that evolved and conditionalized distribution P_{1623}.

If there is a true and complete and fundamental theory of the sort that we have been trying to imagine here, then what Fisher has given will amount to an argument that P_{1623} is indeed going to count it as likely that the preponderance of the babies born in London, to human parents, in each of the 82 years following 1623, will be boys. Period. End of story.

Of course, the business of explicitly <u>calculating</u> P_{1623} from the microscopic laws of motion and the statistical postulate and the past-hypothesis is plainly, permanently, out of the question. But Philip's point was that even if that calculation <u>could</u> be performed, even (as he says) "if we had this "explanation" to hand, and could assimilate all the details, it would still not advance our understanding. For it would not show that Arbuthnot's regularity was anything more than a gigantic coincidence." And this seems just....wrong. And what it misses – I think – is that the fundamental physical laws of the world, merely in order to get the narrowest imaginable construal of their 'work' done, merely in order to get things right (that is) about projectiles and levers and pulleys and tops, are going to have to come equipped, from the word go, with <u>chances</u>.

*

And those chances are going to bring with them – in principle – the complete explanatory apparatus of the special sciences. And <u>more</u> than that: those chances, together with the exact microscopic equations of motion, are going to explain all sorts of things about which all of the special sciences taken together can have <u>nothing whatever to say</u>, they are going to provide us – in principle – with an account of where those sciences <u>come from</u>, and of how they <u>hang together</u>, of how it is that certain particular sets of too-ings and fro-ings of the fundamental constituents of the world can simultaneously instantiate <u>every last one of them</u>, of how each of them separately applies to the world in such a way as to accommodate the fact that the world is a <u>unity</u>.

And so (you see) what gets in the way of explaining things is <u>not at all</u> the conception of science as <u>unified</u>, but the conceit that it can somehow <u>not</u> be.

2.5   The General Business of Legislating Initial Conditions

All of this delicately hangs (of course) on the possibility of making clear metaphysical sense of the assignment of real physical chances to initial conditions.[6]

---

[6] I will be taking it for granted here that a probability-distribution over initial conditions, whatever else it is, is an <u>empirical hypothesis</u> about the way the world <u>contingently happens</u> to be.

But this is by no means the received view of the matter. Indeed, the statistical postulate of Boltzmann and Gibbs seems to have been understood by its inventors as encapsulating something along the lines of an <u>a priori principle of reason</u>, a principle (more particularly) of <u>indifference</u>, which runs something like this: Suppose that the entirety of what you happen to know of a certain system S is that S is X. And let $\{v_i\}_{X,t}$ be the set of the possible exact microconditions of S such that v_i's obtaining at t is compatible with S's being X. Then the principle stipulates that for any two v_j, v_k 0 $\{v_i\}_{X,t}$ the probability of v_j's obtaining at t is equal to the probability of v_k's obtaining at t.

And that (I think) is more or less what the statistical postulate still amounts to in the imaginations of many physicists. And that (to be sure) has a supremely innocent ring to it. It sounds very much, when you first hear it, as if it is instructing you to do nothing more than attend very carefully to what you <u>mean</u>, to what you are <u>saying</u>, when you say that the entirety of what you know of S is that S is X. It sounds very much as if it is doing nothing more than reminding you that what you are saying when you say something like that is that S is X, and (moreover) that for any two v_j, v_k 0 $\{v_i\}_{X,t}$, you have no more reason for believing that v_j obtains at t than you have for believing that v_k obtains at t, that (in so far as you know) nothing <u>favors</u> any particular one of the v_j 0 $\{v_i\}_{X,t}$ over any particular <u>other</u> one of the v_j 0 $\{v_i\}_{X,t}$, that (in other words) the <u>probability</u> of any particular one of those microconditions obtaining at t, given the information you have, is <u>equal</u> to the probability of any particular <u>other</u> one of them obtaining at t.

And this is importantly and spectacularly wrong. And the reasons why it's wrong (of which there are two: a technical one and a more fundamental and less often remarked-upon one too) are worth rehearsing.

And conceptions of chance as anything along the lines of (I don't know) a <u>cause</u> or a <u>pressure</u> or a <u>tendency</u> or a <u>propensity</u> or a <u>pulling</u> or a <u>nudging</u> or an <u>enticing</u> or a <u>cajoling</u> or (more generally) as anything essentially bound up with the way in

---

The technical reason has to do with the fact that the sort of information we can actually <u>have</u> about physical systems – the sort that we can <u>get</u> (that is) by <u>measuring</u> – is invariably compatible with a <u>continuous infinity</u> of the system's <u>microstates</u>. And so the only way of assigning equal probability to all of those states at the time in question will be by assigning each and every one of them the probability <u>zero</u>. And <u>that</u> will of course tell us <u>nothing whatever</u> about how to make our <u>predictions</u>.

And so people took to doing something <u>else</u> – something that looked to them to be very much in the same <u>spirit</u> – <u>instead</u>. They abandoned the idea of assigning probabilities to individual microstates, and took instead to stipulating that the probability assigned to any <u>finite region of the phase space</u> which is entirely compatible with X – under the epistemic circumstances described above – ought to be proportional to the continuous <u>measure</u> of the points <u>within</u> that region.

But there's a trouble with that – or at any rate there's a trouble with the thought that it's <u>innocent</u> – too. The trouble is that there are in general an <u>infinity</u> of equally mathematically legitimate ways of <u>putting</u> measures on infinite sets of points. Think, for example, of the points on the real number line between 0 and 1. There is a way of putting measures on that set of points according to which the measure of the set of points between any two numbers a and b (with a < 1 and b < 1 and b > a) is b – a, and there is <u>another</u> way of putting measures on that set of points according to which the measure of the set of points between any two numbers a and b between (with a < 1 and b < 1 and b > a) is b – a, and according to the first of those two formulae there are "as many" points between 1 and 1/2 as there are between 1/2 and 0, and according to the <u>second</u> of those two formulae there are <u>three times</u> "as many" points between 1 and 1/2 as there are between 1/2 and 0, and there turns out to be no way whatever (or at any rate none that anybody has yet dreamed up) of arguing that either one of these two formulae represents a truer or more reasonable or more compelling measure of the "number" or the "amount" or the "quantity" of points between a and b than the other one does. And there are (moreover) an infinite number of <u>other</u> such possible measures on this interval as well, and this sort of thing (as I mentioned above) is a very general phenomenon.

And anyway, there is a more fundamental problem, which is that the sorts of probabilities being imagined here, probabilities (that is) conjured out of airy nothing, out of pure ignorance, whatever else might be good or bad about them, are obviously and scandalously unfit for the sort of explanatory work that we require of the probabilities of Boltzmann and Gibbs. Forget (then) about all the stuff in the last three paragraphs. Suppose there was no trouble about the measures. Suppose that there were some unique and natural and well-defined way of expressing, by means of a distribution-function, the fact that "nothing in our epistemic situation favors any particular one of the microstates compatible with X over any other particular one of them". So <u>what</u>? Can anybody seriously think that that would somehow <u>explain</u> the fact that the <u>actual microscopic conditions</u> of <u>actual thermodynamic systems</u> are <u>statistically distributed in the way that they are</u>? Can anybody seriously think that it is somehow <u>necessary</u>, that it is somehow <u>a priori</u>, that the particles of which the material world is made up must arrange themselves in accord with <u>what we know</u>, with <u>what we happen to have looked into</u>? Can anybody seriously think that our merely being <u>ignorant</u> of the exact microstates of thermodynamic systems plays some part in <u>bringing it about</u>, in <u>making it the case</u>, that (say) <u>milk dissolves in coffee</u>? How could that <u>be</u>? What can all those guys have been <u>up</u> to? If probabilities have anything whatever to do with how things actually fall out in the <u>world</u> (after all) then knowing nothing whatever about a certain system other than X can in and of itself entail nothing whatever about the relative probabilities of that system's being in one or another of the microstates <u>compatible</u> with X; and if probabilities have <u>nothing</u> whatever to do with how things actually fall out in the world, then they can patently play no role whatever in explaining the behaviors of <u>actual physical systems</u>; and that would seem to be all the options there are to <u>choose</u> from!

which instantaneous states of the world <u>succeed one another in time</u>, is manifestly not going to be up to the job – since the initial condition of the world is (after all) not the temporal successor of anything, and there was (by definition) no historical episode of the world's having been pulled or pressed or nudged or cajoled into this or that particular way of getting started.

Our business here (then) is going to require another understanding of chance. And an understanding of law in general, I think, to go with it. Something Humean. Something wrapped up not with an image of <u>governance</u>, but with an idea of <u>description</u>. Something (as a matter of fact) of the sort that's been worked out, with slow and sure and graceful deliberation, over these past 20 years or so, by David Lewis and Barry Loewer.

*

Here's the idea. You get to have an audience with God. And God promises to tell you whatever you'd like to know. And you ask Him to tell you about the world. And He begins to recite the facts: such-and-such a property (the presence of a particle, say, or some particular value of some particular field) is instantiated at such-and-such a spatial location at such-and-such a time, and such-and-such <u>another</u> property is instantiated at such-and-such <u>another</u> spatial location at such-and-such <u>another</u> time, and so on. And it begins to look as if all this is likely to drag on for a while. And you explain to God that you're actually a bit pressed for time, that this is not all you have to do today, that you are not going to be in a position to hear out the whole story. And you ask if maybe there's something meaty and pithy and helpful and informative and short that He might be able to tell you about the world which (you understand) would not amount to everything, or nearly everything, but would nonetheless still somehow amount to a lot. Something that will serve you well, or reasonably well, or as well as possible, in making your way about in the world.

And what it is to be a law, and <u>all</u> it is to be a law, on this picture of Hume's and Lewis' and Loewer's, is to be an element of the best possible response to precisely this request – to be a member (that is) of that set of true propositions about the world which, alone among all of the sets of true propositions about the world that can be put together, best combines simplicity and informativeness.

On a picture like this, the world, considered as a whole, is merely, purely, <u>there</u>. It isn't the sort of thing that is susceptible of being <u>explained</u> or <u>accounted for</u> or <u>traced back to something else</u>. There isn't anything that it <u>obeys</u>. There is nothing to talk about over and above the totality of the concrete particular facts. And science is the business of producing the most compact and informative possible <u>summary</u> of that totality. And the components of that summary are called <u>laws of nature</u>.[7]

---

[7] This is not <u>at all</u> (of course) to deny that there are such things as scientific explanations! There are <u>all sorts</u> of explanatory relations – on a picture like this one – <u>among</u> the concrete particular facts, and (more frequently) among <u>sets</u> of the concrete particular facts. There are all sorts of things to be said (for example) about how smaller and more local patterns among those facts <u>fit into</u>, or are <u>subsumed under</u>, or are <u>logically necessitated by</u>, larger and more universal ones. But the <u>totality</u> of the concrete particular facts is the point at which – on a view like this one – all explaining necessarily comes to an end.

The world (on this picture) is not what it is in virtue of the laws being what they are, the laws are what they are in virtue of the world's being what it is.

*

Now, different possible worlds – different possible totalities (that is) of concrete particular facts – may turn out to accommodate qualitatively different sorts of maximally compact and informative summaries.

The world might be such that God says: "I have just the thing: The furniture of the universe consists entirely of particles. And the force exerted by any particle on any other particle is equal to the product of the masses of those two particles divided by the square of the distance between them, directed along the line connecting them. And those are all the forces there are. And everywhere, and at every time, the acceleration of every particle in the world is equal to the total force on that particle at that time divided by it's mass. That won't tell you everything. It won't tell you nearly everything. But it will tell you a lot. It will serve you well. And it's the best I can do, it's the most informative I can be, if (as you insist) I keep it short." Worlds like that are called (among other things) Newtonian and particulate and deterministic and non-local and energy-conserving and invariant under Galilean transformations.

Or the world might be different. The world might be such that God says "Look, there turns out not to be anything I can offer you in the way of simple, general, exact, informative, exceptionally true propositions. The world turns out not to accommodate propositions like that. Let's try something else. Global physical situations of type A are followed by global physical situations of type B roughly (but not exactly) 70% of the time, and situations of type A are followed by situations of type C roughly (but not exactly) 30% of the time, and there turns out not to be anything else that's simple to say about which particular instances of A-situations are followed by B-situations and which particular instances of the A-situations are followed by C-situations. That's pithy too. Go fourth. It will serve you well." We speak of worlds like that as being lawful but indeterministic – we speak of them as having real dynamical chances in them.

Or the world might be such that God says: "Sadly, I have nothing whatever of universal scope to offer you – nothing deterministic and nothing chancy either. I'm sorry. But I do have some simple, useful, approximately true rules of thumb about rainbows, and some others about the immune system, and some others about tensile strength, and some others about birds, and some others about interpersonal relationships, and some others about stellar evolution, and so on. It's not elegant. It's not all that concise. But it's all there is. Take it. You'll be glad, in the long run, that you did." We speak of worlds like that – following Nancy Cartwright – as dappled.

Or the world might be such that God has nothing useful to offer us at all. We speak of worlds like that as chaotic – we speak of them as radically unfriendly to the scientific enterprise.

Or the world might (finally) be such that God says: "All of the maximally simple and informative propositions that were true of the Newtonian particulate deterministic non-local energy-conserving Galilean-invariant universe are true of this one too. The furniture of the universe consists entirely of particles. And the force exerted by any particle on any other particle is equal to the product of the masses of those two particles divided by the square of the distance between them, directed along the line connecting them. And those are all the forces there are. And everywhere, and at every time, the acceleration of every particle in the world is equal to the total force on that particle at that time divided by it's mass. But that's not all. I have something more to tell you as well. Something (as per your request) simple and helpful and informative. Something about the initial condition of the world. I can't tell you exactly what that condition was. It's too complicated. It would take too long. It would violate your stipulations. The best I can do by way of a simple and informative description of that condition is to tell you that it was one of those which is typical with respect to a certain particular probability-distribution – the Boltzmann-Gibbs distribution, for example. The best I can do by way of a simple and informative description of that initial condition is to tell you that it was precisely the sort of condition that you would expect, that it was precisely the sort of condition that you would have been rational to bet on, if the initial condition of the world had in fact been selected by means of a genuinely dynamically chancy procedure where the probability of this or that particular condition's being selected is precisely the one given in the probability-distribution of Boltzmann and Gibbs." And this is precisely the world we encounter in classical statistical mechanics. And this is the sought-after technique – or one of them – for making clear metaphysical sense of the assignment of real physical chances to initial conditions. The world has only one microscopic initial condition. Probability-distributions over initial conditions – when they are applicable – are compact and efficient and informative instruments for telling us something about what particular condition that is.[8]

And note that it is of the very essence of this Humean conception of the law that there is nothing whatever metaphysical at stake in the distinctions between deterministic worlds, and chancy ones, and dappled ones, and chaotic ones, and ones of the sort that we encounter in a deterministic statistical mechanics. All of them are nothing whatever over and above totalities of concrete particular facts. They differ only in the particular sorts of compact summaries that they happen – or happen not – to accommodate.

---

[8] The strategy described in footnote 2 – the strategy (that is) of abstaining from the assignment of any particular probability-distribution over those of the possible microconditions of the world which are compatible with its initial macrocondition, has sometimes been presented as a way around the problem, as a way of avoiding the problem, of making clear metaphysical sense of assigning probability-distributions to the initial conditions of the world. But that seems all wrong – for two completely independent reasons. First, the strategy in question makes what looks to me to be ineliminable use of sets of probability-distributions over the possible initial microconditions of the world – and if those distributions themselves can't be made sense of, then (I take it) sets of them can't be made sense of either. Second, the problem of making clear metaphysical sense of the assignment of probability-distributions to the initial microcondition of the world isn't the sort of thing that needs getting around – since (as we have just now been discussing) it can be solved!

2.6 Dynamical Chances

Quantum Mechanics has fundamental chances in it.

And it seems at least worth inquiring whether or not those chances can do us any good. It seems worth inquiring (for example) whether or not those chances are up to the business of guaranteeing that we can safely neglect the possibility of a rock, traveling at constant velocity, through an otherwise empty infinite space, spontaneously disassembling itself into statuettes of the British royal family. And the answer turns out to depend, interestingly, sensitively, on which particular one of the available ways of making sense of Quantum Mechanics as a universal theory, on which particular one of the available ways (that is) of solving the quantum-mechanical measurement problem, turns out to be right.

The sorts of chances that come up in orthodox pictures of the foundations of Quantum Mechanics – the pictures (that is) that have come down to us from the likes of Bohr and von Neumann and Wigner – turn out not to be up to the job. On pictures like those, the chanciness that is so famously characteristic of the behaviors of quantum-mechanical systems enters into the world exclusively in connection with the act of measurement. Everything whatever else – according to these pictures – is fully and perfectly deterministic. And there are almost certainly exact microscopic quantum-mechanical wave-functions of the world which are compatible with there being a rock, traveling at constant velocity, through an otherwise empty infinite space, and which are sitting on deterministic Quantum-Mechanical trajectories along which, a bit later on, if no 'acts of measurement' take place in the interim, that rock spontaneously disassembles itself into statuettes of the British royal family. And it happens to be the case, it happens to be an empirical fact, that the overwhelming tendency of rocks like that not to spontaneously disassemble themselves into statuettes of the British royal family has nothing whatsoever to do with whether or not, at the time in question, they are in the process of being measured!

And the same thing goes (for slightly different reasons) for the chances that come up in more precisely formulable and recognizably scientific theories of the collapse of the wave-function like the one due to Penrose. On Penrose's theory, quantum-mechanical chanciness enters into the evolution of the world not on occasions of 'measurement', but (rather) on occasions when certain particular wave-functions of the world – wave-functions corresponding to superpositions of macroscopically different states of the gravitational field – obtain. But the worry here is that there may be exact microscopic quantum-mechanical wave-functions of the world which are compatible with there being a rock, traveling at constant velocity, through an otherwise empty infinite space, and which are sitting on deterministic Quantum-Mechanical trajectories which scrupulously avoid all of the special collapse-inducing macroscopic superpositions mentioned above, and along which, a bit later on, that rock spontaneously disassembles itself into statuettes of the British royal family.

And the same thing goes (for slightly more different reasons) for the chances that come up in Bohm's theory. The only things that turn out to be chancy, on Bohm's theory, are the initial positions of the particles. The only sort of fundamental chance there is in Bohm's theory is (more particularly) the chance that the initial spatial configuration of all of the particles in the world was such-and-such given that the initial quantum-mechanical wave-function of those particles was so-and-so. And it happens – on Bohm's theory – that those parts of the fundamental physical laws that govern the evolution of the wave-function in time, and those parts of the fundamental physical laws that stipulate precisely how the evolving wave-function drags the particles around, are completely deterministic. And it turns out that there are possible exact wave-functions of the world which are compatible with there initially being a rock, traveling at constant velocity, through an otherwise empty infinite space, which (if those laws are right) will determine, all by themselves, that the probability of that rock's spontaneously disassembling itself into statuettes of the British royal family is overwhelmingly, impossibly, high.

And the long and the short of it is that the same thing goes (for all sorts of different reasons) for the chances that come up in Modal theories, and in the many-worlds interpretation, and in the Ithaca interpretation, and in the transactional picture, and in the relational picture, and in a host of other pictures too.

On every one of those theories, the business of guaranteeing that we can safely neglect the possibility of a rock, traveling at constant velocity, through an otherwise empty infinite space, doing something silly, turns out to require the introduction of another species of chance into the fundamental laws of nature – something over and above and altogether unrelated to the Quantum-Mechanical chances, something (more particularly) along the lines of the non-dynamical un-quantum-mechanical probability-distributions over initial microscopic conditions of the world that we have been discussing throughout the earlier sections of this chapter.

And this seems (I don't know) odd, cluttered, wasteful, sloppy, redundant, perverse.

And there is (perhaps) a way to do better. There is a simple and beautiful and promising theory of the collapse of the quantum-mechanical wave-function due to Ghirardi and Rimini and Weber that puts the quantum-mechanical chanciness in differently.

On the GRW theory – as opposed to (say) Bohm's theory, quantum-mechanical chanciness is dynamical. And on the GRW theory – as opposed to any theory whatever without a collapse of the wave-function in it – quantum-mechanical chanciness turns out to be a chanciness in the time-evolution of the universal wave-function itself. And on the GRW theory – as opposed to theories of the collapse like the one due to Penrose – the intrusion of quantum-mechanical chanciness into the evolution of the wave-function has no trigger; the probability of a collapse per unit time (that is) is fixed, once and for all, by a fundamental constant of nature; the probability of a collapse over the course any particular time-interval (to put it one more way) has nothing whatsoever to do with the physical situation of the world over the course of that interval.

And this is precisely what we want. On the GRW theory – as opposed to any of the other theories mentioned above, or any of the other proposed solutions to the measurement problem of which I am aware – quantum-mechanical chanciness is the sort of thing that there can be no outwitting, and no avoiding, and no shutting off. It insinuates itself everywhere. It intrudes on everything. It seems fit (at last) for all of the jobs we have heretofore needed to assign to probability distributions over initial conditions. If the fundamental dynamics of the world has this sort of chanciness in it, then there will be no microconditions whatsoever – not merely very few, not merely a set of measure zero, but not so much as a single one – which make it likely that a rock, traveling at constant velocity, through an otherwise empty infinite space, will spontaneously disassemble itself into statuettes of the British royal family.[9] And the same thing presumably goes for violations of the second law of thermodynamics, and for violations of the law of the survival of the fittest, and for violations of the law of supply and demand.

And so if something along the lines of the GRW theory should actually turn out to be true, science will apparently be in a position to get along without any probability-distribution whatsoever over possible initial microcinditions.[10] If something along the lines of the GRW theory should actually turn out to be true, then it might imaginably turn out that there is at bottom only a single species of chance in nature. It might imaginably turn out (that is) that all of the robust lawlike statistical regularities there are in the world are at bottom nothing more or less than the probabilities of certain particular GRW collapses hitting certain particular subatomic particles.[11]

Whether or not it does turn out to be true – of course – is a matter for empirical investigation.

---

[9] For details, arguments, clarifications, and any other cognitive requirements to which this sentence may have given birth – see chapter seven of my *Time and Chance*.

[10] It will still be necessary (mind you) to include among the fundamental laws of the world a stipulation to the effect that the world started out in some particular low-entropy macrocondition – but (in the event that something along the lines of GRW should turn out to be true) nothing further, nothing chancy, nothing (that is) along the lines of a probability-distribution over those of the possible microconditions of the world which are compatible with that macrocondition, will be required.

These considerations are spelled out in a great deal more detail in chapter seven of my *Time and Chance*.

[11] The theory we are envisioning here will of course assign no probabilities whatever to possible initial microconditions of the world, and it will consequently assign no perfectly definite probabilities to any of the world's possible conditions – microscopic or otherwise – at any time in its history. What it's going to do – instead – is to assign a perfectly definite probability to every proposition about the physical history of the world given that the initial microcondition of the world was A, and another perfectly definite probability to every proposition about the physical history of the world given that the initial microcondition of the world was B, and so on. But note that the probability that a theory like this is going to assign to any proposition P given that the initial microcondition of the world was A is plausibly going to be very very very very close to the probability that it assigns to P given that the initial microcondition of the world was B – so long as both A and B are compatible with the world's initial macrocondition, and so long as P refers to a time more than (I don't know) a few milliseconds into the world's history.

Chapter 3
Typicality and the Role of the Lebesgue Measure in Statistical Mechanics

Itamar Pitowsky

Abstract Consider a finite collection of marbles. The statement "half the marbles are white" is about counting and not about the probability of drawing a white marble from the collection. The question is whether non-probabilistic counting notions such as *half*, or *vast majority* can make sense, and preserve their meaning when extended to the realm of the continuum. In this paper we argue that the Lebesgue measure provides the proper non-probabilistic extension, which is in a sense uniquely forced, and is as natural as the extension of the concept of cardinal number to infinite sets by Cantor. To accomplish this a different way of constructing the Lebesgue measure is applied.

One important example of a non-probabilistic counting concept is typicality, introduced into statistical physics to explain the approach to equilibrium. A typical property is shared by a vast majority of cases. Typicality is not probabilistic, at least in the sense that it is robust and not dependent on any precise assumptions about the probability distribution. A few dynamical assumptions together with the extended counting concepts do explain the approach to equilibrium. The explanation though is a weak one, and in itself allows for no specific predictions about the behavior of a system within a reasonably bounded time interval.

It is also argued that typicality is too weak a concept and one should stick with the fully fledged Lebesgue measure. We show that typicality is not a logically closed concept. For example, knowing that two ideally infinite data sequences are typical does not guarantee that they make a *typical pair* of sequences whose correlation is well defined. Thus, to explain basic statistical regularities we need an independent concept of typical pair, which cannot be defined without going back to a construction of the Lebesgue measure on the set of pairs. To prevent this and other problems we should hold on to the Lebesgue measure itself as the basic construction.

This chapter was written by Itamar Pitowsky for this volume shortly before his death. As we did not have Itamar's LaTex file, it had to be retyped, and proofread by us. Remaining typos are, therefore, our fault. We thank William Demopoulos for reading the chapter and streamlining some of its formulations.

I. Pitowsky (✉)
Edelstein Center, The Hebrew University, Jerusalem, Israel

Y. Ben-Menahem and M. Hemmo (eds.), *Probability in Physics*, The Frontiers Collection,   41
DOI 10.1007/978-3-642-21329-8_3, © Springer-Verlag Berlin Heidelberg 2012

3.1 Introduction

Consider a finite but large collection of marbles. When one says that a vast majority of the marbles are white one usually means that all the marbles except possibly very few are white. And when one says that half the members are white, one makes a statement about counting, and not about the probability of drawing a white marble from the collection. The question is whether non-probabilistic notions such as *vast majority* or *half* can make sense, and preserve their meaning when extended to the realm of the continuum, especially when the elements of the collection are the possible initial conditions of a large physical system.

A major purpose of this paper is to argue that the task of expanding combinatorial counting concepts to the continuum can be accomplished. In the third section we shall see that counting concepts, which have a straight-forward meaning in the finite realm, also have an extension in the construction of the Lebesgue measure. Moreover, we shall argue that the extension is in a sense uniquely forced, as the famous extension of the concept of cardinal number to infinite sets by Cantor. To accomplish this task a different route to the construction of the Lebesgue measure is taken.

All this relates to the notion of *typicality* [1–9], introduced to statistical physics to explain the approach to equilibrium of thermodynamic systems. This concept has at least three different definitions [9], all entail that a typical property is shared by a vast majority of cases, or almost all cases. Typicality is not a probabilistic concept, this is maintained explicitly [3, 7, 8] or implied, at least in the sense that typicality is robust and "not dependent on any precise assumptions" about the probability distribution [2]. A recent example ([8], page 9):

> "When employing the method of appeal to typicality, one usually uses the language of probability theory. When we do so we do not mean to imply that any of the objects considered is random in reality. What we mean is that certain sets (of wave functions, of orthonormal bases, etc.) have certain sizes (e.g., close to 1) in terms of certain natural measures of size. That is, we describe the behavior that is typical of wave functions, orthonormal bases, etc. However, since the mathematics is equivalent to that of probability theory, it is convenient to adopt that language. For this reason, we do not mean, when using a normalized measure μ, to make an "assumption" of a priori probabilities," even if we use the word "probability".

However, none of the above papers explain in a precise manner why the Lebesgue measure is a "natural measure of size", or what is the connection between the continuum notions of "vast majority of cases" or "typical cases", and the equivalent finite notions which are based on simple counting.

A few modest dynamical assumptions combined with the combinatorial notions do explain the approach to equilibrium. I shall argue that the explanation is a weak one, and in itself allows for no specific predictions about the behavior of the system within a reasonably bounded time interval. Whenever predictions of that kind are made some additional knowledge about the initial condition or dynamics should be added. This is where probability enters the picture. We shall argue this for a finite system in the next section, and consider the infinite case in the 4th section.

Typicality, however, is too weak a concept and it is argued in the last section that one should stick with the full-fledged Lebesgue measure. Typicality does not quite cover measurable subsets whose measure is strictly between zero and one, which we might use in statistical mechanics. Even more seriously, the concept is not logically closed. For example, consider Galton's Board which is a central example in [3, 7]. Knowing that two ideally infinite sequences are typical does not guarantee that they make a *typical pair* of sequences whose correlation is well defined and equal to 0.25. Therefore, the concept of typical sequence cannot be used to explain basic long term statistical regularities. For this we need an independent concept of typical pair, which cannot be defined without going back to a construction of the Lebesgue measure on the set of pairs. Similar observations apply to triples, quadruples, and all k-tuples; in each case typicality cannot be defined just on the basis of the former notions.

3.2 Divine Comedy- the Movie

Consider the set of all possible square arrangements of $1,000 \times 1,000$ black and white pixels. There are 2^{10^6} such arrangements, we shall call each one a picture, and the set of all pictures is our phase space. Imagine that upon his arrival in Hell, a lesser sinner is seated in a movie theater (no air conditioning). The show consists of the following movie:

1. Pictures are projected on the screen at a constant pace of 25 frames a second.
2. The sequence is deterministic, the director has arranged that each picture gives rise to a unique successor. We can assume that the dynamical rule is internal, so that each picture, apart from the first, depends uniquely on the pixel arrangement of its predecessors.
3. The movie goes through all 2^{10^6} pictures, and then starts again. So the show is periodic, but the period is extremely long, more than 10^{301020} years long (compared with the age of the universe which is less than 10^{11} years). The phase space contains all the pictures that were ever shot and will ever be shot, including photo copies of written texts and frames from movies, provided they are cast in the format of a thousand by a thousand black and white pixels. Despite this, the set of pictures that look remotely like regular photographs is very small compared with the totality of pictures. Worse, the set of pictures that contain a large patch of black (or white) pixels is very small. These are just combinatorial facts: the overwhelming majority of pictures look gray, approximately half black and half white, with the black and the white pixels well mixed. The number of pictures with a single color patch of size m decreases exponentially with m.

The conjunction of the three dynamical rules for the movie with this combinatorial observation explains why, in the long run, the movie is extremely boring and looks gray. It also explains why, in the long run, the frequency of the pictures that have more black than white pixels is (a little less than) 0.5.[1] We have to be clear about the meaning of "the long run" here. In the absence of any detail about the dynamics other than rules **1, 2, 3**, we cannot really say how long the long run is. It may be the case that the movie begins with a 50,000 year-long stretch of cinematic masterpieces. However, this cannot last much longer and the movie then settles into almost uniform gray for a vast length of time. Likewise, it is also possible that the director has chosen a dynamics that puts all the pictures with more black than white pixels at the end of the movie. In this case the long run may be very long indeed.

Another way of looking at the long run is to notice that, given the nature of the theater, different spectators arrive at different times. The first picture each new-comer encounters upon arrival can be taken as an "initial condition". So the answer to the question "how long will it take for the movie to settle into almost uniform gray?" depends on the initial condition. Similarly, the number of to frames it takes the time average of an "observable" (a function $f : \{0, 1\}^{10^6} \to \mathbf{R}$) to stabilize depends on the initial condition.

So far nothing has been said about probabilities, it is clear that the frequencies are just proportions in a finite set. The explanation for the frequencies is straight-forward and involves no probabilities. However, the questions that can be answered are limited. On the basis of the three dynamical rules and counting alone we can make no specific forecasts. In the best case we obtain a simple theory which is consistent with what we see.

Probabilistic considerations enter when definite predictions are made, beyond the long run explanations. Given the deterministic nature of the system, probability in this context is invariably epistemic. Consider the claim that the picture to be projected two minutes from now will have more black than white pixels. We can imagine two extreme reactions: A savant spectator (Laplace's demon) may have figured out what the dynamics is, and knowing the present condition, may calculate the pattern of pixels two minutes from now. The probability he assigns his result is one, or very near one allowing for a possible mistake. At the other extreme, where most spectators are, no information beyond the dynamical rules is available. In this case a natural choice of a prior is the uniform distribution, that is, the counting measure represents the probability.[2] The probability assigned to the event is thus (slightly less) than 0.5. It is easy to invent stories where partial information is available, with the consequence that the probability can be anything between zero and one.

Now imagine that upon their arrival in Hell heavier sinners are made to watch a different show. They are seated in front of a large transparent insulated container

---

[1] Note that the number of white pixels in a picture may be considered a "macroscopic" observable, whose measurement requires no detailed knowledge of the pixel distribution. If we assume that white pixels emit light, and black pixels do not, we just measure the light emitted from a picture, and compare it with the all white picture.

[2] This is not a very smart prior, though. It assumes independence, and therefore blocks the possibility of learning from experience.

full of gas at a constant temperature and sulk at it. Nothing much happens of course, and the question is whether we can explain why this is the case on grounds that are similar to the movie story. Here a single picture is analogous to one microscopic state, and the movie as a whole to the continuous trajectory of the microstate in phase space. However, since there is a continuum of microstates it is not clear how to expand the finite concepts to the continuum. In particular, it is not clear what is the meaning of overwhelming majority of microstates, or typical states, or half the microstates, unlike the finite case where we just use the terms with their ordinary meaning. The translation of the dynamical rules **1, 2, 3** to the motion of particles is not obvious either.

Boltzmann had a long and complicated struggle with these issues [10]. In some writings he was clearly attempting to associate combinatorial intuition, finite in origin, with continuous classical dynamics. However, he lacked the appropriate mathematics which had not yet been invented, or at any rate, was not yet widely known among physicists. By the time it became available combinatorial and probabilistic consideration were hopelessly mixed up. The idea of typicality goes a long way to disentangle the two issues.

Putting the dynamical questions aside for a while, the next section is devoted to the extension of the relevant combinatorial concepts to the domain of the continuum. It is therefore a chapter in the philosophy of mathematics.

3.3   The Road Less Travelled to Lebesgue Measure

Our purpose is to extend concepts such as *majority of cases, or one quarter of the cases*, from the finite realm, where their meaning is obvious, to the domain of the continuum. Extensions of mathematical concepts from one realm to a larger domain that contains it are not necessarily unique, and may result in a large variety of quite different creatures [11]. However, in some cases there are very compelling arguments why one particular possible extension is the correct choice, the most important example being Cantor's definition of the cardinality of infinite sets. I shall argue below that the Lebesgue measure plays a similar role in the extension of combinatorial counting concepts.

Usually the Lebesgue measure is introduced as part of the modern theory of integration, the extension of the definition of the integral beyond the limitations of Riemann's construction. This is consistent with the historical development, and answers the requirements of the mathematics curriculum. Here we take another approach altogether. First note that without loss of generality our efforts can concentrate on the interval [0, 1] with the Lebesgue measure on it. The reason is that every (normalized) Lebesgue space is isomorphic to this space, meaning that there is a measure preserving isomorphism between the two spaces.[3] Second, note

---

[3] This general result is due to Caratheodory, see [12, page 16].

that the interval [0, 1] can be replaced with the set of all infinite sequences of zeros and ones $\{0, 1\}^\omega$, when we identify each infinite zero-one sequence $\mathbf{a} = (a_1, a_2, a_3, \ldots)$ with a binary development of a number in [0, 1], that is, $\mathbf{a} \to \sum_{j=1}^\infty a_j 2^{-j}$. This map is not 1–1, but fails to be 1–1 only on the countable set of rational numbers whose denominator is a power of 2 (dynamic numbers), hence a set of measure zero. In sum, our construction of the Lebesgue measure is developed without loss of generality as an extension from sets of finite 0–1 sequences to subsets of $\{0, 1\}^\omega$.

We start with the finite case, where the movie of the previous section is the example we want to generalize. We can represent the movie as the set of sequences of zeros and ones of length one million $\{0, 1\}^{10^6}$, where each picture is an element of that set. Consider more generally the set $\{0, 1\}^n$ where n is any natural number, and $A \subseteq \{0, 1\}^n$. Then the *measure* μ_n of A is defined to be

$$\mu_n(A) = 2^{-n}|A|, \tag{3.1}$$

where $|A|$ is the number of elements of A. So that, for example, if $\mu_n(A) = 0.5$ we can say that half the sequences of $\{0, 1\}^n$ belong to A. The size measure has an important invariance property: If $m > n$ then $\{0, 1\}^m = \{0, 1\}^n \times \{0, 1\}^{m-n}$, we can embed every $A \subseteq \{0, 1\}^n$ in $\{0, 1\}^m$ by the map

$$A \subseteq \{0, 1\}^n \to A' = A \times \{0, 1\}^{m-n} \subseteq \{0, 1\}^m, \tag{3.2}$$

so that $\mu_n(A) = \mu_m(A')$.

With these notations we can formulate the claim made in the movie story, that the overwhelming majority of pictures are approximately half black and half white. Given a sequence $\mathbf{a} = (a_1, a_2, \ldots, a_n) \in \{0, 1\}^n$, let $S_n(\mathbf{a}) = \sum_{j=1}^n a_j$ be the sum of the elements of $\mathbf{a}$, and thus the average number of ones in the sequence is $n^{-1}S_n(\mathbf{a}) = n^{-1}\sum_{j=1}^n a_j$. Therefore, the claim is that for a sufficiently large n the vast majority of sequences satisfy $n^{-1}S_n(\mathbf{a}) \sim 0.5$. Indeed, the weak law of large numbers (LLN) states: *For every $\varepsilon > 0$*

$$\mu_n\left\{\mathbf{a} \in \{0, 1\}^n; \frac{1}{2} - \varepsilon \leq n^{-1}S_n(\mathbf{a}) \leq \frac{1}{2} + \varepsilon\right\} > 1 - \frac{1}{4n^2\varepsilon^4}, \tag{3.3}$$

so that the left hand side tends to 1 as $n \to \infty$.[4]

Students usually encounter this or similar finite versions of LLN in a course on probability and statistics. In rare cases the teachers make it a point to distinguish the

---

[4] The rate of convergence on the right hand side of (3.3) is better than the historical one derived by Bernoulli. We need the stronger result (essentially due to Borel) for later purposes. See [13], page 40.

two meanings of LLN. First there is the familiar one of probability theory concerning, for example, Bernoulli trials with probabilities p and $q = 1 - p$ for the two outcomes. In case the distribution is uniform, $p = q = 0.5$, a formula like (3.3) obtains. The second meaning, the one used here, concerns counting the number of elements in the set between the braces in (3.3), or equivalently, calculating the proportion of such elements in the set of all 0–1 sequences of length n. This combinatorial meaning is much simpler, and is qualitatively apparent by looking at Pascal's Triangle.

The difference between the two meanings of LLN can be better understood when we consider the conditions for their applications. In the probabilistic case we have to describe by which process the digits in the sequence are chosen, for example, by coin tosses with probability p for "heads". Subsequently, we have to justify the assumption that coin flips are independent, and finally to explain that LLN is saying that the probability that the average of "heads" lies close to p is large. By contrast, in the application of the combinatorial theorem there is nothing to explain, the process of counting requires no further analysis. As noted, the distinction between the two meanings of the weak LLN is rarely taught in the class-room or mentioned in text books. Moreover, this distinction is never mentioned at all when it comes to the strong LLN, despite the fact that the strong LLN is a *consequence* of inequality (3.3) and σ-additivity (see below).

Moving to the infinite case, consider the set of all infinite 0–1 sequences $\{0, 1\}^{\omega}$. Given a finite set $A \subseteq \{0, 1\}^n$ we can embed it as a subset of $\{0, 1\}^{\omega}$ using the same method in (3.2) namely

$$A \subseteq \{0, 1\}^n \to F = A \times \{0, 1\} \times \{0, 1\} \times \dots \subseteq \{0, 1\}^{\omega}. \tag{3.4}$$

Call every subset of $\{0, 1\}^{\omega}$ that has the form of F in (3.4) finite. Summarizing, $F \subseteq \{0, 1\}^{\omega}$ is finite if it has the form $F = A \times \{0, 1\} \times \{0, 1\} \times \dots$, with $A \subseteq \{0, 1\}^n$ for some natural number n. Of course F has infinitely many elements, but this does not cause confusion as long as the context is clear. Now, define the measure μ of F to be

$$\mu(F) = \mu_n(A) = 2^{-n}|A|. \tag{3.5}$$

As long as only finite subsets of $\{0, 1\}^{\omega}$ are considered no real expansion of the concept of measure is achieved. Note that the family of all finite subsets is a Boolean algebra, it is closed under complementation and (finite) unions and intersections. The minimal expansion to infinity is achieved by considering countable infinite unions and intersections. Denote the Boolean algebra of finite subsets of $\{0, 1\}^n$ by $\mathcal{F}$. In other words, $F \in \mathcal{F}$ if F has the form $F = A \times \{0, 1\} \times \{0, 1\} \times \dots$ with $A \subseteq \{0, 1\}^n$ for some natural number n. The σ-algebra $\mathcal{B}$ of Borel subsets of $\{0, 1\}^{\omega}$ is defined to be the minimal σ-algebra that contains $\mathcal{F}$. This means that $\mathcal{B}$ is the minimal family of subsets of $\{0, 1\}^{\omega}$ which contains $\mathcal{F}$, and is closed under complementation, and under countable unions and countable intersections of its own elements, to generate $\mathcal{B}$, one takes countable unions of finite sets, then countable intersections of the resulting sets, and so on.[5]

The measure μ is extended from $\mathcal{F}$ to $\mathcal{B}$ using *the σ-additivity rule*: *If* $E_1, E_2, \ldots, E_j, \ldots \in \mathcal{B}$ *is a sequence subsets, disjoint in pairs, i.e.,* $E_i \cap E_j = \phi$ *for* $i \neq j$, *then*

$$\mu \left(\bigcup_{j=1}^{\infty} E_j \right) = \sum_{j=1}^{\infty} \mu \left(E_j \right). \tag{3.6}$$

Usually, one additional "small" step is taken to complete the construction: Given any Borel set $B \in \mathcal{B}$ such that $\mu(B) = 0$ add every such subset of B to the Borel algebra $\mathcal{B}$. The larger σ-algebra which is generated after this addition is the Lebesgue algebra $\mathcal{L}$. The measure μ, which is extended to $\mathcal{L}$ in an obvious way, is the Lebesgue measure.[6]

Why is μ the correct expansion to infinity of the size measure in the finite case? Obviously, the crucial steps in the expansion are the construction of the σ-algebra and the application of σ-additivity. As a consequence new theorems can be formulated and proved, for example, the strong law of large numbers:

$$\mu \left\{ \mathbf{a} \in \{0, 1\}^{\omega}; \lim_{n \to \infty} \left(n^{-1} S_n(\mathbf{a}) \right) = \frac{1}{2} \right\} = 1, \tag{3.7}$$

which says that the set defined within the braces in (3.7) is an element of $\mathcal{L}$ (in fact even $\mathcal{B}$) and its Lebesgue measure is 1; hence in almost every infinite 0–1 sequence half the elements are zero and half one. This is a direct extension of the counting intuition expressed by the weak LLN (3.3). Indeed, the strong LLN (3.7) is a logical consequence of the weak law (3.3) in conjunction with σ-additivity. This means that the finite (3.3) and infinite (3.7) express the same idea, and σ-additivity is a way to translate the cumbersome (3.3) to the compact (3.7). Borel, the author of the strong LLN, actually preferred (3.3), in line with his intuitionistic views. He thought that (3.7) added nothing except for the illusion that infinite sets of infinite sequences made sense.

Similar observations can be made with respect to other limit laws that have familiar infinite formulations in $\mathcal{L}$, but also parallel formulations in $\mathcal{F}$ which together with σ-additivity imply the infinite laws. An important example is the Law of Iterated Logarithm (LIL), a stronger and more subtle law than (3.7), which implies, among other things, that for almost every $\mathbf{a} \in \{0, 1\}^{\omega}$ the sign of n^{-1}

---

[5] The construction of $\mathcal{B}$ is achieved by transfinite induction over the two operations, countable union and then countable intersection, all the way to the first uncountable ordinal.

[6] Further extensions of the Lebesgue measure are possible. The validity of the strong version of the axiom of choice entails the existence of non-measurable sets, that is, $C \subset \{0, 1\}^{\omega}$ such that $C \notin \mathcal{L}$. We can add some of those to $\mathcal{L}$ and extend the measure to them [14]. With this the measure is no longer regular (see below). Moreover, there are models of set theory, with weaker principles of choice, in which every subset of $\{0, 1\}^{\omega}$ is Lebesgue measurable [15].

S_n (a) $-$ 0.5 oscillates infinitely often as $n \to \infty$. Sometimes the infinite law is more easily discovered than its finite parallel which may be even hard to formulate. In any case one can prove the regularity of μ, that every set in $\mathcal{L}$ can be approximated by a set in $\mathcal{F}$ to an arbitrary degree.

Theorem 1. *Let $E \in \mathcal{L}$ be any Lebesgue measurable set and let $\varepsilon > 0$; then there is $F_\varepsilon \in \mathcal{F}$ such that $\mu[(E \setminus F_\varepsilon) \cup (F_\varepsilon \setminus E)] < \varepsilon$.*

The proof is in Appendix 1 (note that the theorem becomes trivial when $\mu(E) = 0$ or $\mu(E) = 1$). Therefore, the expansion of the measure from the finite to the infinite domain conserves the meaning of the counting terms. We can, in principle, replace any set in $E \in \mathcal{L}$ by a finite set $F_\varepsilon \in \mathcal{F}$ which is arbitrarily close to E. If direct counting shows that F_ε comprises 0.75 of the cases, then so does E up to a small error.[7] Moreover, the Lebesgue algebra $\mathcal{L}$ is the maximal extension of $\mathcal{F}$ for which theorem (1) is valid (see footnote 6). This seems to me to be a compelling argument for why $\mathcal{L}$ is the correct extension of $\mathcal{F}$, and why the Lebesgue measure μ on $\mathcal{L}$ is the correct extension of the combinatorial counting measure to infinity. It is also a compelling argument for why the notions of σ-algebra and σ-additivity are the appropriate tools in extending the combinatorial measure to infinity.

Let us come back to the issue of the Lebesgue measure and probability. As noted before only in rare cases do teachers make a point of distinguishing the meanings of weak LLN as a combinatorial and as a probabilistic statement. As for the strong LLN and other similar theorems, teachers and textbooks alike never make the distinction, and invariably interpret the Lebesgue measure in this context as probabilistic. There is no intrinsic reason for this, the application of σ-additivity has no probabilistic qualities. The reason is more sociological: For the pure mathematician there is no difference between the uniform probability distribution and the combinatorial measure, since their formal properties are one and the same. At a certain point in time mathematicians started to use the probabilistic language exclusively, and fellow scientists, physicists in particular, followed in their footsteps. But there is all the difference in the world between the mathematicians who are using the measure probabilistically, as a mere formality, and the physicists who are committing themselves to an application of probability as part of a theory of reality.

This has not always been the case, even for mathematicians! For example, in the struggle to obtain the correct estimation of frequency oscillations (LIL- the law of iterated logarithm), bounds were suggested by Hardy and Littlewood in 1914. They viewed the problem as number theoretic, concerning the binary development of real numbers between zero and one, and related to Diophantine approximation. Even in his final formulation of LIL from 1923 (for the uniform case) Khinchine was using the number-theoretic language, and only a year later switched to probability [16].

Extending the notion of vast majority from the finite to the infinite realm results in typical cases. None of these concepts is intrinsically probabilistic. I believe that

---

[7] Actually $F_\varepsilon = A \times \{0,1\} \times \{0,1\} \times \ldots \subseteq \{0,1\}^\omega$ with $A \subseteq \{0,1\}^n$ for some integer n, and we are counting the elements of A.

this is an important step towards removing the host of problems associated with probability distributions over initial conditions. As an example consider a recent application that does not even involve dynamics. Let a quantum system ("the universe") be associated with a finite dimensional Hilbert space $\mathcal{H}$, with a large dimension D. Now, consider a small subsystem of dimension $d \ll D$ that corresponds to a subspace $\mathcal{H}_1$. We can write $\mathcal{H} = \mathcal{H}_1 \otimes \mathcal{H}_2$, where $\mathcal{H}_2$ is the Hilbert space of the environment with a large dimension $d^{-1}D$. The set of pure states in $\mathcal{H}$ is the unit sphere of $\mathcal{H}$; let μ be the normalized Lebesuge measure on it. Each pure state induces a mixed relative state on the small subsystem. The following recent result was proved independently in [17, 18]: *Almost all pure states in $\mathcal{H}$ induce on $\mathcal{H}_1$ a relative state which is very close, in the trace norm, to the maximally mixed state on $\mathcal{H}_1$, that is, $d^{-1}I_d$ with I_d the unit operator on $\mathcal{H}_1$.*

One possible reading is that with probability one the state of the large system induces the near uniform state on the subsystem.[8] A natural question is, "What does probability mean in this context?" Assume the large system is a model of the universe; it began in one pure state, and after time t it is again in one particular pure state. This state has been deterministically developed from the initial condition by the unitary time transformation. So the question is, "What do we mean by saying that the initial condition of the universe was picked from a uniform rather than some other probability distribution?" The only sensible answer is that this statement represents the epistemic probability of an agent who has no knowledge at all about the initial condition. However, this agent cannot be a physicist, who usually knows something about the present and earlier (macroscopic) states of the universe.

In the typicality approach, by contrast, the result simply means that the vast majority of pure states of the big system have the property in question, a combinatorial claim. This claim gives rise to a weak, but still informative conditional statement: If the universe began from a typical state then equilibrium should be a widespread phenomenon. A simple assumption (typicality) explains a large set of observations.

3.4 Dynamics

Our aim is to discuss the dynamical conditions that are the infinite parallels of the constraints **1**, **2**, **3** we have imposed on the movie. To fix notations let Γ denote the energy hypersurface of the closed system under consideration. If $x_0 \in \Gamma$ is a point, it can be considered as a possible initial condition, let $x_0(t)$ denote the trajectory starting from this point in Γ. Alternatively, if t is fixed $x_0(t)$ is the point to which x_0 travels after time t. The Lebesgue measure on Γ will be denoted by μ, and we assume it is normalized (we ignore the difficulties arising from a non compact Γ, which are settled by known techniques). The σ-algebra of the Lebesgue measurable

---

[8] See [17]. In a later important dynamical extension of this result the authors adopt the typicality point of view [19].

sets will again be denoted by $\mathcal{L}$. If $E \in \mathcal{L}$, define E_t to be the time translation of E, that is, $E_t = \{x_0(t); x_0 \in E,\}$ for $0 \leq t < \infty$.

The second assumption **2** corresponds to the determinism inherent in classical mechanics and already reflected in the notation. The classical dynamical rule closest to assumption **1** is the conservation of energy. In the case of an ideal gas the velocities of the individual particles are varied but the average (square of the) particle's speeds remains constant (by analogy, the pace of the movie is constant). Energy conservation, that is, the Hamiltonian character of the system, also guarantees that the dynamics is measure preserving: $\mu(E) = \mu(E_t)$. In the movie case measure preservation is trivial.

Condition **3** corresponds to ergodicity. Historically, a major difficulty was associated with the formulation of this condition, Boltzmann mistakenly thought that a path can fill the whole energy hypersurface in phase space, so that every state will be visited. However, this requirement contradicts basic topological facts.[9] It took a long struggle until the modern version of the ergodic condition was formulated, and the ergodic theorems subsequently proved [16]. Instead of referring to individual points visited by the path, the condition takes (measurable) set of points, and puts a constraint on the way the set fills up the space. Let E be a measurable subset of the energy hypersurface in phase space. Then E is *invariant* if for some $t > 0$ we have $E_t \subseteq E$. The system is ergodic if all invariant sets have measure zero or one.

In the finite case the dynamical rules provide an explanation why, in the long run, the movie is extremely boring and looks almost always gray. They also explain why, in the long run, the frequency of the pictures that have more black than white pixels is (a little less than) 0.5. This corresponds, in the infinite case, to the identity of the long run averages and the phase space averages of thermodynamic observables, a highly non-trivial fact which is the content of the ergodic theorems. In both cases the long run may be very long, in the infinite case there is no a priori bound on its length. This is the explanation why the system is at maximal entropy most of the time, or why about half the time the pressure in the left half of the container is less (even very slightly so) than in the right half.

However, there seems to be a difference between the finite and infinite case here. Given a thermodynamic observable, only typical initial conditions result in the identity of its phase space and long time averages. This may seem like a major difference from the finite case in which all initial conditions behave properly. However, a small amendment to the movie story can lead us to the conclusion that the movie satisfies condition **3** only for a vast majority of initial conditions, not all. To see this imagine that the set of pictures is divided into two disjoint subsets, one very small containing 20 pictures and the other containing the rest. When a movie begins with a picture in the small subset it goes through a small loop,

---

[9] Dimension is a topological invariant, as proved by Brouwer in 1911. Partial results concerning the non-existence of a homeomorphism between the real line and higher dimensional real spaces existed in Boltzmann's time. For example, Lüroth in 1878.

visiting all 20 pictures and starts again. Similarly for an initial condition in the second set, but then it covers all the pictures except 20. In both cases determinism is satisfied. We can say that for the vast majority of initial conditions the time and space averages of "thermodynamic observables", functions $f:\{0,1\}^{10^6} \to \mathbf{R}$, are (very nearly) the same.

It must be emphasized that the sense of explanation obtained in this manner is significant but limited. As a result of the unbounded nature of the long run, and in the absence of more information, there is no way we can combine the dynamical rules with the combinatorial facts to yield a definite prediction, for example, about what will take place 2 days from now. The kind of explanation we do have is weaker, and has the conditional form: "If the initial condition is typical, then... " The assumption of typicality explains why the (calculated) space averages of observables are the same as the measured long time averages (which stabilize quickly in practice). Thus, assuming we are on a typical trajectory, one of a vast majority, explains much of what we actually see.

So far the explanation relies on the dynamical rules and the observations derived from the combinatorial nature of the Lebesgue measure. One may object to the latter point on the ground that the measure here does not seem to be "the same" as the measure on the set of infinite 0-1 sequences, being a Lebesgue measure on a Euclidean manifold of high dimension. This objection can be answered on two levels, the first is purely formal. As indicated before, all Lebesgue spaces which are defined on compact subsets of real or complex Euclidean spaces are isomorphic (after normalization of the measure) to the interval [0,1] with the Lebesgue measure on it. Therefore, they are also isomorphic to the space of all 0-1 sequences, and every measurable set $E \subseteq \Gamma$ corresponds to a measurable set $\hat{E} \subseteq \{0,1\}^\omega$, with the same measure, and $\hat{E}$ can be approximated by a finite set $F \in \mathcal{F}$ as indicated in theorem 1.

On a deeper level there often exists a connection between ergodic systems and the sequence space when we apply a mapping of the ergodic system, including its dynamics, to the set two sided infinite 0-1 sequences [12, page 274]. This space, denoted by $\{0,1\}^z$, is equipped with the (uniform) Lebesgue measure, and its elements can be written as $\mathbf{a} = (\ldots, a_{-2}, a_{-1}, a_0, a_1, a_2, \ldots)$, with $a_i \in \{0,1\}$, $i = 0, \pm1, \pm2, \ldots$. To perform the mapping between the thermodynamic system and this space one has to replace the continuous time variable by a discrete parameter. It turns out that many important ergodic systems, including the few physically realistic systems for which ergodicity was actually proved, are isomorphic as dynamical systems to the Bernoulli shift on $\{0,1\}^{\mathbf{Z}}$, defined by[10] $(S\mathbf{a})_i = a_{i-1}$. These results were proved in a sequence of papers, mainly by Orenstein and his collaborators [12]. Ergodic systems with this property include the standard model of the ideal gas (hard-sphere molecules in a rectangular box),

---

[10] The reason why the double sided sequence space is used is to make the Bernoulli shift well defined and invertible.

Brownian motion in a rectangular region with reflecting boundary, geodesic flows in hyperbolic and many other spaces.

The connection with the combinatorial character of the measure is even more transparent in this case. For example, the Ergodic theorem for $\{0, 1\}^{\mathbb{Z}}$ with the shift entails the strong LLN. To see this let $A \subseteq \{0, 1\}^{\mathbb{Z}}$ be a measurable set then the ergodic theorem for the Bernoulli shift states,

$$\mu\left\{\mathbf{a} \in \{0, 1\}^{\mathbb{Z}}; \lim_{n \to \infty} \frac{1}{n} \sum_{i=1}^{n} \chi_A(S^i(\mathbf{a})) = \mu(A)\right\} = 1. \tag{3.8}$$

Here χ_A is the indicator function of A, so that $\chi_A(\mathbf{a}) = 1$ if $a \in A$, and $\chi_A(\mathbf{a}) = 0$ otherwise. Now take $A = \{a \in \{0, 1\}^{\mathbb{Z}}; a_0 = 1\}$, then $\mu(A) = 0.5$ and $\sum_{i=1}^{n} \chi_A(S^i(\mathbf{a})) = \sum_{i=1}^{n} a_i$, and we obtain the strong LLN as a special case.

Probabilistic considerations enter when definite predictions are made, beyond the weaker long term explanations that are possible on the basis of ergodicity. Given the deterministic nature of the system we shall take probability in this context to be epistemic, although this may be disputed [7, 20]. The assignments of probabilities are based on knowledge about the system that may go beyond the simple rules we have considered. Some-times, in the absence of any knowledge about the initial condition and the dynamics beyond ergodicity, the uniform Lebesgue measure can serve as the degree of knowledge regarding the system. Often more knowledge is available, which can be theoretical, but frequently concerns the initial condition and is based on experience. For example, we may know something about the rate with which the dynamics is moving to mix the molecules. Usually the rate cannot be derived directly on the basis of the interactions between the particles. Higher theories such as fluid dynamics may be involved, together with experimental data. If a gas is prepared in a container with a divider, and the pressure on the left hand side much higher than the pressure on the right, then upon removing the divider the pressures will equalize very swiftly. By contrast, when we drop ink into water we know that it will take much longer to mix uniformly with the medium. Therefore, if we where to bet whether the pressures on both sides will equalize 20 s from now, the answer will be yes with probability close to 1, but the probability that the ink will be well mixed within 20 s is near zero. This does not follow from ergodicity which just explains why the system will eventually arrive at equilibrium and stay there most of the time.

We also know that in all recorded human history the reverse of these processes has never seen reported. Consequently, the probability assigned to a spontaneous large pressure differences occurring within the next week (or month, or year...) is zero or very nearly so. This observation too cannot be derived logically from the dynamical and combinatorial rules. Given ergodicity, almost all initial conditions will take the system arbitrarily near every possible state. How do we know that the creation of a spontaneous large pressure difference is not around the corner?

We do know from combinatorial considerations that non equilibrium states are very rare, but this condition is insufficient to derive the probabilistic conclusion, because we do not know what the trajectory is, and have no clue about the way rare states are distributed on it. The movie analog is a photograph of the Empire State Building appearing suddenly in the midst of gray pictures. This photograph must appear sometime, but in the absence of detailed knowledge of the dynamics one cannot tell when. However, after sitting 10^{10} years and watching gray pictures one may assign the sudden pop up of the Empire State Building in the next week a very small probability. This would not be the case after a long stretch of pictures of buildings. By analogy, we assign zero probability to the creation next week of a spontaneous large pressure differential because this has never happened, and not just because we know abstractly that this is an atypical event.[11]

3.5   Troubles with Typicality

The problem is that typicality is too restrictive a notion, and the reasons are twofold, physical and logical. Physically, there are good reasons to deal with measurable sets of intermediate size. For example, the set of micro states for which the pressure in the left half of the container is equal or less than the pressure in the right comprise 0.5 of all the states. Logically, we shall see that the concept of typicality lacks closure. For example, even after typical points have been "fixed" one cannot use this stipulation to define typical pairs of points, that is, *a pair of typical points is not necessarily a typical pair of points*. To define the latter, one has to go back to the Lebesgue measure on the set of pairs (which is defined in terms of the Lebesgue measure on the set of singletons) and redefine typicality for pairs.

As for the physical restriction, one important case is that of smooth classical Hamiltonian systems which are not ergodic, but only measure preserving. By Birkhoff's theorem convergence of the time average of a thermodynamic observable for typical initial conditions is guaranteed, but the result is not identical to the space average. In this case the phase space is partitioned into invariant sets of positive measure, such that the restriction of the dynamics to each element in the partition is ergodic (after a suitable renormalization). By KAM's theorem many Hamiltonian systems are not ergodic, although the partition is often composed of one large invariant set and other much smaller elements. (For such systems the notion of ε ergodicity has been introduced [22]). Even in this case one has to say something about sets of initial conditions with measure smaller than 1, which cannot even be formulated without the full Lebesgue measure.

The logical point is that exchanging the full Lebesgue measure for the weaker notion of typicality does not even accomplish the task of explaining the long run statistical regularities. In order to provide such an explanation one has to introduce an infinite sequence of logically independent concepts of typicality, none of which

---

[11] A similar point about the role of induction in statistical mechanics is made in [21].

are definable in terms of the former. Consider Galton's board, which serves as a central example in the papers by Dürr [3] and Maudlin [8]. The first notion introduced is that of a typical initial condition, which explains, e.g., the stability of relative frequencies of going left and going right. Next, we must introduce a new notion of typical pairs of initial conditions to explain the stability of the frequency of the correlated sequence obtained from two runs of the board, then we have to introduce a new notion of typical triples to explain the stability of triple correlated sequences obtained from three runs, and so on. Each one of these notions is logically independent of the former notions, that is, none of them can be defined on the basis of the previous concepts of typicality. In each case one has to reintroduce the fully fledged Lebesgue measure (respectively, on the interval of initial conditions, the Cartesian product of the interval by itself, the three–fold Cartesian product, and so on), and only then, in each case separately, throw away the ladder as it were, and introduce the new notion of typicality in the manner described by Maudlin for the singleton case.

One consequence of this state of affairs is that being typical is not an intrinsic property of a point even for a single dynamical system, but is a property induced by its relations to other points. Moving to the system comprising the whole universe (which after all has only one initial state) does not solve the problem. In this case it also arises in the context of the typicality of idealized sequences of empirical observations, the correlations or independence of two such sequences, and of triples, etc. Even if we observe only one (ideally infinite) typical sequence, the problem arises with respect to its subsequences and their relations.

To see this consider a pair of $\mathbf{a}, \mathbf{b} \in \{\mathbf{0}, \mathbf{1}\}^\omega$ and denote $\mathbf{a} \cdot \mathbf{b} = (a_1 b_1, a_2 b_2, a_2 b_2, \ldots)$. We know that typically $\mathbf{a} \cdot \mathbf{b}$ is a sequence whose averages satisfy $\frac{1}{n} \times \sum_{i=1}^n a_i b_i \rightarrow 0.25$. But does this fact follow if we assume that $\mathbf{a}$ and $\mathbf{b}$ are typical? The negative answer follows from

Theorem 2. *Let $A \subset \{0, 1\}^\omega$ be any measurable set with $\mu(A) > \frac{1}{2}$; then there are $\mathbf{a}, \mathbf{b} \in A$ such that $\mathbf{a} \cdot \mathbf{b}$ has a divergent sequence of averages.*

The proof is in Appendix 2. This means that no matter what the set of typical sequences is, there will always be pairs of typical sequences whose correlation is not even defined. One might object on the ground that the set of such bad pairs has measure zero, and the set of typical pairs has measure one. However, this refers to the measure *on the Lebesgue space of pairs*. The set of typical pairs does not have the form $A \times A$ with $A \in \mathcal{L}$, and $\mu(A) = 1$. By theorem 2 any set of the form $A \times A$ contained in the set of typical pairs has at most measure $\mu(A) \leq 0.5$. Therefore, to be able to speak about typical pairs one has to construct first the Lebesgue measure on the set of pairs $\{0, 1\}^\omega \times \{0, 1\}^\omega$, or alternatively $[0, 1] \times [0\ 1]$, and only then define typicality for pairs. One cannot do it by relying on the already established set of typical points. This observation can be extended to triple, quadruple correlations, and so forth. In the case of triples the equivalent theorem applies when $\mu(A) > \frac{1}{3}$, and so on, for k-tuples when $\mu(A) > \frac{1}{k}$ In all these cases the notion of typicality cannot be derived from the lower dimensional ones.

As noted this also means the being typical is not an intrinsic property of an initial condition, not even for a single fixed system, but depends on the relation between the point and other possible initial conditions. The way suggested here to avoid this difficulty is to use the fully fledged Lebesgue measure, in its combinatorial interpretation. In this case subsets of measure one are just special cases. I think all the advantages of the concept of typicality that were pointed out in the literature are preserved, but the difficulties are avoided.

Acknowledgement *I would like to thank Meir Hemmo and Orly Shenker for their valuable advice. This research is supported by the Israel Science Foundation, grant number 744/07.*

Appendix 1: Proof of Theorem 1

Theorem 1. *Let $E \in \mathcal{L}$ be any Lebesgue measurable set and let $\varepsilon > 0$; then there is $F_\varepsilon \in \mathcal{F}$ such that $\mu[(E \backslash F_\varepsilon) \cup (F_\varepsilon \backslash E)] < \varepsilon$.*

Proof. Consider first a Lebesgue measurable subset $E \subseteq [0, 1]$. By the regularity of the Lebesgue measure (see [23], page 230) given any $\varepsilon > 0$ there is an open set U, with $E \subseteq U$ and $\mu(U \backslash E) < \frac{\varepsilon}{2}$. The family of open intervals with dyadic endpoints forms a basis for the usual topology on $[0, 1]$ (recall that a dyadic number is a rational whose denominator is a power of 2). Thus, we can represent U as a disjoint countable union $U = \cup_{j=1}^{\infty}(c_j, d_j)$, where c_j and d_j are dyadic, and $\mu(U) = \sum_{j=1}^{\infty}(d_j - c_j)$. By choosing a sufficiently large natural number N we can make sure that $U' = \cup_{j=1}^{N}(c_j, d_j) \subseteq U$ satisfies $\mu(U') > \mu(U) - \frac{\varepsilon}{2}$. Now define U'' to be the set obtained from U' by adding the endpoints of each interval: $U'' = \cup_{j=1}^{N}(c_j, d_j)$. Since we have added just finitely many points the measure of U'' is the same as that of U', and therefore, $\mu[(E \backslash U'') \cup (U'' \backslash E)] < \varepsilon$.

Now apply the map $\sum_{j=1}^{\infty} a_j 2^{-j} \rightarrow (a_1, a_2, a_3, \ldots)$ which takes real numbers in $[0, 1]$ to their sequence of binary coefficients in $\{0, 1\}^{\omega}$. Dyadic rationals have two binary developments, one ending with an infinite sequence of zeroes, and the other ending with an infinite sequence of ones. Adopt the convention that in case of a dyadic rational d, the map takes d to its two binary sequences. Since the set of dyadic numbers has measure zero the map is measure preserving. The set E is then mapped to a subset of $\{0, 1\}^{\omega}$ which we shall also denote by E. The set U'' is mapped to a *finite* subset of $\{0, 1\}^{\omega}$ which we will denote by $F_\varepsilon \in \mathcal{F}$. The reason is that every closed interval with dyadic endpoints is mapped to a finite set, for example, $[\frac{1}{4}, \frac{5}{8}] \rightarrow \{(0, 1, 0), (0, 1, 1), (1, 0, 0)\} \times \{0, 1\} \times \{0, 1\} \times \ldots \subseteq \{0, 1\}^{\omega}$, . and U'' is a finite union of such intervals. This completes the proof.

Appendix 2: Proof of Theorem 2

This theorem and proof appeared first in [24] as part of a criticism of the frequency interpretation of probability.

Theorem 2. *Let $A \subset \{0, 1\}^\omega$ be any measurable set with $\mu(A) > \frac{1}{2}$, then there are* **a**, **b** $\in A$ *such that* **a·b** *has a divergent sequence of averages.*

Proof. Denote by **a** $\oplus$ **b** the XOR of the elements **a** and **b**, in other words $(\mathbf{a} \oplus \mathbf{b})_i = a_i + b_i \pmod 2$. We first show that $\mu(A) > \frac{1}{2}$ implies that $A \oplus A = \{\mathbf{a} \oplus \mathbf{b}; \mathbf{a}, \mathbf{b} \in A\} = \{0, 1\}^\omega$. Indeed if $\mathbf{c} \notin A \oplus A$, then $(\mathbf{c} \oplus A) \cap A = \phi$, where $\mathbf{c} \oplus A = \{\mathbf{c} \oplus \mathbf{a}; \mathbf{a} \in A\}$. Otherwise, if $\mathbf{d} \in (\mathbf{c} \oplus A) \cap A$ then $\mathbf{d} \in A$ and $\mathbf{d} = (\mathbf{c} \oplus \mathbf{a})$ for some $\mathbf{a} \in A$. Hence $\mathbf{c} = (\mathbf{d} \oplus \mathbf{a}) \in A \oplus A$, contradiction. Therefore, $(\mathbf{c} \oplus A) \cap A = \phi$, but this also leads to a contradiction since $\mu(\mathbf{c} \oplus A) = \mu(A) > \frac{1}{2}$, hence $A \oplus A = \{0, 1\}^\omega$.

We can assume without loss of generality that all elements of A have a convergent sequence of averages. This is the case because the set of elements of $\{0, 1\}^\omega$ whose averages diverge has measure zero. Let $\mathbf{c} \in \{0, 1\}^\omega$ be some sequence with a divergent sequence of averages. Then by the above argument there are **a**, **b** $\in A$ such that $\mathbf{c} = (\mathbf{a} \oplus \mathbf{b})$, that is $c_i = a_i + b_i \pmod 2 = a_i + b_i - 2a_ib_i$ and therefore

$$\frac{1}{n} \sum_{i=1}^{n} c_i = \frac{1}{n} \sum_{i=1}^{n} a_i + \frac{1}{n} \sum_{i=1}^{n} b_i - \frac{2}{n} \sum_{i=1}^{n} a_ib_i.$$

The sequence on the left diverges, and the first two sequences on the right converge. Hence, $n^{-1} \sum_{i=1}^{n} a_ib_i$ diverges. This completes the proof.

References

1. Lebowitz, J.L.: Boltzmann's entropy and time's arrow. Phys. Today **46**, 32–38 (1993)
2. Lebowitz, J.L.: Macroscopic laws, microscopic dynamics, time's arrow and Boltzmann's entropy. Physica A **194**, 1–27 (1993)
3. Dürr, D.: Über den Zufall in der Physik. http://www.mathematik.uni-muenchen.de/~duerr/ Zufall/zufall.html (1998)
4. Goldstein, S.: Boltzmann's approach to statistical mechanics. In: Jean, B., Dürr, D., Galavotti, M.C., Ghirardi, G.C., Petruccione, F., Zangh, N. (eds.) Chance in Physics: Foundations and Perspectives, pp. 39–54. Springer, Berlin and New York (2001)
5. Goldstein, S., Lebowitz, J.L.: On the (Boltzmann) entropy of non-equilibrium systems. Physica D **193**, 53–66 (2004)
6. Lavis, D.: Boltzmann and Gibbs: an attempted reconciliation. Stud. Hist. Philos. Mod. Phys. **36**, 245–273 (2005)
7. Maudlin, T.: What could be objective about probabilities? Stud. Hist. Philos. Mod. Phys. **38**, 275–291 (2007)
8. Sheldon, G., Lebowitz, J.L.: Mastrodonato Christian, Tumulka Roderich, Zanghi Nino. Normal Typicality and von Neumann's Quantum Ergodic Theorem. Proc. R. Soc. A 466(2123), 3203–3224 (2009)

9. Frigg, R.: Typicality and the approach to equilibrium in Boltzmannian statistical mechanics. In: Ernst, G., Huttemann, A. (eds.) Time Chance and Reduction. Philosophical Aspects of Statistical Mechanics, pp. 92–118. Cambridge University Press, Cambridge (2010)

10. Uffink, J.: Compendium of the foundations of classical statistical physics. In: Butterfield, J., Earman, J. (eds.) Handbook for Philsophy of Physics, pp. 923–1047. North Holland, Amsterdam (2006)

11. Buzaglo, M.: The Logic of Concept Expansion. Cambridge University Press, Cambridge (2002)

12. Patersen, K.: Ergodic Theory. Cambridge University Press, Cambridge (1983)

13. Chow, Y.S., Teicher, H.: Probability Theory Independence, Interchangeability, Martingales. Springer, Berlin and New York (1978)

14. Hewitt, E., Ross, K.A.: Abstract Harmonic Analysis. Springer, Berlin and New York (1994)

15. Solovay, R.M.: A model of set-theory in which every set of reals is Lebesgue measurable. Ann. Math. **92**, 1 (1970)

16. von Plato, J.: Creating Modern Probability. Cambridge University Press, Cambridge (1994)

17. Sandu, P., Short, A.J.: Winter Andreas. The foundations of statistical mechanics from entanglement: individual states vs. averages. http://arxiv.org/abs/quantph/0511225 (2005)

18. Sheldon, G., Lebowitz, J.L., Tumulka, R., Zanghi, N.: Canonical typicality. Phys. Rev. Lett. **96**, 050403 (2006)

19. Noah, L., Sandu, P., Short, A.J.: Winter Andreas. Quantum mechanical evolution towards thermal equilibrium. http://arxiv.org/abs/0812.2385 (2008)

20. Loewer, B.: Determinism and chance. Stud. Hist. Philos. Mod. Phys. **32**, 609–629 (2001)

21. Hemmo, M., Orly, S.: Measures over initial conditions, in Y. Ben-Menahem and M. Hemmo (eds.), Probability in Physics, pp. 87–98. The Frontiers Collection, Springer-Verlag Berlin Heidelberg (2011)

22. Vranas, P.B.M.: Epsilon-ergodicity and the success of equilibrium statistical mechanics. Philos. Sci. **65**, 688–708 (1998)

23. Rudin, W.: Principles of Mathematical Analysis. McGraw Hill, New York (1964)

24. Bub, J., Pitowsky, I.: Critical notice on Popper's postscript to the logic of scientific discovery. Can. J. Philos. **15**, 539–552 (1986)

Chapter 4
Typicality and Notions of Probability in Physics

Sheldon Goldstein

Abstract A variety of notions of probability, playing different roles, are relevant in physics. One crucial notion, typicality, while not genuinely probabilistic at all, is arguably the mother of them all.

4.1 Introduction

There are lots of different words for probability. Here are some: chance, likelihood, distribution, measure. There are also a variety of different notions of probability:

- Subjective chance (Bayesian?)
- Objective chance (propensity?)
- Relative frequency, empirical (pattern)
- A mathematical structure providing a measure of the size of sets (Kolmogorov)

Sometimes these are presented as competing notions. That's not my intention here. I wish only to emphasize at this point that when one speaks of probability it is a good idea to be clear about which notion one has in mind.

My main concern in this paper, however, is with typicality, a notion that, while extremely important for understanding probability, is not really a notion of probability at all. The logic of typicality is this. Many important phenomena, in physics and beyond, while they cannot be shown to hold without exception, can be shown to

I am very pleased to dedicate this article to the memory of Itamar Pitowsky. Itamar was an exceptionally creative philosopher and scientist, and one with a sophisticated understanding of mathematics. In recent years he had become very interested in typicality, a subject on which he was working at the time of his death.

S. Goldstein (✉)
Departments of Mathematics, Physics, and Philosophy – Hill Center Rutgers,
The State University of New Jersey, Piscataway, NJ, USA
e-mail: oldstein@math.rutgers.edu

Y. Ben-Menahem and M. Hemmo (eds.), *Probability in Physics*, The Frontiers Collection,   59
DOI 10.1007/978-3-642-21329-8_4, © Springer-Verlag Berlin Heidelberg 2012

hold with very rare exception, suitably understood. Such phenomena are said to hold typically; a proof that they do so is a typicality proof.

Regarded as mathematics, such results can be very interesting, with prizes awarded for their achievement. Of course the practical relevance of such results is that if some observed behavior has been shown to hold with rare exception, one should not be surprised if no exceptions are seen and one will tend to feel justified in regarding the behavior as explained.

It must be admitted, however, that as a matter of logic such practical conclusions don't follow. If exceptions exist there is nothing that would preclude the exceptional cases from being the only cases we ever encounter. Nonetheless, science could make little if any progress without invoking appeals to typicality, at least implicitly.

Here is an important example of a typicality statement:

> One should not forget that the Maxwell distribution is not a state in which each molecule has a definite position and velocity, and which is thereby attained when the position and velocity of each molecule approach these definite values asymptotically. . . . It is in no way a special singular distribution which is to be contrasted to infinitely many more non-Maxwellian distributions; rather it is characterized by the fact that by far the largest number of possible velocity distributions have the characteristic properties of the Maxwell distribution, and compared to these there are only a relatively small number of possible distributions that deviate significantly from Maxwell's. Whereas Zermelo says that the number of states that finally lead to the Maxwellian state is small compared to all possible states, I assert on the contrary that by far the largest number of possible states are "Maxwellian" and that the number that deviate from the Maxwellian state is vanishingly small. (Ludwig Boltzmann, 1896 [1])

Notice that this statement of Boltzmann involves probability ("distribution") and typicality. Boltzmann is saying here that states with Maxwellian probabilities are typical ("by far the largest number of possible states are 'Maxwellian' . . . the number that deviate from the Maxwellian state is vanishingly small"). This illustrates an important source of confusion in this business: that typicality statements often concern probabilities, making it all too easy to conflate typicality and probability.

4.2 History

There has been a revival of interest in typicality among physicists and philosophers in recent years. However the recognition of the importance of the notion is not new. That goes back to the very beginnings of probability theory in the eighteenth century. What I shall describe here of the relevant ancient history I've learned from Glenn Shafer [2, 3]. Notice in what follows how some of the founding fathers of probability theory struggled to finesse the gap between an event having extremely small size as measured in some natural way and the event being impossible, or certain to fail.

4.2.1   Ancient History (<1950)

- Jakob Bernoulli, in his great work Ars Conjectandi (1713), writes that "Because it is only rarely possible to obtain full certainty, necessity and custom demand that what is merely morally certain be taken as certain."
- Antoine Cournot (1843) writes that "A physically impossible event is one whose probability is infinitely small. This remark alone gives substance—an objective and phenomenological value—to the mathematical theory of probability." This later became known as *Cournot's principle*.
- According to Paul Levy (≈1919), Cournot's principle is the only connection between probability and the empirical world. He calls it "the principle of the very unlikely event."
- Hadamard refers instead to "the principle of the negligible event."
- Kolmogorov, in his Foundations of Probability (1933), Chapter 1, §2, The Relation to Experimental Data, writes that "Only Cournot's principle connects the mathematical formalism with the real world."
- Similarly Borel (≈1948) writes that "The principle that an event with very small probability will not happen is the only law of chance."

4.2.2   Modern History (>1950)

Notice that while the probablists did not refer to "typical" or "typicality," that notion, or something very much in its vicinity, is what they had in mind. In more recent years the "t"-word has been used quite frequently, most often, curiously, in connection with probability in quantum mechanics. I hope the following quotations help convey the idea of the method of appeal to typicality.

> In order to establish quantitative results, we must put some sort of measure (weighting) on the elements of a final superposition. This is necessary to be able to make assertions which hold for almost all of the observer states described by elements of the superposition. We wish to make quantitative statements about the relative frequencies of the different possible results of observation—which are recorded in the memory—for a typical observer state; but to accomplish this we must have a method for selecting a typical element from a superposition of orthogonal states. . . .
>
> The situation here is fully analogous to that of classical statistical mechanics, where one puts a measure on trajectories of systems in the phase space by placing a measure on the phase space itself, and then making assertions . . . which hold for "almost all" trajectories. This notion of "almost all" depends here also upon the choice of measure, which is in this case taken to be the Lebesgue measure on the phase space. . . . Nevertheless the choice of Lebesgue measure on the phase space can be justified by the fact that it is the only choice for which the "conservation of probability" holds, (Liouville's theorem) and hence the only choice which makes possible any reasonable statistical deductions at all. (Hugh Everett, III, 1957 [4], page 460)
>
> Then for instantaneous macroscopic configurations the pilot-wave theory gives the same distribution as the orthodox theory, insofar as the latter is unambiguous. However,

this question arises: what is the good of *either* theory, giving distributions over a hypothetical ensemble (of worlds!) when we have only one world.

...a single configuration of the world will show statistical distributions over its different parts. Suppose, for example, this world contains an actual ensemble of similar experimental set-ups. ... it follows from the theory that the 'typical' world will approximately realize quantum mechanical distributions over such approximately independent components. The role of the hypothetical ensemble is precisely to permit definition of the word 'typical.' (John S. Bell, 1981 [5], page 129)

4.3 Typicality in Statistical Mechanics

If there is a branch of physics in which typicality is most prominently used it is probably statistical mechanics. And the most famous use of typicality in statistical mechanics concerns Boltzmann's equation. Moreover one could scarcely have a better illustration of the point of and the need for a typicality argument than in the transition from Boltzmann's presentation of 1872 to his presentation in 1877. Boltzmann (1872) claimed that (at low density) the state of a gas *must* evolve in accord with his equation. Boltzmann (1877) claimed, in effect, only that it would typically do so. Here are some details.

Boltzmann's equation is an evolution equation for a function $f(\mathbf{q}, \mathbf{v}, t)$, where $\mathbf{q}$ is a point in physical space, $\mathbf{v}$ is a velocity, and t of course is time. Boltzmann analyzed the behavior of a certain function of $\mathbf{q}$ and $\mathbf{v}$ that provides an efficient summary of the most important details of the state of a gas, namely the empirical one-partical distribution $\rho_{emp}(\mathbf{q}, \mathbf{v}) \equiv f_X(\mathbf{q}, \mathbf{v})$, giving basically the density of particles of the gas that are at or near $\mathbf{q}$ with velocity more or less $\mathbf{v}$.

Here, for an N-particle system, $X = (\mathbf{q}_1, \mathbf{v}_1, \ldots, \mathbf{q}_N, \mathbf{v}_N)$ is the point in the N-particle phase space describing the detailed state of the gas. The subscripts "emp" and X on ρ and f are to emphasize that f_X is indeed an empirical distribution, determined by the phase point X, and not a probability distribution that describes a random system or a hypothetical ensemble of systems. As the phase point $X(t)$ evolves according to the Hamiltonian dynamics for the system, ρ_{emp} evolves accordingly: $\rho_{emp}(\mathbf{q}, \mathbf{v}, t) \equiv f_{X(t)}(\mathbf{q}, \mathbf{v})$.

What Boltzmann claimed to have shown in 1872 is that for a low density gas it must be the case that $f_{X(t)}(\mathbf{q}, \mathbf{v})$ is well approximated by a solution $f(\mathbf{q}, \mathbf{v}, t)$ to Boltzmann's equation. On the basis of an analysis of that equation using his H-function

$$H(f(\mathbf{q}, \mathbf{v}, t)) = \int f(\mathbf{q}, \mathbf{v}, t) \log f(\mathbf{q}, \mathbf{v}, t) \, \mathbf{dq} \mathbf{dv}$$

Boltzmann then argued that for large times $t, f(\mathbf{q}, \mathbf{v}, t)$—and hence also $f_{X(t)}(\mathbf{q}, \mathbf{v})$—will approach the distribution that minimizes H, namely the equilibrium distribution—the Maxwellian distribution—$f_{eq}(\mathbf{q}, \mathbf{v}) \propto e^{-\frac{1}{2}m\mathbf{v}^2/kT}$, where k is Boltzmann's constant and T is the temperature of the gas.

Because of Loschmidt's reversibilty objection, by 1877 Boltzmann had realized that his earlier claim could not be right. He concluded that he had shown, not that $f_{X(t)}$ (**q**, **v**) is, approximately, a solution to Boltzmann's equation for all initial phase points $X(0)$, but only for most of them. More precisely, he concluded that he had shown that given any distribution function f (**q**, **v**), even one that is non-Maxwellian and that does not correspond to equilibrium, $f_{X(t)}$ will approximate a solution to Boltzmann's equation for the overwhelming majority, suitably understood, of initial phase points for which $f_{X(0)}$ is (approximately) f—the overwhelming majority of phase points in the macrostate defined by f. In other words, in 1877 Boltzmann argued that the evolution of a gas in accord with Boltzmann's equation, while not inevitable, is typical. (Boltzmann's proof was not rigorous. Almost a century later, a rigorous typicality proof, valid only for short times, was found by Oscar Lanford [6].)

More important for our understanding of the origin of thermodynamics, in 1877 Boltzmann arrived at a far deeper appreciation of why a gas will tend to approach a state of equilibrium, in which nothing seems to change. Crucial to this understanding is the notion of macrostate, alluded to above. The macrostate $\Gamma_f = \{X \in \Gamma_E \mid f_X (\mathbf{q}, \mathbf{v}) \approx f (\mathbf{q}, \mathbf{v})\}$ corresponding to f is the set of phase points, in the energy surface Γ_E of phase points having energy E, that are all macroscopically like f—in the sense that the macro-variable f_X is approximately f. The phase points in the same macrostate are thus very similar from a macroscopic perspective.

The most important fact about these macrostates, recognized by Boltzmann, concerns their sizes as measured using the natural volume measure on the phase space, Lebesgue or Liouville measure. It is, in fact, this natural volume measure that provides a sufficiently precise notion of "overwhelming majority" for his typicality claim.

Here are two depictions of the partition of Γ_E into macrostates (corresponding to different choices of f):

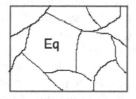 vs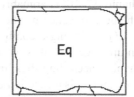

One special macrostate is singled out here by "Eq," indicating the equilibrium macrostate $\Gamma_{f_{eq}}$, which is larger than all the others. The crucial fact is that the depiction on the left is utterly misleading, giving a very wrong sense of the relative sizes of the macrostates.

The depiction on the right is much better. But in fact the equilibrium macrostate is so very much larger than the other macrostates that no picture could adequately depict the difference in sizes. In fact, as Boltzmann showed, at low density $|\Gamma_f| \sim e^{-NH(f)}$. For a macroscopic system, with particle number $N \sim 10^{20}$ or greater, this means that the overwhelming majority of the points of Γ_E are in $\Gamma_{f_{eq}}$, the ratio of the size of a non-equilibrium macrostate to that of the equilibrium macrostate being ridiculously small, of order $10^{-10^{20}}$.

The depiction on the right illustrates another one of Boltzmann's typicality results—that equilibrium is typical: the overwhelming majority of the phase points X in the energy surface Γ_E correspond to a gas having equilibrium properties, in the sense that f_X is approximately f_{eq}. This typicality result, which is easy, should not be confused with the typicality result for Boltzmann's equation, which is very hard. For the former, "overwhelming majority" is relative to the entire energy surface, while for the latter it is mainly relative to the incredibly small non-equilibrium macrostates.

Be that as it may, the estimates associated with the depiction do in fact provide us with a good rule of thumb for the relative size of atypical events of all sorts in statistical mechanics: corresponding to the ratio $10^{-10^{20}}$. You should at least not be surprised when events corresponding to sets of possibilities that are so small aren't observed.

Besides the two typicality results that I've mentioned there are many others in statistical mechanics—either proven already or awaiting a rigorous proof. Some examples are the second law of thermodynamics, the derivation of hydrodynamic equations, approach to equilibrium in quantum mechanics, and the universality of the canonical ensemble in quantum mechanics (canonical typicality). And outside of statistical mechanics there is, for example, the origin of quantum randomness in Bohmian mechanics, to which I now turn.

4.4   Bohmian Mechanics

Bohmian mechanics [7–10] is arguably the simplest formulation of non-relativistic quantum mechanics. It concerns the dynamics of a system of particles, with positions $\mathbf{Q}_1,\ldots,\mathbf{Q}_N$, defining a configuration Q. This dynamics is determined by the usual quantum mechanical wave function ψ, itself evolving, as in standard quantum mechanics, according to Schrödinger's equation. In the simplest case, of particles without spin, ψ is a function $\psi(\mathbf{q}_1,\ldots,\mathbf{q}_N)$ of the possible positions of the particles. The joint evolution of ψ and Q is deterministic. Nonetheless, as a consequence of a typicality analysis, the usual quantum probabilities, given by $|\psi(q)|^2$, govern the results of observations in a Bohmian universe.

Quantum equilibrium, corresponding to the quantum equilibrium distribution $\rho_{qe}(q) = |\psi(q)|^2$, should be thought of, in this regard, as roughly analogous to thermodynamic equilibrium, corresponding to the Maxwellian $f_{eq} \propto e^{-\frac{1}{2}mv^2/kT}$. A proper understanding of quantum equilibrium probabilities and of thermodynamic equilibrium probabilities both require that we appreciate that there are a variety of conceptually different probablistic objects relevant to the analysis, as I shall explain later. They also require that we appreciate that there are, in both cases, two different sorts of systems to be dealt with: a large system, for thermodynamics a gas in a box, and for Bohmian mechanics the entire universe; and a small subsystem of the large system, which in both cases we will take here, for simplicity, to be a single particle.

4.4.1 The Wave Function of a Subsystem

Consider a 1-particle subsystem of an N-particle Bohmian universe. Let's denote by $\mathbf{Q}$ the position of the particle and by Q_{env} the configuration of the rest of the particles of our universe—the configuration of the *environment*. Let Ψ be the wave function of the universe. It is a function $\Psi(\mathbf{q}, q_{env})$. (Here we've used, as is common in Bohmian mechanics, lower case letters to indicate possible values, generic values, as opposed to the actual values, denoted with capital letters.)

The appropriate notion for the wave function of our subsystem is given by the *conditional wave function*

$$\psi(\mathbf{q}) = \Psi(\mathbf{q}, Q_{env}),$$

a function of the generic position of our particle obtained by plugging the actual configuration of its environment into the wave function of the universe. Note that ψ need not be normalized—its absolute square integral over all of space need not be 1. Whenever ψ appears as part of a probability formula it should be regarded as having been normalized via multiplication by the appropriate positive real number. Note also that because of the dependence on the actual configuration of the environment, which inherits its own typically complicated evolution from the Bohmian evolution of the configuration of the universe, the wave function of our particle depends on time in a somewhat complicated way:

$$\psi_t(\mathbf{q}) = \Psi_t(\mathbf{q}, Q_{env}(t)),$$

with Ψ itself, as a solution of Schrödinger's equation, depending on t. As a consequence of this evolution, the wave function of a subsystem in Bohmian mechanics can evolve in a variety of ways. In particular it will evolve according to Schrödinger's equation when the system is suitably decoupled from its environment, and will undergo collapse of the wave packet in the appropriate measurement situations.

For our purposes here, the most important fact about the conditional wave function is that it provides us with the probability distribution of our subsystem—in fact in a variety of senses. The most basic sense in which it does so is expressed in the following simple mathematical fact:

$$P(\mathbf{Q}(t) \in d\mathbf{q} \mid Q_{env}(t)) = |\psi_t(\mathbf{q})|^2 d\mathbf{q} \tag{4.1}$$

Here P is the probability distribution on initial configurations of the universe (at a time, say, shortly after the big bang) given by $|\Psi(q)|^2$. This *fundamental conditional probability formula* of Bohmian mechanics says that for such a random universe the conditional distribution of the position of a particle at any time, given its environment at that time, depends only on its conditional wave function at that time, and does so via the usual Born, quantum equilibrium, probability formula.

As a consequence of this formula it follows via a typicality analysis [10] that for the overwhelming majority—in the sense of the measure P—of initial configurations of a Bohmian universe, the empirical distribution for the positions of particles (and for larger subsystems) in suitable real world ensembles of systems having given conditional wave function ψ is (approximately) the quantum equilibrium distribution $|\psi|^2$. In short, quantum equilibrium is typical.

4.5 Probability and Typicality

In a typical typicality analysis in physics—and arguably in any serious application of probability to the real world—probability structures play several quite different roles, the most important of which are the following:

- ρ_{emp}: empirical distribution (relative frequency)
- ρ_{th}: theoretical distribution (idealization, $N \to \infty$)
- P: measure for typicality

It is the empirical distribution that describes a real world pattern of events that is responsible for what we observe. The theoretical distribution is an idealization providing a good approximation to the empirical distribution, $\rho_{emp} \approx \rho_{th}$, in the limit of large ensembles of subsystems. P is a probability distribution on the big system containing the subsystems. It is via a law of large numbers kind of analysis using P that one can show that it typically happens that $\rho_{emp} \approx \rho_{th}$, with typicality defined in terms of P.

Many different probability distributions P define the same sense of typicality. This is because, insofar as typicality is concerned, the detailed probability of a set is not relevant; all that matters is which sets have very large measure and which very small. Nonetheless, it is often the case that a particular choice of P is special. It is for such a choice that one in fact can most efficiently carry out the relevant analysis. For this special P the theoretical distribution will be in some sense a conspicuous part of P, meaning that:

$$\rho_{th}(\mathbf{x})d\mathbf{x} = \rho^P(\mathbf{x})d\mathbf{x} = {}''P(\mathbf{X} \in d\mathbf{x}){}''.$$

Here I use $\mathbf{X}$ and $\mathbf{x}$ for the subsystem variables. (I shall use X and x—without bold—to refer to the variables for the big system.) ρ^P is a sort of marginal distribution of the subsystem, arising from the distribution P of the big system, and the quotation marks are to indicate that ρ^P is often only "sort of" a marginal, and not always an actual marginal. For example the relevant ρ^P in Bohmian mechanics is a conditional marginal, just as is suggested by the fundamental conditional probability Formula 4.1.

For the second typicality result in statistical mechanics mentioned above, that "equilibrium is typical," $P(dx)$ is the microcanonical ensemble on Γ_E, the uniform distribution over the energy surface. However, to illustrate the points I wish to make here it would be better to make a different but physically equivalent choice for P, namely the canonical ensemble, given by

$$P(dx) \propto e^{-H(x)/kT} dx.$$

And we shall assume we are dealing with the simplest case, that of noninteracting particles, with $H = \sum_i \frac{1}{2} m v_i^2$. In this case $P = \prod_i f_{eq}(v_i)$ is simply the product over all the particles of the equilibrium distribution for each particle.

Then [with $\mathbf{x} = (\mathbf{q}, \mathbf{v})$] we have that

$$\rho_{emp}(\mathbf{x}) \equiv \rho_{emp}^{(X)}(\mathbf{x}) = f_X(\mathbf{q}, \mathbf{v})$$

(the precise definition of ρ_{emp} is $\rho_{emp}^{(X)}(\mathbf{q}, \mathbf{v}) = \frac{1}{N} \sum_{(\mathbf{q}_i, \mathbf{v}_i) \in X} \delta(\mathbf{q} - \mathbf{q}_i) \delta(\mathbf{v} - \mathbf{v}_i))$ and that

$$\rho_{th}(\mathbf{x}) \equiv f_{eq}(\mathbf{x}) = \rho^P(\mathbf{x}) \propto e^{-\frac{1}{2} m v^2 / kT}.$$

In particular the theoretical distribution here is a factor of P—a piece, as it were, of the measure for typicality.

In Bohmian mechanics (writing x for q, etc.) we have that

$$P(dx) = |\Psi(x)|^2 dx$$
$$\rho_{emp}(\mathbf{x}) \equiv \rho_{emp}^X(\mathbf{x}) = \frac{1}{N} \sum_{\mathbf{x}_i \in X} \delta(\mathbf{x} - \mathbf{x}_i)$$

and

$$\rho_{th}(\mathbf{x}) \equiv \rho_{qe}(\mathbf{x}) = \rho^P(\mathbf{x}) = |\psi(x)|^2.$$

Here too the theoretical distribution is sort of a piece of the measure for typicality.

A typicality analysis binds tightly together these three very different probablistic objects. This is particularly so for the special choice of P, a choice for which P has some nice properties—more on these shortly—the simplest such being that P be a product measure as above. When P is thus "nice" one can show via a law of large numbers type analysis that (when N is large) $\rho_{emp} \approx \rho^P$, P-typically—that for the P-overwhelming majority of points X, ρ_{emp}^X is approximately the theoretical distribution, with the latter itself being a quasi-marginal of the measure for typicality P We shall say in this situation that P is *statistically transparent*.

4.5.1 Ergodicity and Statistical Transparency

Statistical transparency is closely connected to the notion of ergodicity [11]. That is because it is often the case that ρ_{emp} is more or less a time-average or space-average—the sorts of things with which ergodicity is concerned. The ergodicity of P (under either space or time translation) implies that these averages agree with the phase averages, i.e., with the theoretical distribution ρ^P arising from P. Thus we can more or less identify statistical transparency with the ergodicity of P.

This does, however, have to be taken with a grain of salt, since the space averages relevant to ergodicity would be infinite system averages (so $N = \infty$) or for time averages, infinite time averages ($T = \infty$), idealizations that might not exactly match the typicality analysis under consideration. We shall however ignore this point, abusing mathematics a bit, and simply pretend without qualification that statistical transparency can be identified with ergodicity—that it is the ergodicity of P that makes it special, so that we have statistical transparency.

4.5.2 Symmetry and Statistical Transparency

There is another way in which P might be special: among all measures defining the same sense of typicality, it might be one that is symmetric, and the only one that is.

The relevant symmetry here depends upon whether ρ_{emp} involves space or time averages. In the former case the symmetry is that of spatial-translation invariance (remember we are pretending that our system is suitably idealized, and thus spatially infinite if necessary), in the latter case that of time-translation invariance. Suppose P is in this sense symmetric. The set of measures $\tilde{P}$ defining the same sense of typicality as P are those of the form $\tilde{P}(dx) = g(x)P(dx)$ obtained from P by multiplying it by a positive function g that is bounded above and away from zero below. And if, as we are pretending, ρ_{emp} involves infinite space or time averages, the set of probability measures equivalent to P in the sense of typicality could now be taken to correspond to the requirement that g be positive, with integral with respect to P equal to 1—i.e., to the set of probability measures equivalent to P in the sense of measure theory.

The connection between symmetry and statistical transparency is then this: P is the only symmetric probability measure in the class of equivalent ones precisely in case there is statistical transparency. That's because of the connection between ergodicity and statistical transparency just discussed together with the fact that a system is ergodic precisely in the case of there being a unique symmetrical P in the equivalence class.

4.5.3 Predictive Typicality and Ergodicity

The discussion has so far taken for granted that we have unambiguous (space or time) averages, i.e., that ρ_{emp} is unambiguous. This *predictive typicality*, we should remark, is a property of the typicality class itself, and not a characterization of a special member of that class. Predictive typicality is more or less equivalent to the requirement that the typicality equivalence class has an ergodic member P. This will of course be the member of the class that most directly expresses the observed probabilities ρ_{emp}.

4.5.4 The Good, the Bad, and the Ugly

To summarize, the three probability measures, ρ_{emp}, $\rho_{th} = \rho^P$, and P, involved in the usual typicality analysis are intimately related: We have that

$$\rho_{emp} \leftrightarrow \rho^P \leftrightarrow P,$$

conveying that the observed probability distribution ρ_{emp}, which of course varies from trial to trial of the same experiment, is well approximated by the theoretical distribution ρ^P (which of course is the same for all trials of an experiment), the latter being a conspicuous part of the measure for typicality P.

This is both good and bad. It is good, because it suggests a nice simplicity, inasmuch as it means that for many practical purposes one need worry about just one probability measure and not three. At the same time it is bad, because the simplicity is a misleading simplicity, since the three probability measures are conceptually of entirely different natures, despite their closeness for practical purposes. And the consequences of the conflation of three such very different notions—the discussions and analyses in which crucial distinctions between very different objects are not properly recognized—can be quite ugly.

The confusion is probably greater still with regard to the typicality analysis for Bohmian mechanics, which can be summarized like so:

$$\rho_{emp} \leftrightarrow \rho_{th}^{\psi} \leftrightarrow P^{\Psi}$$

Here $\rho_{th}^{\psi} = |\psi|^2$ and $P^{\Psi} = |\Psi|^2$ are given by the very same formula, with the only difference being ψ versus Ψ. Unless one appreciates the great difference between the wave function ψ of a subsystem and that of the universe Ψ, this can make it difficult to accept that the probabilistic objects involved are so very different.

4.6   Two Directions for Typicality Research

I have discussed here the method of appeal to typicality and given some examples.
I have indicated that, while typicality is fundamentally not a version of probability,
it can nonetheless easily seem to be one. But we have not attempted here to justify
the conclusions that scientists arrive at by appealing to typicality. In particular
I have not explained why what is typical should be expected to happen.

Nor shall I do so here: a systematic analysis would require that we deal with some
of the most fundamental issues in the philosophy of science, such as the meaning and
nature of scientific explanation. I do feel, however, that a comprehensive philosophical analysis of scientific explanation and the logic of appeal to typicality would be
most welcome. (Some gestures in this direction can be found in Sect. 6 of [12].)

4.6.1   Types of Typicality

I would, however, like to mention here three distinctions between types of typicality
that are relevant to how strongly typicality seems to compel our expectations:
(i) natural versus axiomatic, (ii) continuum versus finite, and (iii) hypothetical
versus actual. The measure of typicality might be natural, like the uniform distribution over the space of possibilities, naturally expressed; or it might be merely
stipulated axiomatically. The set of possibilities might be finite, or it might be
a continuum (it of course might also be infinite but not a continuum). The possibilities might be merely possibilities—they might be hypothetical—or they might,
as with many-worlds, be all actual.

Other things being equal, typicality corresponding to the first type of each pair
seems to more strongly compel our expectations. For example, the notion of most
elements of a finite set seems entirely unambiguous, corresponding to counting measure,
whereas with a continuum one might be able to argue that there are a variety of
reasonable senses of most. The worst case in this regard is that of axiomatic typicality
with a finite set of actualities. (It might well be that the only way typicality can be
persuasively applied to the case in which the possibilities are in fact actual is within the
Human approach to law and probability advocated by Barry Lower [13]; however,
the goal in this approach is more modest: description rather than explanation.)

4.6.2   Typicality Not Given by Probability

I will conclude by putting on the table a possibility afforded by the recognition that
typicality is not probability. While it is usually the case that typicality is defined
using a probability measure, a different way of deciding which sets are large and
which small, for example one that is given by a set function that violates the axioms

of probability, is feasible. Such a wider notion of typicality could be used for the formulation of new types of physical theories.

Along such lines not much has yet been done. But Murray Gell-Mann and James Hartle [14] have noted with regard to their decoherent histories version of quantum mechanics that insofar as their decoherence functional fundamentally is used to define, in effect, typicality (though they don't use that word) the fact that it may end up violating the axioms of probability in a limited sort of way need not concern us. And Bruno Galvan [15] has proposed a trajectory based version of quantum mechanics that, unlike Bohmian mechanics, is defined solely in terms of a typicality that is not based on probability. While Galvan's theory is a bit odd, it does have the virtue of seeming to exploit only traditional quantum mechanical structure.

The possibility of a typicality liberated from probability might be a great source of inspiration for theory formation. I think that this possibility would have pleased Itamar very much.

Acknowledgements This work was supported in part by NSF Grant DMS–0504504.

References

1. Boltzmann, L.: Annalen der Physik **57**, 773 (1896); reprinted and translated as Chapter 8 in Brush, S.G: Kinetic Theory. Pergamon, Oxford (1966)
2. Shafer, G.: Why did Cournot's principle disappear? Talk at Ecole des Hautes Etudes en Sciences Sociales. Paris (2006). http://www.glennshafer.com/assets/downloads/disappear.pdf
3. Shafer, G., Vovk, V.: The sources of Kolmogorov's Grundbegriffe. Stat. Sci. **21**, 70–98 (2006)
4. Everett III, H.: "Relative state" formulation of quantum mechanics. Rev. Mod. Phys. **29**, 454–462 (1957)
5. Bell, J.S.: Speakable and Unspeakable in Quantum Mechanics. Cambridge University Press, Cambridge (1987)
6. Lanford III, O.E.: Time evolution of large classical systems. In: Moser, J. (ed.) Dynamical Systems, Theory and Applications. Lecture Notes in Physics, vol. 38, pp. 1–111. Springer, Berlin (1975)
7. Goldstein, S.: Bohmian Mechanics. In: Zalta, E.N. (ed.) Stanford Encyclopedia of Philosophy, published online by Stanford University (2001). http://plato.stanford.edu/entries/qm-bohm/
8. Bohm, D.: Phys. Rev. **85**, 166 (1952)
9. Bohm, D.: Phys. Rev. **85**, 180 (1952)
10. Dürr, D., Goldstein, S., Zanghì, N.: J. Stat. Phys. **67**, 843 (1992). http://arxiv.org/abs/quant-ph/0308039
11. Arnold, V.I., Avez, A.: Ergodic Problems of Classical Mechanics. Addison-Wesley, Reading (1989)
12. Goldstein, S., Lebowitz, J.L., Tumulka, R., Zanghì, N.: Long-time behavior of macroscopic quantum systems: commentary accompanying the English translation of John von Neumann's 1929 article on the quantum ergodic theorem. Eur. Phys. J. H: Hist. Perspect. Contemp. Phys. **35**, 173–200 (2010). http://arxiv.org/abs/1003.2129v1
13. Loewer, B.: David Lewis' theory of objective chance. Philos. Sci. **71**, 1115–1125 (2004)
14. Gell-Mann, M., Hartle, J.B.: Quantum mechanics in the light of quantum cosmology. In: Zurek, W. (ed.) Complexity, Entropy, and the Physics of Information. Addison-Wesley, Reading (1990)
15. Galvan, B.: Typicality vs probability in trajectory-based formulations of quantum mechanics. Found. Phys. **37**, 1540–1562 (2007)

Chapter 5
Deterministic Laws and Epistemic Chances

Wayne C. Myrvold

Abstract In this paper, a concept of chance is introduced that is compatible with deterministic physical laws, yet does justice to our use of chance-talk in connection with typical games of chance, and in classical statistical mechanics. We take our cue from what Poincaré called "the method of arbitrary functions," and elaborate upon a suggestion made by Savage in connection with this. Comparison is made between this notion of chance, and David Lewis' conception.

5.1 Probability, Chance, and Credence: A Brief History

As has been often pointed out, the word "probability" has been used in at least two distinct senses.[1] One sense, the *epistemic* sense, has to do with degrees of belief of a rational agent. The other sense, which Hacking calls the *aleatory* sense, is the concept appropriate to games of chance; this is the sense in which one speaks, for example, of the probability (whether known by anyone or not) of rolling at least one pair of sixes, in 24 throws of a pair of fair dice.

A particularly clear statement that there are two concepts that need to be distinguished is found in Poisson's book of 1837.

> In ordinary language, the words *chance* and probability are nearly synonymous. Quite often we employ the one or the other indifferently, but when it is necessary to distinguish between their senses, we will, in this work, relate the word chance to events in themselves, independently of our knowledge of them, and we will reserve for the word probability the previous [epistemic] definition. Thus, an event will have, by its nature, a greater or less

---

[1] See Hacking [1] for a masterful overview of the history.

W.C. Myrvold (✉)
Department of Philosophy, The University of Western Ontario, London, ON, Canada
e-mail: wmyrvold@uwo.ca

Y. Ben-Menahem and M. Hemmo (eds.), *Probability in Physics*, The Frontiers Collection, 73
DOI 10.1007/978-3-642-21329-8_5, © Springer-Verlag Berlin Heidelberg 2012

chance, known or unknown; and its probability will be relative to the knowledge we have, in regard to it.

For example, in the game of *heads* and *tails*,[2] the chance of getting *heads*, and that of getting *tails*, results from the constitution of the coin that one tosses; one can regard it as physically impossible that the chance of one be equal to that of the other; nevertheless, if the constitution of the coin being tossed is unknown to us, and if we have not already subjected it to trials, the probability of getting *heads* is, for us, absolutely the same as that of getting *tails*; we have, in effect, no reason to believe more in one than the other of the two events [2, p. 31].[3]

Note that Poisson's use of "chance" refers to single events, and the chance of heads on a coin toss is a matter of the physical constitution of the chance set-up (he says "the constitution of the coin," but clearly it matters also how the coin is tossed). This is not a frequency interpretation.

Something happened on the way to the twentieth century: the notion of objective chance—that is, single-case probability thought of as a feature of a physical situation—largely dropped out of discussions. There remained that the idea that "probability" can be taken in either an objective or an epistemic sense, but the objective sense became identified with a frequency interpretation.

Thus we find de Finetti rejecting objective notions of probability on the basis of a rejection of Laplacean and frequency conceptions ([3], pp. 16ff; [4], pp. 71ff). In a similar vein, Savage [5, p. 4] identifies an objective conception of probability with a frequency interpretation, and takes it as a virtue of the subjectivist view that it is applicable to single cases. Absence of the notion of objective chance in so many discussions of the foundations of probability led Popper [6, 7] to conclude that he had an entirely new idea in single-case objective probabilities, which he called "propensities."

We also find some historians of probability have projecting a frequency conception onto writers of previous centuries, in place of the notion of chance. For example, though Hacking [1, p. 12] initially characterizes this duality much as we

---

[2] Poisson says "*croix* et *pile*"; heads and tails is our equivalent.

[3] Dans la langage ordinaire, les mots *chance* et probabilité sont á peu prés synonymes. Le plus souvent nous emploierons indifféremment l'un et l'autre; mais lorsqu'il sera nécessaire de mettre une différence entre leurs acceptions, on rapportera, dans cet ouvrage, le mot chances aux événements en eux-mêmes et indépendamment de la connaissance que nous en avons, et l'on conservera au mot probabilité sa définition précédente. Ainsi, un événement aura, par sa nature, une chance plus ou moins grande, connue ou inconnue; et saprobabilité sera relative á nos connaissances, en ce qui le concerne.

Par exemple, au jeu de *croix* et *pile*, la chance de l'arrivée de *croix* et celle de l'arrivée de *pile*, résultent de la constitution de la pièce que l'on projette ; on peut regarder comme physiquement impossible que l'une de ces chances soit égale à l'autre; cependant, si la constitution du projectile nous est inconnue, et si nous ne l'avons pas déjà soumis à des épreuves, la probabilité de l'arrivée de croix est, pour nous, absolument la même que celle de l'arrivée de pile: nous n'avons, en effet, aucune raison de croire plutôt à l'un qu'à l'autre de ces deux événements. I'l n'en est plus de même, quand la pièce a été projetée plusieurs fois: la chance propre à chaque face ne change pas pendant les épreuves; mais, pour quelqu'un qui en connaît le résultat, la probabilité de l'arrivée future de *croix* ou de *pile*, varie avec les nombres de fois ces deux faces se sont déjà présentées.

have here, later in the book the aleatory concept of probability identified with a frequency interpretation: "On the one hand it [probability] is epistemological, having to do with support by evidence. On the other hand it is statistical, having to do with stable frequencies" (p. 43). Hacking is far from alone in this. For example, Daston [8, p. 191] attributes a frequency notion to A.A. Cournot [9], who distinguishes between probability and what he calls degrees of physical possibility in much the same way that Poisson distinguished between probability and chance. Howie [10, p. 36] attributes a frequency concept of chance to Poisson.

It's not entirely clear what the reason is for the disappearance of chance from discussions of probability. A plausible conjecture is that it had to do with increasing acceptance that the laws that govern the physical world are deterministic, together with the notion that objective chance and determinism are incompatible. Laplace famously began his *Philosophical Essay on Probabilities* (1841) with a discussion of determinism. A being that had complete knowledge of the laws of nature and the state of the world at some time, and was able to perform the requisite calculations, would have no need of the calculus of probabilities, according to Laplace. It is only because our abilities depart from those of such a being that we employ probabilities. Laplace then proceeds to characterize probability in epistemic terms; "[t]he theory of chance consists in reducing all the events of the same kind to a certain number of cases equally possible," where "equally possible" means that *we* are equally undecided about which of the cases obtain [11, p. 6]. It is not clear that Laplace is consistent in maintaining an epistemic view throughout; there are passages in the Essay that suggest that it is a matter of fact, which we can investigate empirically, whether events are truly equipossible. Nevertheless, the official view, for Laplace, is an epistemic one.[4]

Though it may be that emphasis on determinism led to the decay of the notion of objective chance, there is nevertheless a tension between the idea that chance and determinism are incompatible, and the way that we talk about chances. We talk about the chance of heads on a coin toss, casino owners worry about whether their roulette wheels show discernible bias, and these seem to be matters that can subjected to experimental test, by doing multiple trials and performing a statistical analysis on the results.[5] Yet we also think that, at least to the level of approximation required, these systems can be adequately modelled by deterministic, classical physics.[6] Nor does it seem that Poisson was supposing any departure from determinism.

Of course, such talk might simply be deeply ill-conceived. David Lewis declared that "[t]o the question how chance can be reconciled with determinism, ... my answer is: *it can't be done*" [13, p. 118], and Schaffer [14] has provided arguments for this conclusion. It is the purpose of this paper to present a notion that may

---

[4] See Hacking [12; 1, Ch 14] for a lucid discussion.

[5] Though perhaps this should go without saying, it should be emphasized that taking frequency data as evidence about chances is not tantamount to holding a frequency *interpretation* of chance.

[6] Though, it must be noted, Lewis [13, p. 119] has suggested that this is false.

appropriately be called *chance*, which fits quite nicely with deterministic evolution of a certain kind, and which is suited to play the role of chance in our chance-talk. Schaffer [14] distinguishes between genuinely objective chances and what he calls "epistemic chances," while noting that the latter can be objectively informed. "Epistemic chance" is an apt name for the notion that we will be introducing, as it highlights the fact that we will be interweaving epistemic and physical considerations.

In what follows, we will use the word "chance" for the aleatory concept; a chance is an objective, single-case probability. When we are speaking of the belief states of a (possibly idealized) agent, we will use the word "credence." "Probability" will be used when we want to be noncommittal.

5.2 Learning About Chances

The chance of heads on a coin toss, if it is regarded as an objective feature of the set-up, is *ipso facto* the sort of thing that we can have beliefs about, beliefs that may be correct or incorrect, better or worse informed. Under certain conditions, we can learn about the values of chances.

Particularly conducive to learning about chances are cases in which we have available (or can create) a series of events that we take to be similar in all aspects relevant to their chances, that are, moreover, independent of each other, in the sense that occurrence of one does not affect the chance of the others. The paradigm cases are the occurrence of heads on multiple tosses of the same coin, occurrence of a six on multiple throws of the same die, and the like. Consider a sequence of N coin tosses. If, on each toss, the chance of heads is λ, then the chance of any given sequence of results is

$$\lambda^m (1 - \lambda)^{N-m},$$

where m is the number of heads in the sequence. Considered as a function of λ, this is peaked at the observed relative frequency m/N, and becomes more sharply peaked, as N is increased.

Let E be the proposition that expresses the sequence of results these N tosses, and, for any λ, let H_λ be the proposition that the chance of heads on each toss is equal to λ. Consider an agent who has some prior credences about the chance of heads, and updates them by Bayesian conditionalization:

$$cr(H_\lambda) \;\Rightarrow\; cr(H_\lambda|E) = \frac{cr(E|H_\lambda)\, cr(H_\lambda)}{cr(E)}.$$

It seems natural to suppose—and, indeed, in the statistical literature this is typically assumed without explicit mention—that our agent's credences set $cr(E|H_\lambda)$

equal to the chance of E according to H_λ, as is required by what Lewis [26] has dubbed the *Principal Principle*. This has the consequence that our agent's credence in chance-values close to the observed relative frequency is boosted, and her credence in other values, diminished. Moreover, since the likelihood function $\lambda^m(1 - \lambda)^{N-m}$ is more sharply peaked, the larger the number of trials, relative frequency data becomes more valuable for narrowing credence about chances as the number of trials is increased.

Note that there are three distinct concepts at play here: chance, credence, and relative frequency in repeated trials. None of these three is to be identified with any of the others. They do, however, connect in a significant way: relative frequency data furnish evidence on which we update credences about chances.

5.3 "Almost Objective" Chances

We take our cue from what Poincaré called the *method of arbitrary functions* (see [15] for the history of this). Poincaré's analysis leaves some crucial questions unanswered, and, in particular, leaves it unclear what notion of probability might be in play. Savage [16] argued that subjective credence has a role to play. And so it does, but it is not a radical subjectivism that is needed, but a tempered personalism that distinguishes between reasonable and unreasonable credence-functions.

5.3.1 Tempered Personalism About Credences

Objective Bayesians hold that there are, for any body of background knowledge, unique credences that would be the degrees of belief of an ideally rational agent. At the opposite extreme would be *radical subjectivists* (if there are any), who hold that the only constraint on credences is the requirement of coherence, that is, that they satisfy the axioms of probability; within the class of coherent credence functions, there can be no grounds for judging one better or worse than another.[7]

The attitude that Abner Shimony [17] has dubbed *tempered personalism* steers between these extremes, finding cognitive virtue between two opposing vices. Without supposing that there are *uniquely* rational credences, not all are equally acceptable; excessively dogmatic credences are to be eschewed as obstacles to learning about the world. In what follows, we will assume that there is a class of credences, perhaps imprecisely defined, that represent the possible credences of a reasonable agent.

---

[7] The parenthetical qualification is due to the fact that, though, in some passages de Finetti sounds like a radical subjectivist, there are others that indicate a more moderate position.

As an example relevant to the sorts of cases we are discussing, suppose that a gambler in a casino becomes convinced that the next spin of the roulette wheel will be 23, to the extent that he is willing to bet his life savings on it.[8] Not because he thinks the game is rigged; he takes it to be an ordinary roulette wheel, being spun in the ordinary way, with the ordinary sorts of causal influences on the outcome. Though the gambler's credences need not violate coherence, we would nevertheless take such a conviction to be unreasonable. If, further, our gambler claimed that his conviction was based on a belief that the current state of everything causally relevant to the outcome was such as to lead, via the unfolding of deterministic laws of physics, to the result 23, we would, even if we shared his conviction in determinism, nevertheless regard it as ludicrous to pretend to knowledge to the degree of precision that such a conviction would require. In what follows, I will simply take it that the reader shares this judgment, with the issue of what justification we might have for this to be left for another occasion.

5.4   Poincaré and the Method of Arbitrary Functions

Poincaré [18, 19] considered a simple, roulette-like game, in which a wheel, divided into a large number n of sectors of equal size, alternately colored red and black, is spun, and eventually comes to rest due to friction. Bets are to be placed on whether a pointer affixed to the wheel's mount will point to a red or a black sector when the wheel comes to rest. The set-up is such that small differences in the initial impulse, too small to be perceived or controlled, can make a difference between the outcome being red or black.

Poincaré supposes the probabilities of initial impulses to be given by a density function ϕ, a function that is "entirely unknown" [18, p. 148]. This function yields, via the dynamics of the set-up, a density function f over the angle θ at which the wheel comes to rest. Suppose, now, that the function f is continuous and differentiable, and that the derivative is bounded, so that, for some M, $|f'(\theta)| < M$ for all θ. If the angle (in radians) subtended by each sector is ε, then the difference between the probability of red and the probability of black is at most $M\pi\varepsilon$. This goes to zero as ε goes to zero.

There are a number of questions left open by Poincaré's discussion. First is the status of the limit $\varepsilon \to 0$. We are, after all, considering the probability of a red or black outcome on a spin of a particular wheel, with fixed number of sectors, not a sequence of wheels with an ever-increasing number of sectors. We need not take Poincaré's limit-talk literally. What matters is that f vary slowly enough over intervals of size ε that the probability of landing in any red sector be approximately equal to the probability of landing in the adjacent black sector, and that any differences between probabilities associated with successive sectors be small enough that they remain negligible when $n/2$ of them are summed.

---

[8] This example is inspired by the behaviour of the character played by James Garner in the comic western *Support Your Local Gunfighter* (1971).

For our purposes, a more serious issue is the status of the function ϕ, which yields probabilities over initial conditions. Poincaré calls ϕ an unknown function, which suggests that there is a matter of fact about what function it actually is. In a discussion of the game of roulette, Poincaré writes,

> What is the probability that the impulse has this or that value? About this I know nothing, but it is difficult for me not to admit that the probability is represented by a continuous analytic function [18, p. 12].[9]

In a parallel discussion in *Science and Hypothesis*, he writes,[10]

> I do not know what is the probability that the ball is spun with such a force that this angle should lie between and $\theta + d\theta$, but I can make a convention.
>
> I can suppose that this probability is $\phi(\theta)$. As for the function $\phi(\theta)$, I can choose it in an entirely arbitrary manner. I have nothing to guide me in my choice, but I am naturally induced to suppose the function to be continuous [19, p. 201].

This, it must be admitted, is puzzling. Poincaré alternates between treating the probability as something objective but unknown, and treating it as something that we can make arbitrary choices about.

Suppose, now, that we take the function ϕ to represent an agent's degrees of belief about the impulse imparted to the wheel. Applying the dynamics of the system to this credence function yields a probability density f over the orientation of the wheel after it has come to rest, at some later time t_1. The probability function yielded by f might not represent our agent's degrees of belief about the final angle, if she doesn't know the dynamics of the set-up, or is unable to perform the requisite calculation; this will be important in Sect. 5.3, below.

Suppose, now, that small changes in the initial conditions—too small to be controlled or noticed by our agent—yield differences in the final angle that are large compared to the width of a single sector. It is reasonable to suppose that an agent's credences would not vary much over such small scales; a credence function that changed appreciably when shifted by an imperceptible amount would represent more detailed knowledge of initial conditions than would be available to an agent in the epistemic situation we are imagining. Then application of the system's dynamics to the agent's credences will yield roughly equally probabilities of red and black outcomes. Moreover, this conclusion does not depend sensitively on the function ϕ. Though, *pace* Poincaré, it is not true that an *arbitrary* probability density over initial conditions, or even an arbitrary density with bounded variation, yields equal probabilities for red and black, it is true that a mild constraint on the density function ϕ—moreover, a constraint that arguably any reasonable credence should satisfy—suffices to entail that f yield approximately equal probabilities for red and

---

[9] Quelle est la probabilité pour que cette impulsion ait telle ou telle valeur? Je n'en sais rien, mais il m'est difficile de ne pas admettre que cette probabilité est representée par une fonction analytique continue.

[10] Note that there is a shift of notation between *Calcul des Probabilités* and *Science and Hypothesis*; ϕ is here a density function over the final angle, that is, the function we have been calling f.

black. The dynamics of the set-up ensure that any reasonable credences about states of affairs at one time yield approximately the same probabilities for certain coarse-grained propositions about a later state of the system.

Generalizing, the situations of interest to us are ones in which we have a physical set-up such that, for some proposition A (or class of propositions) about the outcome of an observation undertaken at time t_1, the dynamics of the set-up are such that any reasonable credences about states of affairs at time t_0 yield, as a result of the evolution of the system, approximately the same value for the probability of A. There are four interacting components at play here. One is a limitation on the knowledge of the system available to the agent. Although it remains true that a being in possession of precise knowledge of initial conditions and able to do the requisite calculation would be in a position to have precise knowledge about the outcome, we suppose limits to the precision of the knowledge available to our agents. Second is a judgment about what sorts of credences are reasonable, given the knowledge available to our agents; we are neither supposing uniquely reason-able credences, nor are we supposing that coherence is the only criterion of reasonableness. Third is a limitation of attention to certain macroscopic propositions about the system's state at a later time. Lastly, and crucially, it should be a feature of the dynamics of the system that any differences between reasonable credence-functions wash out; any reasonable credences about initial conditions lead to approximately the same credence in the proposition A.

In situations like this—in which the dynamics of the system lead all reasonable credences about the state of affairs at t_0 to effectively the same probability for some proposition A about states of affairs at a later time t_1—we are justified, I think, in calling this common probability the *chance* of A. What value these chances have depends on the physics of the set-up, and, moreover, it makes sense to talk of unknown chances, or in cases of disagreement about what the chances are, about one value being more correct than another. The limitations on knowledge might be *in principle* limitations. For an example that may or may not apply to the real world, consider the de Broglie-Bohm hidden-variable interpretation of quantum mechanics. There, it is provably impossible for an agent (who must interact with a system via physical means to gain information about it) to gain enough information about corpuscle positions to make betting at other than the quantum-mechanical probabilities reason-able. In other cases, even if it might in principle be possible to gain further knowledge, obtaining such knowledge is so far beyond feasibility that it might as well be impossible in principle. Consider a real roulette wheel, to be spun by a human croupier, and take t_0 to be some time *before* the spin.[11] Even if it is possible, in principle, to gain sufficient information about the croupier's physical state and all influences on it that would be sufficient to make it reasonable to bet at other than the values we are calling chances, this matters little to the credences of actual agents. Quantities of this sort have been called "almost objective" [20]. On the issue of terminology, see Sect. 5.5, below.

---

[11] *After* the ball has been released is another matter.

5.4.1 Learning About Chances, Revisited

The function φ is meant to represent an agent's credences about states of affairs at t_0; f, the result of applying the system's dynamics to this function. These dynamics will often be imperfectly known to the agent, who might be unsure, say, whether the wheel is biased in some way. Even if the dynamics are known, the requisite computation might be intractable.

Nonetheless, our agent might believe that there is some value, unknown to her, that gives the probability assigned to a proposition A by time-evolving, not only her current credences, but those of any reasonable agent. This value is the degree of belief in A that a reasonable agent would have if she knew the dynamics of the set-up and could do the calculation, and in this sense represents credence that makes optimal use of information available. Our agent can have credences about what this value is. For any real number λ, let H_λ be the proposition that this value is equal to λ. It is a reasonable constraint on our agent's credences that they satisfy

$$cr(A|H_\lambda) = \lambda,$$

and that, moreover, if E is any proposition whose truth-value could be ascertained by the agent at t_0,

$$cr(A|H_\lambda \& E) = \lambda.$$

To see that this is a reasonable condition on an agent's credences, recall that, if λ is the chance of A, and our agent's credences in A is not equal to λ, this is due to the agent's imperfect knowledge of the dynamical laws governing the system, or else to her inability to apply these laws. Her conditional credence, conditional on the supposition that her credence would be λ were these limitations lifted, is required to be λ.

Our constraint on credence suffices for our agent to learn about the chances of a series of events that are regarded as having equal and independent chances, in the manner outlined in Sect. 5.2, above. The constraint is a cousin of Lewis' Principal Principle. The chief difference is that, in the Principal Principle, E may be any *admissible* proposition, and, though Lewis does not explicitly define admissibility, he takes all statements about the past to be admissible. This would be unjustified on our treatment.

There is considerable literature on the justification of the Principal Principle; to some it appears a mysterious constraint.[12] About our constraint there is no mystery.

---

[12] For some references to this literature, and skepticism about the possibility of a cogent justification, see Strevens [21].

5.5 Chances in Statistical Mechanics

An appropriate physical set-up can wash out very considerable differences in credences about initial conditions. Consider a gas in a box with a partition down the middle. Alice believes that at t_0 the gas is initially in the left side of the box; Bob, that it is in the right. The partition is removed, and, a few minutes later, at t_1, some measurements are to be performed on the gas. Let $\phi^A(t_0)$ and $\phi^B(t_0)$ be Alice and Bob's credences about the state of the system at t_0, and let $\phi^A(t_1)$ and $\phi^B(t_1)$ be the result of applying the actual dynamical evolution of the system to these credence-functions. That is, the probability assigned by $\phi^A(t_1)$ to a region Δ of the system's phase space is the probability that $\phi^A(t_0)$ assigns to the system being at t_0 in some state that will evolve into a state in Δ.[13] Though this would be difficult to prove rigorously for anything like a realistic gas, there is good reason to believe that, provided that their initial credences don't vary too rapidly within the respective regions of phase space on which they are non-zero, the probability functions that result from applying the dynamics of the system to Alice's and Bob's credences about initial conditions will yield virtually the same probabilities for the results of any feasible measurements (that is, there is no feasible experiment to be performed on the gas that will be informative about whether, a few minutes earlier, the gas had been in the left or the right side of the box). This is true even though, in one sense, the time-evolved credences are as different as the original ones: if there is no overlap between the regions of phase space that Bob and Alice believe the gas to be in at t_0, there will be no overlap in the regions assigned non-zero probability by the time-evolved credence functions. Nevertheless, these two regions will be finely intertwined in the phase space of the system, and macroscopic regions will contain roughly equal proportions of both, so that the two probability functions will agree closely on the probabilities of outcomes of macroscopic observations.

Here we see a role to play for the equilibrium probability measures used in statistical mechanics. Provided the relaxation to the new equilibrium proceeds as we think it does, Alice's and Bob's time-evolved credences about measurements will not only agree with each other, but with those of a third agent, Charles, who believes that both sides of the box initially contained gas of the same temperature and pressure. Charles will take the removal of the partition to effect no change in macroscopically observable properties of the gas, and his credences may be represented by an equilibrium distribution, a probability measure that is invariant under dynamical evolution. If Alice and Bob are convinced that this distribution yields the same probabilities for results of macroscopic measurements as would their own credences, applied to the system, then they may use the equilibrium distribution—which will typically be much more tractable mathematically—as a surrogate for their own credences.

---

[13] Once again, these time-evolved credence functions might not be Alice and Bob's credences about the states of affairs at t_1, if they don't know the dynamics of the system, or are unable to do the requisite calculation.

Taking statistical mechanical probabilities in this way removes some of the puzzles that have been associated with them. The equilibrium distribution distinguishes no direction in time. If we took it to represent Alice or Bob's credences about the state of the system at t_1, this would clash with their beliefs about the state of the system at t_0, as the equilibrium distribution at t_1 renders it overwhelmingly probable that the gas was spread evenly over the box at t_0. But this is not how it is being used; Alice and Bob are using it as a *surrogate* for the more complicated functions $\phi^A(t_1)$ and $\phi^B(t_1)$, and the justification for doing so is that the equilibrium distribution yields what are effectively the same probabilities for the results of measurements performed after t_1. There is no justification for applying this distribution to past events, and hence we do not encounter the disastrous retrodictions that prompt David Albert [22] to introduce his Past Hypothesis.

5.6 Chances, Real or Counterfeit?

As mentioned, some philosophers might be willing to accept all the substantive claims made in this paper, yet resist the use of the word "chance" for the quantities we have discussed. Lewis himself might be among these; with reference to ideas advanced by Jeffrey [23, Sect. 12.7] and Skyrms [24, 25], Lewis speaks of a "kind of counterfeit chance" [13, p. 120].

There is an argument, stemming from the Principal Principle, for the incompatibility of non-extremal chances with deterministic laws of nature (see [14], pp. 128–129). Recall that the PP says that a reasonable agent's credences should have it that, for any proposition A, any real number λ in [0, 1], and any admissible information E,

$$cr(A|E\&ch(A) = \lambda) = \lambda.$$

Lewis does not offer a definition of admissibility, but he does declare that all propositions about past events and present states of affairs are admissible, "every detail—no matter how hard it might be to discover—of the structure of the coin, the tosser, other parts of the setup, and even anything nearby that might somehow intervene" [26, p. 272]. If the laws of nature are deterministic, then these laws, together with sufficient information about events to the past of A, entail either A or its negation. Suppose that laws of nature are always admissible. This means that, for a suitable choice of admissible E, probabilistic coherence requires

$$cr(A|E) = 0 \text{ or } 1.$$

This in turn entails that our agent must assign zero credence to any proposition that asserts that the value of a chance lies in an interval not containing 0 or 1. An agent whose credences satisfy Lewis' PP must be certain that, in a deterministic world, there are no non-extremal chances (that is, the agent must assign zero credence to the conjunction of some proposition that entails that the laws are deterministic and some proposition that entails that there are nonextremal chances). If we add the further condition, as Lewis [26, p. 267] does, that the agent assign zero credence only to the empty proposition, true at no possible world, then the incompatibility of determinism and non-extremal chances follows.[14]

If it is part of our notion of chance that reasonable credences must satisfy Lewis' Principle Principle, with laws of nature and all propositions about the past of an event counted as admissible, then Lewis is right; determinism and chance are irreconcilable. This is a symptom of the fact that Lewis' notion of chance differs from the conception we are trying to capture in this paper. On Lewis' notion, chance requires chancy laws. The notion we are trying to capture stems from the idea that, in the face of unavoidable (or, perhaps, unavoidable for all practical purposes) limitations on accessible information about the world, there might be some credences that are optimal for an agent who makes maximal use of available information and dynamical features of the systems involved. There is nothing incoherent in Lewis' notion; indeed, as suggested by quantum mechanics, the fundamental laws of physics probably *are* chancy. We need not leave the notion of chance behind, however, when we emerge from the domain in which quantum chanciness predominates and enter into the realm of systems whose behaviour can be adequately modelled by classical mechanics. There is a useful conception of chance that is compatible with determinism.

A terminological distinction between the two notions is in order. Schaffer's term "epistemic chance" seems to be an apt one, as a term that combines epistemic and objective connotations. Lewisian metaphysicians may, if they choose, call such chances "counterfeit" chances, but we should not let this obscure the value that lies in the concept, nor should we let it dissuade us from accepting such chances as valid currency when appropriate.

Acknowledgement This work is supported by a grant from the Social Sciences and Humanities Research Council of Canada.

References

1. Hacking, I.: The Emergence of Probability. Cambridge University Press, Cambridge (1975)
2. Poisson, S.-D.: Recherches sur la Probabilité des Jugements en Matiére Criminelle et en Matiére Civile, précédées des régles générales du calcul des probabilités. Bachelier, Imprimeur-Libraire, Paris (1837)
3. de Finetti, B.: La prevision: ses lois logiques, ses sources subjectives. Ann. Inst. Henri Poincaré **7**, 1–68 (1937). English translation in [4]
4. de Finetti, B.: Foresight: its logical laws, its subjective sources. In: Kyburg, H.E., Smokler, H.E. (eds.) Studies in Subjective Probability. Krieger, Huntington (1980). Translation of [3]
5. Savage, L.J.: The Foundations of Statistics. Dover, New York ([1954] 1972)

---

[14] Note that, without this condition on credences, which Lewis calls *regularity*, nothing can follow about what chances actually are like from the PP, which is a condition on the credences of a reasonable agent, and hence can only tell us what a reasonable agent must believe.

6. Popper, K.R.: The propensity interpretation of the calculus of probability, and the quantum theory. In: Körner, S. (ed.) Observation and Interpretation: A Symposium of Philosophers and Physicists, pp. 65–70. Butterworths, London (1957)
7. Popper, K.R.: The propensity interpretation of probability. Br. J. Philos. Sci. **37**, 25–42 (1959)
8. Daston, L.: Classical Probability in the Enlightenment. Princeton University Press, Princeton (1988)
9. Cournot, A.A.: Exposition de la Théorie des Chances et des Probabilités. Librairie de L. Hachette, Paris (1843)
10. Howie, D.: Interpreting Probability: Controversies and Developments in the Early Twentieth Century. Cambridge University Press, Cambridge (2002)
11. Laplace, P.-S.: A Philosophical Essay on Probabilities. Dover, New York ([1902] 1951) [27]
12. Hacking, I.: Equipossibility theories of probability. Br. J. Philos. Sci. **22**, 339–355 (1971)
13. Lewis, D.: Postscripts to "A subjectivist's guide to objective chance". In: Philosophical Papers, IIth edn, Vol. II. Oxford University Press, Oxford (1986)
14. Schaffer, J.: Deterministic chance? Br. J. Philos. Sci. **58**, 113–140 (2007)
15. von Plato, J.: The method of arbitrary functions. Br. J. Philos. Sci. **34**, 37–47 (1983)
16. Savage, L.J.: Probability in science: a personalistic account. In: Suppes, P. (ed.) Logic Methodology, and Philosophy of Science IV, pp. 417–428. North-Holland, Amsterdam (1973)
17. Shimony, A.: Scientific inference. In: Colodny, R. (ed.) The Nature and Function of Scientific Theories, pp. 79–172. Pittsburgh University Press, Pittsburgh (1971). Reprinted in [28]
18. Poincaré, H.: Calcul des probabilités. Gauthier-Villars, Paris (1912)
19. Poincaré, H.: Science and Hypothesis. Dover, New York ([1905] 1952)
20. Machina, M.J.: Almost-objective uncertainty. Econ. Theory **24**, 1–54 (2004)
21. Strevens, M.: Objective probability as a guide to the world. Philos. Stud. **95**, 243–275 (1999)
22. Albert, D.: Time and Chance. Harvard University Press, Cambridge (2000)
23. Jeffrey, R.C.: The Logic of Decision. McGraw-Hill, New York (1965)
24. Skyrms, B.: Resiliency, propensities, and causal necessity. J. Philos. **74**, 704–713 (1977)
25. Skyrms, B.: Causal Necessity. Yale University Press, New Haven (1980)
26. Lewis, D.: A subjectivist's guide to objective chance. In: Jeffrey, R.C. (ed.) Studies in Inductive Logic and Probability, Vol. II, pp. 263–293. University of California Press, Berkeley (1980)
27. Laplace, P.-S.: Essai Philosophique sur les Probabilités. Courcier, Paris (1814). English translation in Laplace (1951)
28. Shimony, A.: Search for a Naturalistic World View. Scientific Method and Epistemology, vol. I. Cambridge University Press, Cambridge (1993)

Chapter 6
Measures over Initial Conditions

Meir Hemmo and Orly Shenker

Abstract This paper concerns the meaning of the idea of typicality in classical statistical mechanics and how typicality is related to the notion of probability.

6.1 Introduction

This paper concerns the meaning of the idea of typicality in classical statistical mechanics and how typicality is related to the notion of probability. Our thoughts about these issues have been greatly influenced along the years by numerous conversations with Itamar Pitowsky. In his last paper [1] which he devoted to the issue of typicality, he writes:

> Consider a finite but large collection of marbles. When one says that a vast majority of the marbles are white one usually means that all the marbles except possibly very few are white. And when one says that half the marbles are white, one makes a statement about counting, and not about the probability of drawing a white marble from the collection.

Here Itamar is making a sharp distinction between the *size* of a set of outcomes of an experiment and the probability of these outcomes. The size of a set of outcomes is fixed by a measure defined on the event space. In the discrete case, the size of the set is fixed by counting the number of outcomes that belong to it. Itamar thought that in the discrete case the measure obtained by counting is natural, and therefore he thought that it is worthwhile to generalize this measure to the continuous case. In his paper (ibid.) he argues that the Lebesgue measure in the continuous case is

M. Hemmo (✉)
Department of Philosophy, University of Haifa, Haifa, Israel
e-mail: meir@research.haifa.ac.il

O. Shenker
Department of Philosophy, The Hebrew University of Jerusalem, Jerusalem, Israel
e-mail: orlyshenker@pluto.mscc.huji.ac.il

Y. Ben-Menahem and M. Hemmo (eds.), *Probability in Physics*, The Frontiers Collection,   87
DOI 10.1007/978-3-642-21329-8_6, © Springer-Verlag Berlin Heidelberg 2012

the natural extension of the counting measure in the discrete case, and he takes this result to establish a preference for the Lebesgue measure in the continuous case. This means that in classical statistical mechanics, for example, the Lebesgue measure is the natural measure to determine sizes of sets in the state space. If this is right, the problem of justifying the choice of measure in classical statistical mechanics is partially solved. The reason why it is only partially solved is that on the standard way of thinking about statistical mechanics, the problem concerns the justification of the statistical mechanical probabilities, and as Itamar himself stresses (in the quotation above) the measure of sets is not enough to determine probability.

Despite the distinction between measure and probability, Itamar thought (see ibid.) that the Lebesgue measure in statistical mechanics plays some role, admittedly weak, in the explanation of thermodynamic behavior. In this paper we examine this question. Our starting point is similar to Itamar's that measure is indeed different from probability, but while Itamar thought that the Lebesgue measure is natural in some *a priori* sense, it seems to us that the choice of measure in physics is guided by experience, which in turn guides our choice of probabilistic laws.

The structure of the paper is as follows. We begin in Sect. 6.2 by describing the so-called typicality approach (as it is usually framed in the context of deterministic theories in physics). In Sect. 6.3 we describe the way in which probabilistic statements in classical statistical mechanics ought to be understood. In Sect. 6.4 we examine arguments based on the classical dynamics to the effect that the Lebesgue measure is natural in statistical mechanics. In Sect. 6.5 we analyze the significance of Lanford's theorem in classical statistical mechanics, and we explain how the theorem ought to be understood without appealing to typicality. Section 6.6 is the conclusion.

6.2 Typicality

In classical statistical mechanics the standard way of understanding the thermodynamic behavior of systems around us appeals to a *probability* distribution over the initial microstates of the systems (compatible with the initial thermodynamic macrostate). On the standard way of thinking one says that given the uniform probability distribution (relative to the Lebesgue measure) over the initial macrostate, it is highly *probable* that the system will, for example, approach equilibrium after some designated time. In this way, the behavior of the system is explained by the fact that its actual microstate is highly likely to sit on a trajectory, which will take it to equilibrium at the time in question. Here the high probability pertains to subsystems of the universe, and it is assumed further that the trajectory of the whole universe that gives rise to this high probability itself sits on an initial condition which has high probability. Note that here there are two notions of probability: a probability distribution over the initial macrostate (i.e. the microstates compatible with the macrostate at some present time) of subsystems of the universe, and a probability distribution over the initial conditions of the universe.

Another important example of the central role played by the measure in explaining physical behavior in statistical mechanics is in Einstein's [2] account of Brownian motion, as developed by Wiener (see [3]). As is well known, Wiener has proved that the so-called Wiener measure of trajectories in the phase space of a Brownian particle which are continuous but nowhere differentiable is one. The explanation of the actual behavior of Brownian particles is based on the assumption that their actual trajectories belong to this measure one set. Avogadro's number is derived from this assumption.

A question that immediately arises concerning this understanding is what could a probability distribution over the *initial* conditions of the universe possibly mean. A probability distribution suggests some sort of a random sampling of an initial condition out of the set of all possible conditions. But with respect to the initial conditions of the universe any such sampling (if it is to be physical) would be external to the universe, and therefore this seems to suggest an empirically meaningless fairy tale. This problem does not arise with respect to subsystems of the universe, since one can ground a probability distribution over initial conditions in experience (as we show in Sect. 6.3). Moreover, probability in physical theories is usually conceived as involving (or as being tested by) repetitions of experiments, which in the case of the initial conditions of the universe are trivially impossible.

We understand the typicality approach[1] as an attempt to solve these problems by appealing to a certain natural measure over initial conditions, where the measure is *not* understood as a probability measure (see [5] for a similar construal).

Here is an example of how the distinction between typicality and a probability distribution over initial conditions is made:

> When employing the method of appeal to typicality, one usually uses the language of probability theory. When we do so we do not mean to imply that any of the objects considered is random in reality. What we mean is that certain sets (of wave functions, of orthonormal bases, etc.) have certain sizes (e.g., close to one) in terms of certain natural measures of size. That is, we describe the behavior that is typical of wave functions, orthonormal bases, etc. However, since the mathematics is equivalent to that of probability theory, it is convenient to adopt that language. For this reason, we do not mean, when using a normalized measure μ, to make an "assumption of *a priori* probabilities," even if we use the word "probability." Rather, we have in mind that, if a condition is true of most D, or most H, this fact may suggest that the condition is also true of a concrete given system, unless we have reasons to expect otherwise. [7].

And in another place [8], they say:

> When we express that something is true for most H or most ψ relative to some normalized measure μ, it is often convenient to use the language of probability theory and speak of a random H or ψ chosen with distribution μ. However, by this we do not mean to imply that the actual H or ψ in a concrete physical situation is random, nor that one would obtain, in repetitions of the experiment or in a class of similar experiments, different H's or ψ's whose empirical distribution is close to μ. That would be a misinterpretation of the measure μ, one

---

[1] For various formulations and extensive discussions of the typicality approach, see Dürr et al. [4], Maudlin [5], Callender [6].

that suggests the question whether perhaps the actual distribution in reality could be non-uniform. This question misses the point, as there need not be any actual distribution in reality. Rather, Theorem 1 means that the set of "bad" Hamiltonians has very small measure μ.

There are three different statements made here about the idea of typicality:

(1) The set of initial conditions compatible with the initial macrostate of the universe is divided into two subsets, T1 and T2 such that all the microstates in T1 but not in T2 give rise to some property F. The property F may be for example the approach to equilibrium in statistical mechanics, or the Born rule in Bohmian mechanics.
(2) There is some natural (normalized) measure μ over the initial conditions such that $\mu(T1)$ is close to one (and $\mu(T2)$ is close to zero). In this sense, *most* initial conditions, as determined by μ, are in T1 (and are called typical).
(3) In a given experiment, the actual initial microstate of the universe belongs to T1.

Let us explain these three statements in turn. The statement in (1) above expresses a contingent fact about the dynamics, namely a fact about how the initial conditions are mapped by the equations of motion into microstates at later times. There are various theorems in classical statistical mechanics that demonstrate that special cases of (1) hold under some conditions with some appropriate property F. Examples are Lanford's theorem in which F is (roughly) entropy increase and the Birkhoff-von Neumann theorem in which F is the so-called pointwise ergodic theorem, which we discuss below. Statement (1) is not controversial in our discussion.

The notion of *most* in statement (2) above requires a measure over the phase space. That is, there are infinitely many ways to determine the size of subsets of a continuous set of points. The question is on what grounds one can justify the choice of measure, or the choice of some class of measures. Usually, in classical statistical mechanics the measure chosen is the Lebesgue measure (or the class of measures absolutely continuous with the Lebesgue measure), and in quantum mechanics the measure is given by the absolute square of the wavefunction. The grounds for these choices are that each of these measures has a preferred dynamical status in the theory.

Statement (3), as stated above, seems as expressing the brute fact, without further reasoning, that the microstate of the universe invariably (in every experimental set up) belongs to T1. But since there are microstates of the universe that don't belong to T1 this fact calls for a justification. It is evident that (2) is taken in the typicality approach to completely justify (3), that is if T1 were to contain only a small fraction of the microstates of the universe, one would not see (3) as justified. It is important to stress that in this approach the justification of (3) makes no appeal to probability. Rather, it is the measure of T1 that is supposed to do the whole work. This implies that, lacking reasons to expect otherwise, microstates of the universe that belong to T2 are not realized.

In short, there are two questions that need be answered in the context of typicality: what justifies the choice of measure in (2), and what justifies the passage from (2) to (3). In particular, the question we consider is whether there are grounds that justify the choice of measures in a way that *explains* the observed behavior of physical systems. If such grounds could be spelled out the problems concerning the meaning of probability distributions over the initial conditions of the universe would obviously evaporate together with the probability distribution itself. In the subsequent sections we attempt to answer these two questions. We will see that statements (2) and (3) are both wanting. Again, statement (1) is not controversial in the context of typicality. Our analysis will lead us to reject the typicality approach.

6.3 Probability in Classical Statistical Mechanics

In order to set the stage we need to go into some detail concerning the way in which probability statements arise in classical statistical mechanics and how precisely the choice of measure over the state space is carried out.

Consider the paradigmatic case of an ideal gas S, which is initially confined by a partition to the left half of a container, and then, by removing the partition, is allowed to expand. Finally, the gas fills out the entire container. Suppose that we set up a very large number of such gases $S_1 \ldots S_k$, all of which are prepared in the same initial macrostate M_0 in which the gas is confined to the left half of the container by a partition. We then remove the partitions and follow the spontaneous macroscopic evolution of these gases for a certain time interval Δt, and we see by simple counting that the overwhelming majority of the gases $S_1 \ldots S_k$ quickly reach and then remain in macrostate M_1 in which they fill up the entire container. We now wish to predict the evolution of another system, call it S_{k+1}, which is prepared in the same initial macrostate as $S_1 \ldots S_k$. We know that the dynamical equation of motion that governs the evolution of S_{k+1} is the same as the ones governing $S_1 \ldots S_k$, but we do not know the details of this dynamics, nor do we know the exact initial microscopic conditions of S_{k+1} and therefore all we can rely in this prediction is the above experiment.

Can we infer from the experiment with $S_1 \ldots S_k$ that S_{k+1} is highly likely to end up in macrostate M_1? That is, can we use the experiment with $S_1 \ldots S_k$ in order to come up with a probabilistic law, on which we can base our bets regarding the evolution of S_{k+1}? The answer is, of course, yes, we can infer the probabilities from the finite observed relative frequencies.[2] This inference is valid just to the extent that we can infer from experience any other physical law or prediction, such as $F = ma$. However, the way in which our probabilistic predictions can be *justified*, and the *extent* to which they can be justified – are not always clear in the literature, as we show later.

To see how to understand probabilistic statements in statistical mechanics let us describe the above experiment in the phase space of the gas. Classical mechanics

---

[2] This inference is a subtle issue which depends on how probability is understood. We don't address this question here.

tells us that the universe consists of microscopic particles, and that our experience is an *effect* of the *microstate* of the universe, which is the state of those particles. However, it is a physical fact that our senses are too coarse to reflect the full details of the microscopic structure of the universe; we can only perceive some of its general features. In this sense our experience is *macroscopic*. In the above experiment, we can only observe relative frequencies of *transitions* between macrostates of the gas. Let us see how these transitions are described in the phase space, and then how these relative frequencies are accounted for in the phase space.

The phase space of a system (in our example, of any of the systems S_i) is partitioned into sets of microstates, which are indistinguishable by an observer; these sets are called macrostates. The phase space regions corresponding to the macrostates express the observer's maximal observational capability, and therefore while the observer can tell which macrostate contains the actual microstate of the system at the time of observation, it cannot tell which part of the macrostate contains that microstate.

We now formulate what we take to be the essential way for calculating transition probabilities in statistical mechanics. Suppose that at time t_0 an observer O finds the system S in macrostate M_0 (as for example in our experiment above; see Fig. 6.1). Suppose also that O knows the laws of classical mechanics, which govern S's evolution in time. If O knows the Hamiltonian of S, that is: if O knows the equations of motion of S, then O can (in principle) calculate the evolution of all the trajectory segments that start out in the microstates contained in M_0 and find out the end points of these trajectory segments after the time interval Δt. These end points make up a set, which we call the *dynamical blob* $B(t_0 + \Delta t)$ of S at $t_0 + \Delta t$ given that it was in M_0 at t_0. In general, the region covered by $B(t_0 + \Delta t)$ overlaps with several macrostate regions; for instance, it may partially overlap with M_1 (in which the gas fills out the entire container), and with some other macrostates, such as M_2 or M_3 in which the macrostate of the gas is different. If the system S, which started out in M_0 at t_0, is observed to be (say) in macrostate M_1 at $t_0 + \Delta t$, then this means that the microstate of S is actually in the region of *overlap* between the region of macrostate M_1 and the region of the dynamical blob $B(t_0 + \Delta t)$. Now, in our above experiment, O carries out the experiment k times (or on k identical systems). In some of these experiments – actually in most of them (in our story) – at $t_0 + \Delta t$

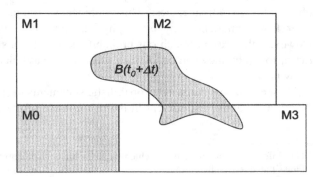

Fig. 6.1 The time evolved blob B(t) spreads over different macrostate

the system S is observed to be in M_1 and in other fewer experiments it is found in M_2 or M_3, or more precisely in the regions of *overlap* of the dynamical blob $B(t_0 + \Delta t)$ with these macrostates, with some relative frequencies F_1, F_2 and F_3 respectively. These relative frequencies are the empirical basis on which the probabilistic statements of the theory can be based, and on the basis of which these statements can be tested or justified.

The next step towards constructing or justifying the probabilistic theory is as follows. Given the above experimental outcomes, we have the relative frequencies with which systems of type S that start out in M_0 at t_0 are found in the macrostates M_1, M_2 or M_3. We conclude that the phase points of our k systems evolved into the regions of overlap of the dynamical blob $B(t_0 + \Delta t)$ with the macrostates M_1, M_2 or M_3. We then conjecture on the basis of our experience that this statistical behavior will be repeated (more or less) in the future. Since any of the microstates in M_0 is a possible initial condition of S_{k+1} and since the phase space is continuous, such a generalization of our experience requires that we impose a measure on the phase space. We identify the set of probability measures that, if applied to the continuous phase space of S, yield a measure of the regions of overlap of the blob $B(t_0 + \Delta t)$ with the macrostates M_1, M_2 or M_3 that are (to a satisfactory approximation) identical with the relative frequencies F_1, F_2 and F_3, respectively. There are many – possibly infinitely many – such measures, and all of them are *empirically adequate*. Among them we *choose* one measure, using *pragmatic* criteria such as simplicity, convenience, meshing with other theories, etc. Call this measure μ. The (normalized) measures of the regions of overlap are then given by $\mu(B(t) \cap M_1) \approx F_1$, $\mu(B(t) \cap M_2) \approx F_2$, $\mu(B(t) \cap M_3) \approx F_3$. This measure μ is imposed *over the blob* $B(t_0 + \Delta t)$ and provides the basis for predicting the evolution of system S_{k+1} in terms of transition probability (roughly) as follows:

(*) The transition probability that S_{k+1} will evolve to macrostate M_i at $t_0 + \Delta t$ given that it was in macrostate M_0 at t_0, is equal to

$$\mu(B(t_0 + \Delta t) \cap M_i | B(t_0) = M_0) \approx F_i.$$

That is, the transition probability from the macrostate M_0 at t_0 to M_i at $t_0 + \Delta t$ is equal to the (normalized) measure of the region of overlap of the blob $B(t_0 + \Delta t)$ with the macrostates M_i. This is the basis of our probabilistic theory.

Note that in general $\mu(M_i)/\mu(M_j)$ need not be equal to $\mu(B(t) \cap M_i)/\mu(B(t) \cap M_j)$.[3] Note further that despite the deterministic dynamics these transition probabilities between *macro*states are physically objective provided the partition to the macrostates is objective.[4]

What is the significance of taking the μ (normalized) measure over the blob $B(t_0 + \Delta t)$ as underwriting our probabilities for measurement outcomes? It is crucial

---

[3] This implies that the transition to a given macrostate need not be equal to the entropy of that macrostate, even if both are measured by the same measure μ.

[4] This last condition needs to be flashed out; we skip this here.

to see that the probabilistic statements are about *transitions* from M_0 at t_0 to any one of the macrostates M_i. We don't distribute probabilities relative to the μ measure over the initial macrostate M_0 at t_0. Of course, if the measure μ is invariant under the classical dynamics, e.g. if it happens to be the Lebesgue measure , then one can map, backwards (as it were), the measure of regions over the blob at later times to the corresponding regions over the *initial* mactostate. That is, in this case the measure of a set of points in M_0 is equal to the measure of the time evolved set of points to which it is mapped by the dynamics. Once the (normalized) measure is fixed (by the probabilities) one can distribute uniform probabilities relative to the Lebesgue measure over the initial macrostate. But note that this interpretative move is derivative. In general, whether or not the measure that best fits our observations is the Lebesgue measure, or more generally a measure that is invariant under the dynamics, is a contingent matter.

We can now see what justifies the choice of measure and what justifies probabilistic statements in classical statistical mechanics, and moreover how these two issues are related. First, probabilistic statements are grounded in the experience of relative frequencies in the way stated above. Second, the choice of measure is dictated inductively (not uniquely) by the observed relative frequencies. That is, the measure is implied by the probabilities rather than the other way around. We can only justify empirically transition probabilities as sketched in (*) above rather than distributions over initial conditions.

The implications of this analysis for the typicality approach are as follows.

1. The probability measure μ is applicable only to subsystems of the universe. Of course, if the dynamics is deterministic, each microstate of all the subsystems of the universe can be mapped backwards to the initial conditions and the measure over the initial conditions will depend on the measure at the later times. But in this way the justification of the choice of the measure over the initial conditions is grounded in experience, and therefore it cannot be taken to explain (non-circularly) experience. Note that this argument applies to the question of the choice of the measure regardless of whether the measure is understood as determining the typical set of initial conditions (as in the typicality approach) or as a probability measure over the initial conditions of the universe (as in standard approaches to statistical mechanics).

2. This strategy of grounding the measure over the initial conditions of the universe in experience can hold only with respect to a fraction of all possible initial conditions of the universe (compatible with the initial macrostate). It excludes by construction initial conditions that lead to a universe at the later times which is macroscopically different from what we see.

3. Our ignorance about the initial microstate of S_{k+1} is often illustrated by appealing to some random sampling of a point out of M_0. Of course, this idea need not be taken too seriously (as describing a fairy tale about some mechanism of selection). However, the point to be stressed here is the following. A random sampling is a sampling that depends only on the measure. The measure with respect to which the sampling is random need only be the measure that fits the observed relative frequencies in experience. In particular the measure need not

be the Lebesgue measure, and may not even be conserved under the dynamics. By appealing to the *probability* measure we can now justify statements about the probability of randomly sampling initial conditions for subsystems of the universe. Here unlike the statement (3) of the typicality approach, the sampling is described in terms of probability rather than typicality. The role of the measure in our approach is derivative rather than fundamental and is patently probabilistic.

6.4   Are There Natural Measures?

In the literature there are attempts to justify the choice of the measure (in the typicality approach and in other approaches) on the basis of dynamical considerations.

An argument sometime given for preferring the Lebesgue measure as 'natural' on the basis of the classical dynamics is the invariance of the Lebesgue measure under the dynamics as expressed by Liouville's theorem. If a measure is invariant under the dynamics it means that the measure of a given set of points in the state space is equal to the measure of the set to which it is mapped by the time evolution equations for all times. Of course this feature has very attractive properties (simplicity, elegance, etc.) but it is unclear why this fact is relevant at all to the issue at stake, namely the explanation and prediction of physical behavior.

A similar argument is sometimes given in the case of ergodic dynamics. Obviously, the ergodic theorem gives a preferred status to the Lebesgue measure (or to any measure absolutely continuous with the Lebesgue measure) since it shows that the relative frequency of any macrostate M along an infinite trajectory is equal to the Lebesgue measure of M for a Lebesgue measure one of points in the phase space of the system. There are various senses in which the preferred status of the Lebesgue measure here is irrelevant for the issue at stake. First, the ergodic theorem yields no predictions concerning finite times, and therefore strictly speaking the theorem is not empirically testable. For example, it is extremely difficult to distinguish empirically between an ergodic system and a system with KAM dynamics (see [9]). Second, even if the dynamics of the universe is granted to be ergodic and even if one accepts the fairy tale about an initial random sampling, this does not imply that the sampling is random relative to the Lebesgue measure. One can say metaphorically that God could have used a non-Lebesgue sort of die in sampling at random the initial condition of the universe even if the universe were ergodic. Third, and with respect to the typicality approach. Consider again statement (3) in Sect. 6.1. Here the idea is that the fact that T1 has measure (close to) one suggests that the initial condition of the universe belongs to T1. Since the measure is not to be understood as a probability measure, this seems to mean that the measure zero set is excluded as impossible in some sense. But the measure zero set belongs to the initial macrostate of the universe and we don't see what justifies this exclusion.

Finally, it is important to stress in this context that in understanding the ergodic theorem as a theorem about probability one must identify from the outset that a set of Lebesgue measure zero (one) has zero (one) probability. Although the theorem is usually understood in probabilistic terms, it should be stressed that this identification is *not* part of von Neumann's and Birkhoff's ergodic theorem. Whether or not the Lebesgue measure may be interpreted as the right *probability* measure for thermodynamic systems depends on whether it satisfies our probability rule (*).

Another argument sometimes given for taking the Lebesgue measure as the natural measure in statistical mechanics is that the Lebesgue measure of a macrostate corresponds to the thermodynamic entropy of that macrostate. However, this correspondence is true *only if* the Second Law of thermodynamics (even in its probabilistic version) is true. But as we argued elsewhere (see our [7, 8, 10], Chap. 5) the Second Law of thermodynamics is not universally true in statistical mechanics.

6.5 Lanford's Theorem

The above conclusion has implications for the significance of measure one theorems in statistical mechanics. We focus here as an example on Lanford's theorem.[5] Lanford proved on the basis of the classical equations of motion, that, roughly, given some specific initial macrostate, and some specific kind of Hamiltonian, a *Lebesgue* measure one of the microstates in that macrostate will evolve to a macrostate with larger entropy, after a certain short time.[6] Can such a theorem endow the Lebesgue measure with a status that is stronger than that of an empirical generalization (as sketched in (*) above)?

In terms of our transition probabilities Lanford's theorem proves that the Lebesgue measure of the overlap between the blob $B(t_0 + \Delta t)$ and the macrostate E of equilibrium (or some other high entropy macrostate) is 1. Of course, as we said above, since the Lebesgue measure is conserved under the dynamics, one may interpret Lanford's theorem as referring to the Lebesgue measure of subsets of the *initial* macrostate M_0 at t_0. However, inferring anything about the measure of subsets of the initial macrostate is an artifact of the contingent fact that the Lebesgue measure matches the observed relative frequencies.

Another crucial point in this context is the following. There are two different and logically independent ways of understanding the role of the Lebesgue measure in Lanford's theorem. (**A**) The size of the overlap between the blob $B(t_0 + \Delta t)$ and the macrostate E, as determined by the Lebesgue measure, is 1; (**B**) Upon a random sampling of a point out of the blob $B(t_0 + \Delta t)$, one is highly likely to pick out a point

---

[5] For details concerning Lanford's theorem see Uffink [11].

[6] The fact that a Maxwellian Demon is compatible with classical statistical mechanics demonstrates that there can be no theorem in mechanics that implies a universal entropy increase. See Albert [10, Chap. 5] and Hemmo and Shenker [12, 13].

from the overlap of the blob with E. The distinction between (**A**)-type statements about sizes of sets and (**B**)-type statements about probabilities is general.

Lanford's theorem is about the size of the overlap with E, that is, it is only an (**A**)-type theorem, whereas in order to make predictions about the future behavior of S-type systems (such as our S_{k+1} in the above example) one needs to add a (**B**)-type statement, *which is not proven by Lanford's theorem*. In other words, assuming that we already know from experience that the Lebesgue measure of the overlap regions (of the blob with the macrostates) matches the relative frequencies of the macrostates, Lanford's theorem provides possible mechanical conditions, which underwrite these observations.

To appreciate this point, note that if the measure μ that matches our experience were not the Lebesgue measure, but some other measure (that may not be absolutely continuous with Lebesgue) then Lanford's theorem would have a completely different significance: for instance, it could happen that by the measure μ the number of systems that go to equilibrium given Lanford's Hamiltonian would be small. The theorem that a set of Lebesgue measure one of points has a certain property (such as approaching equilibrium after some finite time interval) would be empirically insignificant – unless this fact is supplemented by the additional fact that the Lebesgue measure happens to correspond (to a useful approximation) to the observed relative frequencies.

The general structure of Lanford's theorem is that it proves a certain statement about the dynamics of the form of (1) in the typicality approach (see Sect. 6.2). That is, Lanford's theorem shows that a certain subset of micrsostates T1 share some property F (entropy increase, for example), such that all the points in T1 are mapped by the dynamics to points in T1*. Moreover, the theorem shows that the subset T1 has Lebesgue measure one. But nothing in this theorem justifies the choice of the measure. In particular, the fact that T1 has Lebesgue measure one does not constitute such a justification. What's important in Lanford's theorem is that it identifies two sets T1 and T1* and proves that T1 evolves to T1* under the dynamics. That is, the theorem is about the structure of trajectories. The fact that T1 has Lebesgue measure one is important only if there are independent reasons for preferring the Lebesgue measure. As we saw in Sect. 6.3 such reasons can be grounded essentially only in experience.

6.6 Conclusion

In this paper we showed that one can understand the full scope of classical statistical mechanics by appealing to the notion of *transition* probabilities between macrostates, without resorting to probability distributions over initial conditions or to typicality considerations.

Acknowledgement This research is supported by the Israel Science Foundation, grant numbers 240/06 and 713/10.

References

1. Pitowsky, I.: Typicality and the role of the Lebesgue measure in statistical mechanics, In Y. Ben-Menahem and M. Hemmo (eds.), Probability in Physics, pp. 41–58. The Frontiers Collection, Springer-Verlag Berlin Heidelberg (2011)
2. Einstein, A.: Über die von der molekularkinetischen Theorie der Wärme geforderte Bewegung von in ruhenden Flüssigkeiten suspendierten Teilchen. Annal. Phys. 17, 549–560 (1905). English translation in: Furth, R. (ed.) Einstein, A., Investigations on the Theory of Brownian Motion. Dover, New York (1926)
3. Pitowsky, I.: Why does physics need mathematics? A comment. In: Ulmann-Margalit, E. (ed.) The Scientific Enterprise, pp. 163–167. Kluwer, Dordrecht (1992)
4. Dürr, D., Goldstein, S., Zanghi, N.: Quantum equilibrium and the origin of absolute uncertainty. J. Stat. Phys. 67(5/6), 843–907 (1992)
5. Mauldin, T.: What could be objective about probabilities? Stud. Hist. Philos. Mod. Phys. 38, 275–291 (2007)
6. Callender, C.: The emergence and interpretation of probability in Bohmian mechanics. Stud. Hist. Philos. Mod. Phys. 38, 351–370 (2007)
7. Goldstein, S., Lebowitz, J., Mastrodonato, C., Tumulka, R., Zanghi, N.: Normal typicality and von Neumann's quantum ergodic theorem. Proc. R. Soc. A 466(2123), 3203–3224 (2010)
8. Goldstein, S., Lebowitz, J., Mastrodonato, C., Tumulka, R., Zanghi, N.: Approach to thermal equilibrium of macroscopic quantum systems. Phys. Rev. E 81, 011109 (2010)
9. Earman, J., Redei, M.: Why ergodic theory does not explain the success of equilibrium statistical mechanics. Br. J. Philos. Sci. 47, 63–78 (1996)
10. Albert, D.: Time and Chance. Harvard University Press, Cambridge (2000)
11. Uffink, J.: Compendium to the foundations of classical statistical physics. In: Butterfield, J., Earman, J. (eds.) Handbook for the Philosophy of Physics, Part B, pp. 923–1074. (2007)
12. Hemmo, M., Shenker, O.: Maxwell's demon. J. Philos. 107(8), 389–411 (2010)
13. Hemmo, M., Shenker, O.: Szilard's perpetuum mobile. Philosophy of Science 78, 264–283 (2011)

Chapter 7
A New Approach to the Approach
to Equilibrium

Roman Frigg and Charlotte Werndl

Abstract Consider a gas confined to the left half of a container. Then remove the
wall separating the two parts. The gas will start spreading and soon be evenly
distributed over the entire available space. The gas has approached equilibrium.
Why does the gas behave in this way? The canonical answer to this question,
originally proffered by Boltzmann, is that the system has to be ergodic for the
approach to equilibrium to take place. This answer has been criticised on different
grounds and is now widely regarded as flawed. In this paper we argue that these
criticisms have dismissed Boltzmann's answer too quickly and that something
almost like Boltzmann's answer is true: the approach to equilibrium takes place if
the system is epsilon-ergodic, i.e. ergodic on the entire accessible phase space
except for a small region of measure epsilon. We introduce epsilon-ergodicity and
argue that relevant systems in statistical mechanics are indeed espsilon-ergodic.

7.1 Introduction

Let us begin with a paradigmatic example. A gas is confined to the left half of a
container by a dividing wall. We now remove the wall, and as a result the gas spreads
uniformly across the entire container. It reaches equilibrium. Thermodynamics (TD),
via its Second Law, regards this process as uniform and irreversible: once the wall is
removed, the entropy increases until it reaches its maximum which it will thereafter

Authors are listed alphabetically. This work is fully collaborative.

R. Frigg (✉)
Department of Philosophy, Logic and Scientific Method and Centre for Philosophy of Natural and
Social Science, London School of Economics, London, UK
e-mail: r.p.frigg@lse.ac.uk

C. Werndl
Department of Philosophy, Logic and Scientific Method, London School of Economics, London, UK
e-mail: c.s.werndl@lse.ac.uk

Y. Ben-Menahem and M. Hemmo (eds.), *Probability in Physics*, The Frontiers Collection,   99
DOI 10.1007/978-3-642-21329-8_7, © Springer-Verlag Berlin Heidelberg 2012

never leave. Statistical mechanics (SM) tries to understand this manifest macroscopic behaviour in terms of the dynamics of the micro-constituents of the system.

One might expect SM to provide a justification of the *exact* laws of TD, in our case a justification of why systems invariably exhibit monotonic and irreversible entropy increase. This is asking for too much. In fact we have to rest content with less in two respects. First, classical Hamiltonian systems are time-reversal invariant and show Poincaré recurrence; it is therefore impossible for the entropy of such a system to increase irreversibly: sooner or later the system will move out of equilibrium again. Thermodynamics is an approximation, which, echoing Callender's [1] memorable phrase, we should not take too seriously.[1] Instead of trying to derive irreversible behaviour *stricto sensu* we should aim to show that systems in SM exhibit *thermodynamic-like behaviour* (TD-like behaviour): the entropy of the evolving system is most of the time close to its maximum value, from which it exhibits frequent small and rare large fluctuations [3, p. 255]. Second, the Second Law of TD does not allow for exceptions. However, no statistical theory can ever justify an exceptionless law. The best one could hope for is to show that something happens with probability equal to one (but even then zero-probability-events are not ruled out because zero probability is not impossibility!). But even that is a tall order since probability zero results are usually unattainable. What we have to aim for instead is showing that the desired behaviour is *very likely* [4].

These considerations suggest a new approach to the approach to equilibrium: rather than trying to derive monotonic and exceptionless entropy increase, we ought to aim to show that systems in SM are very likely to exhibit TD-like behaviour. The aim of this paper is to propose a response to this challenge. But before turning to our proposal, let us briefly comment on a recent approach which offers an explanation of TD-like behaviour in terms of the notion of typicality (see, for instance, [5]) and without explicit reference to dynamical properties of the system. In our view, such an explanation is either flawed or incomplete.[2]

TD-like behaviour is a dynamical phenomenon. SM is a reductionist enterprise in that its constitutive assumption is that the behaviour of large systems is determined by the behaviour of its constituents. In the case of the initial example this means that the behaviour of the gas is determined by the behaviour of the gas molecules; that is, the gas spreads because the individual molecules bounce around in such a way that they fill the space evenly and that their velocities obey the Maxwell-Boltzmann distribution. So the question is: what kind of motion do the molecules have to carry out for the gas as a whole to show TD-like behaviour? The motion of molecules is governed by the laws of mechanics, which we assume to be the laws of classical Hamiltonian mechanics. What kind of motion a Hamiltonian system carries out is determined by the Hamiltonian of the system. The question then becomes: what dynamical properties does the Hamiltonian have to possess for the system to show TD-like behaviour?

---

[1] In passing we would like to mention that deriving the exact laws of TD from SM is also not a requirement for a successful reduction (see 2).

[2] For a detailed discussion of this approach see Frigg [6, 7].

An answer to this question is an essential ingredient of an explanation of the approach to equilibrium. For one, the only way to deny that the dynamics of molecules matters is to deny reduction, but this amounts to pulling the rug from underneath SM altogether. For another, the answer to the question about dynamics is non-trivial because there are Hamiltonians under which systems do *not* show TD-like behaviour (for instance, quadratic Hamiltonians). So we need to know what properties a Hamiltonian must have for TD-like behaviour to take place. And this question must be answered in a non-trivial way. Just saying the relevant Hamiltonians possess the dynamical property of TD-likeness has no explanatory power—it would be a pseudo-explanation of the *vis dormitiva variety*. *The challenge is to identify in a non-question-begging* way a dynamical property (or, indeed, properties) that those Hamiltionians whose flow is TD-like have.

The traditional answer to this question (which can be traced back to Boltzmann) is that the system has to be ergodic. In recent discussions this answer has fallen out of favour. After introducing the formalism of Boltzmannian Statistical Mechanics (Sect. 7.2), we briefly discuss the Boltzmannian justification of TD-like behaviour along with the criticisms levelled against it (Sect. 7.3). There is indeed a serious question whether the original proposal is workable (although, rife prejudice notwithstanding, there is no *proof* that it fails). For this reason is seems sensible to look for a less uncertain solution. We point out that to justify TD-like behaviour it suffices that a system be almost ergodic, where being almost ergodicis is explained in terms of epsilon-ergodicity (Sect. 7.4). The most important criticism of the ergodic programme is that relevant systems in SM are, as a matter of fact, not ergodic. We review the two most powerful arguments for this conclusion—based on the so-called KAM-Theorem and Markus-Meyer-Theorem, respectively—and argue that they have no force against epsilon-ergodicity (Sect. 7.5). Not only do these arguments have no force against epsilon-ergodicty, there are good reasons to believe that relevant systems in SM are epsilon-ergodic (Sect. 7.6). We end with a summary of our results (Sect. 7.7).

7.2 Boltzmannian Statistical Mechanics

We consider Boltzmannian SM and set Gibbsian SM aside, and we restrict attention to gases.[3] Furthermore we assume systems to be classical;[4] a discussion of quantum SM can found in Emch and Liu [10].

Consider a system of n particles moving in three-dimensional physical space. The system's *microstate* is specified by a point x in its $6n$-dimensional phase

---

[3] The explanation of TD-like behaviour in liquids and solids demands conceptual resources we cannot discuss here. Let us just mention that an explanation of thermodynamic-like behaviour in liquids and solids might well differ from an explanation of thermodynamic-like behaviour in gases. In other words, we see no reason why for systems as different as gases and solids there has to be one single dynamical property that explains thermodynamic-like behaviour.

[4] For a discussion of Gibbsian SM see Frigg [8] and Uffink [9].

space Γ. This space is endowed with the standard Lebesgue measure μ. The time evolution of the system is governed by Hamilton's equations, and the function s_x : $\mathbb{R} \to \Gamma_E$, $s_x(t) = \phi_t(x)$ is the solution originating in x. Because the energy is conserved, the motion of the system is confined to a $6n - 1$ dimensional energy hypersurface Γ_E, where E is the value of the energy of the system. The measure μ is preserved under the dynamics of the system, and so is its restriction to Γ_E, μ_E. If normalised, μ_E is a probability measure on Γ_E. From now on we assume that μ_E be normalised. The triple $(\Gamma_E, \mu_E, \phi_t)$ is a *measure-preserving dynamical system*, where $\phi_t : \Gamma_E \to \Gamma_E (t \in \mathbb{R})$ is a family a one-to-one measurable mappings such that $\phi_{t+s} = \phi_t(\phi_s)$ for all $t, s \in \mathbb{R}$, $\phi_t(x)$ is jointly measurable in (x, t), and $\mu_E(R) = \mu_E(\phi_t(R))$ for all measurable $R \subseteq \Gamma_E$ and all $t \in \mathbb{R}$ (which is the condition of measure-preservation).

From a macroscopic perspective the system is characterised by a set of *macrostates M_i, $i = 1, \ldots, m$*. To each macrostate corresponds a macro-region Γ_{M_i} consisting of all $x \in \Gamma_E$ for which the system is in M_i. The Γ_{M_i} form a partition of Γ_E, meaning that they do not overlap and jointly cover Γ_E. The Boltzmann entropy of a macrostate M_i is $S_B(M_i) := k_B \log[\mu(\Gamma_{M_i})]$ (where k_B is the Boltzmann constant), and the Boltzmann entropy of a *system* at time t, $S_B(t)$, is the entropy of the macrostate of the system at t: $S_B(t) := S_B(M_{x(t)})$, where $x(t)$ is the microstate at t and $M_{x(t)}$ is the macrostate corresponding to $x(t)$ (cf. [11]). The equilibrium state, M_{eq}, and the macrostate at the beginning of the process, M_p, also referred to as the 'past state', are particularly important. For gases $\Gamma_{M_{eq}}$ is vastly larger (with respect to μ_E) than any other macro-region, a fact also known as the 'dominance of the equilibrium macrostate' (we briefly return to this in the conclusion); in fact Γ_E is almost entirely taken up by equilibrium microstates (see, for instance, [5], p. 45).[5] For this reason the equilibrium state has maximum entropy. The past state is, by assumption, a low entropy state. The Boltzmann entropy is the quantity that is expected to show TD-like behaviour.

7.3 The Ergodic Programme

We now introduce the notion of ergodicity and discuss the problems that attach to it when used in the context of Boltzmannian SM. The *time-average* of the phase flow ϕ_t relative to a measurable set A of Γ_E of a solution starting at $x \in \Gamma_E$ is

$$L_A(x) = \lim_{t \to \infty} \frac{1}{t} \int_0^t \chi_A(\phi_\tau(x)) d\tau, \tag{7.1}$$

---

[5] As Lavis [3, pp. 255–258] has pointed out, some care is needed here. Non-equilibrium states can be degenerate and *together* can take up a large part of Γ_E. However, those non-equilibrium states that occupy most of the non-equilibrium area have close to equilibrium entropy values and so one can then lump together equilibrium and close-to-equilibrium states and get an 'equilibrium or almost equilibrium' region, which indeed takes up most of Γ_E. The approach to equilibrium can then be understood as the approach to this 'equilibrium or almost equilibrium' state.

where the measure on the time axis is the Lebesgue measure and $\chi_A(x)$ is the characteristic function of A: $\chi_A(x) = 1$ for $x \in A$ and 0 otherwise. Birkhoff [12] could prove that $L_A(x)$ exists except for a set of measure zero, i.e., except for a set B in Γ_E with $\mu_E(B) = 0$. A system is ergodic (on the energy hypersurface) if and only if (iff) for all measurable A in Γ_E

$$L_A(x) = \mu_E(A) \tag{7.2}$$

for all $x \in \Gamma_E$ except for a set of measure zero.

Ergodic systems exhibit TD-like behaviour. Setting $A = \Gamma_{M_{eq}}$ and taking into account the dominance of the equilibrium macro region, it follows immediately that almost all initial conditions lie on solutions that spend most the time in equilibrium and only show relatively short fluctuations away from it (because non-equilibrium regions are small compared to $\Gamma_{M_{eq}}$). Therefore, the Boltzmann entropy is maximal most of the time and fluctuates away from its maximum only occasionally: the system behaves TD-like.[6]

In passing we would like mention that neither TD itself, nor TD-like behaviour as defined above, make any statement about how quickly a system approaches equilibrium; that is, they remain silent about relaxation times. The same holds true of ergodicity, which is also silent about how long it takes a system to reach equilibrium. This is no drawback: it is unlikely that one can say much about the speed of convergence in general because this will depend on the system under consideration. However, it is true that many gases approach equilibrium fairly quickly, and a full justification of the macroscopic behaviour of systems has to show that the relevant dynamical systems show realistic relaxation times. For want of space we do not pursue this issue further.

The two main arguments levelled against the ergodic approach are the measure zero problem and the irrelevancy charge. The measure zero problem is that $L_A(x) = \mu_E(A)$ holds only 'almost everywhere', i.e. except, perhaps, for initial conditions of a set of measure zero. This is seen as a problem because sets of measure zero can be rather 'big' (for instance, the rational numbers have measure zero within the real numbers) and because sets of measure zero need not be negligible if sets are compared with respect to properties other than their measures (see, for instance, [16], pp. 182–188).

What lies in the background of this criticism is the quest for a justification of a strict version of the Second Law. However, as we have pointed out in the introduction, this is an impossible goal. At best SM can show that TD-like behaviour is very likely, and there is no way to rule out that there are initial conditions for which this is not the case. As long as the probability for this to happen is low, this is no threat to the programme. In fact, our explanation (in the next section) for why systems behave TD-like is even more permissive than the traditional ergodic programme:

---

[6] We can then interpret μ_E as a time-average. For a discussion of this interpretation see Frigg [13], Lavis [14] and Werndl [15].

it allows for sets of 'bad' initial conditions that have finite (yet very small) measure.[7]

The second objection, the irrelevancy challenge, is that ergodicity is irrelevant to SM because real systems are not ergodic. In effect, by appealing to ergodicity we are like the proverbial fool who searches for his lost wallet underneath the lantern. This is a serious objection, and the aim of this paper is to develop a response to it. Our response departs from the observation that less than full-fledged ergodicity is sufficient to explain why systems behave TD-like most of the time. We introduce epsilon-ergodicity and then argue that epsilon-ergodicity gives us what we need. We then revisit the main two arguments for the conclusion that SM systems are not ergodic and show that they have no force against epsilon-ergodicity (or ergodicity).

7.4 Epsilon-Ergodicity and Thermodynamic-Like Behaviour

Roughly speaking, a system is epsilon-ergodic if it is ergodic on the entire energy hypersurface except, perhaps, on a set of measure ε, where ε is very small or zero.[8] In order to eventually introduce epsilon-ergodicity, we first define the (different!) notion of ε-ergodicity. The latter captures the idea that a system is ergodic on a set of measure $1 - \varepsilon$: $(\Gamma_E, \mu_E, \phi_t)$ is ε-ergodic, $\varepsilon \in \mathbb{R}, 0 \leq \varepsilon < 1$, iff there is a set $Z \subset \Gamma_E$, $\mu(Z) = \varepsilon$, with $\phi_t(\hat{\Gamma}_E) \subseteq \hat{\Gamma}_E$ for all $t \in \mathbb{R}$, where $\hat{\Gamma}_E := \Gamma_E \setminus Z$, such that the system $(\hat{\Gamma}_E, \mu_{\hat{\Gamma}_E}, \phi_t^{\hat{\Gamma}_E})$ is ergodic, where $\mu_{\hat{\Gamma}_E}(\cdot) := \mu_E(\cdot)/\mu_E(\hat{\Gamma}_E)$ for any measurable set in $\hat{\Gamma}_E$ and $\phi_t^{\hat{\Gamma}_E}$ is ϕ_t restricted to $\hat{\Gamma}_E$. Trivially, a 0-ergodic system is simply an ergodic system. We now say that a dynamical system $(\Gamma_E, \mu_E, \phi_t)$ is epsilon-ergodic iff there exists a very small ε (i.e. $\varepsilon \ll 1$) for which the system is ε-ergodic.

An epsilon-ergodic system $(\Gamma_E, \mu_E, \phi_t)$ is ergodic on $\Gamma_E \setminus Z$, and, therefore, it shows thermodynamic-like behaviour for the initial conditions in $\Gamma_E \setminus Z$. If ε is very small compared to $\mu_E(\Gamma_{M_p})$, then the system will behave TD-like for most initial conditions (i.e. for all initial conditions except, perhaps, ones that form a set of measure ε).[9] If we now interpret μ_E as a probability density (which we are free to do because it has the formal properties of a probability measure),[10] then it follows that

---

[7] This solution (or rather: dissolution) of the measure zero problem presupposes that the initial conditions are measured with respect to the Lebesgue measure. Justifying this choice is a well-known and thorny problem which we cannot address here. In what follows we assume that such a justification can be given and that the Lebesgue measure is the right measure to use in these cases.

[8] Epsilon-ergodicity has been introduced into the foundations of SM by Vranas [17]. However, Vranas uses it to justify Gibbsian equilibrium theory, while we use it within Boltzmannian SM. For a discussion of Vranas' views, see Frigg [8, pp. 149–151].

[9] A weaker antecedent still warrants the consequent: $\mu_E(\Gamma_{M_p} \setminus Z)/\mu_E(\Gamma_{M_p})$ has to be close to one. This is trivially true if $\mu_E(Z)$ is small compared to $\mu_E(\Gamma_{M_p})$, but it can also be true if $\mu_E(Z)$ is larger (but substantial parts of Z come to lie in other macro-regions).

[10] For a discussion of how to interpret these probabilities see Frigg and Hoefer [18].

the system is overwhelmingly likely to behave TD-like. Therefore, we find that if a system is epsilon-ergodic, then it is overwhelmingly likely to behave TD-like. This is the sought after result.

7.5   Threats from the Sidelines

This result is relevant only if real systems are actually epsilon-ergodic. In this section we discuss two general mathematical theorems that are often marshaled against ergodicity and argue that these arguments are based on a misinterpretation of the theorems. In the next section we look at some important systems in SM and provide evidence (both mathematical and numerical) that they are indeed epsilon ergodic.

The Kolmogorov-Arnold-Moser theorem (KAM-Theorem). Physically speaking, a first integral of a dynamical system is constant of motion. Formally, a function G is a *first integral* of a dynamical system with Hamiltonian H just in case the Poisson bracket $\{H, G\}$ equals zero. A dynamical system with n degrees of freedom is called integrable (in the sense of Liouville) just in case there are n independent first integrals G_i which are in involution (the G_i are in involution iff $\{G_i, G_j\} = 0$ for all $i, j, 1 \le i, j \le n$). Iff a dynamical system is not integrable, it is called *nonintegrable*. For an integrable system the energy hypersurface is foliated into tori, and on each torus there is either periodic motion or quasi-periodic motion with a specific frequency [19, 20].

The KAM-theorem gives an answer to the question of what happens when an integrable system is perturbed by a small perturbation which is nonintegrable. According to the KAM-theorem, under certain conditions,[11] there are two kinds of motion on the hypersurface of constant energy. Namely, first, there is the motion on tori with sufficiently irrational frequencies; the solutions on these tori behave like the ones in the integrable case, meaning that there is quasi-periodic motion (these tori are said to "survive the perturbation"). Second, between the surviving tori the motion is irregular and unpredictable. As the perturbation decreases, the measure of the tori which survive the perturbation goes to one. Thus the hypersurface of constant energy splits into two regions invariant under the dynamics: the region where the tori survive and the region where this is not the case; moreover, the measure of the former goes to one as the perturbation goes to zero. The motion on the region where the tori survive cannot be ergodic or epsilon-ergodic because the solutions are confined to tori. Consequently, dynamical systems to which the KAM-Theorem applies fail to be ergodic, and for a small enough perturbations they also fail to be epsilon-ergodic (cf. [20]).

---

[11] It is required that (i) one of the frequencies never vanishes, and (ii) that the ratios of the non-vanishing frequency to the remaining $n - 1$ frequencies are functionally independent on the entire energy hypersurface (this means that the ratios depend on the action) [20, pp. 182–183].

This consequence of the KAM-theorem is often taken to show that many, or even all, systems in SM fail to be ergodic. Consider, for instance, the following quotations:

[T]he evidence against the applicability [of ergodicity in SM] is strong. The KAM-Theorem leads one to expect that for systems where the interactions among the molecules are non-singular, the phase space will contain islands of stability where the flow is non-ergodic. [21, p. 70]

Actually, demonstrating that the conditions sufficient for the regions of KAM-stability to exist can only be done for simple cases. But there is strong reason to suspect that the case of a gas of molecules interacting by typical intermolecular potential forces will meet the conditions for the KAM result to hold. [...] So there is plausible theoretical reason to believe that more realistic models of typical systems discussed in statistical mechanics will fail to be ergodic. [16, p. 72]

First appearances notwithstanding, these claims are unfounded. The KAM-Theorem does not show that gases in SM fail to be ergodic (and hence does not show that they fail to be epsilon-ergodic). The crucial point, which is often ignored, is that the KAM-theorem only applies to *extremely small* perturbations of integrable systems. For systems in SM it has been found that the largest admissible perturbation parameter rapidly converges toward zero as the number of degrees of freedom n goes to infinity [22, 23]. Consequently, as Pettini points out, "for large n-systems – which are dealt with in statistical mechanics – the admissible perturbation amplitudes for the KAM-theorem to apply drop down to exceedingly tiny values of no physical meaning" [22, p. 60]. For larger perturbations the surviving tori disappear and the motion can be epsilon-ergodic or even ergodic. Thus the KAM-theorem is simply irrelevant because it does not apply to gases in SM.

Moreover, it is at best unclear whether systems in SM can be represented as integrable systems plus a small perturbation.[12] Hamiltonians of that kind are extremely special. And not only is there no reason to believe that SM systems are of this special kind; the systems commonly studied in SM are not (as becomes clear in the next section). Hence, once again, the KAM-Theorem is just irrelevant to the question of whether or not systems in SM are epsilon-ergodic (or ergodic), and dismissals of the ergodic approach based on the KAM-Theorem are misguided.

The Markus-Meyer Theorem (MM-Theorem). The MM-theorem is about the class of infinitely differentiable Hamiltonians on a compact manifold. It says that in this class nonergodic systems are generic in a topological sense (of first Baire category) [24]. Furthermore, when studying the proof of the MM-Theorem, one sees that the proof implies that the set of Hamiltonians which are not epsilon-ergodic are also generic. Here an *ergodic Hamiltonian (epsilon-ergodic Hamiltonian)* (as opposed to a dynamical system) is defined to be a Hamiltonian which is ergodic (epsilon-ergodic) on the energy hypersurface for a dense set of energy values. So is the MM-Theorem a threat to the claim that all gases in SM are epsilon-ergodic?

---

[12] Thanks to Pierre Lochack for making us aware of this.

We do not think so for two reasons. First, the proof of the MM-theorem shows that those Hamiltonians which are generic are not epsilon-ergodic because there is exactly one minimum value of the energy (where the motion is a general elliptic equilibrium point). And for energy values which are arbitrarily close to this minimum the motion on the energy hypersurface is not epsilon-ergodic. However, these very low energy values are of no relevance to gases in SM. Either for very low energy values quantum effects come in, rendering these energy values irrelevant for SM. Or these low energy values do not correspond to gases but to glasses or solids [17, 25–28].

Second, the MM-Theorem only holds for compact phase spaces. However, for systems considered in SM the phase space is usually not compact (see, e.g., [19]).[13] The proof of the MM-Theorem cannot be easily transferred to noncompact phase spaces; but this is exactly what would be needed. For these reasons, also the MM-Theorem is no threat to the claim that all gases in SM are epsilon-ergodic (or ergodic).

7.6 Relevant Cases

A different line of attack draws attention to particular systems that fail to be ergodic and yet behave TD-like, from which it is concluded that ergodicity cannot explain TD-like behaviour. We will argue that these examples are besides the point and that there are good reasons to believe that gases in SM are epsilon-ergodic.

Common counterexamples to the ergodic programme are the following. First, solids show thermodynamic-like behaviour; however, in a solid the molecules oscillate around fixed positions in a lattice, implying that a state can only access a small part of the energy hypersurface [9, p. 1017]. Second, a system of n uncoupled anharmonic oscillators of identical mass shows TD-like behaviour, but it is not ergodic [29]. Third, the Kac Ring Model is known not to be ergodic, but it still shows TD-like behaviour (ibid.). Fourth, a system of non-interacting point particles is not ergodic, yet it is still often studied in SM [30, p. 381].

None of these examples threatens our claim that gases in SM are epsilon-ergodic. Clearly, solids are not gases and hence can be set aside. Similarly, uncoupled harmonic oscillators and the Kac-ring model are irrelevant because they seem to have nothing to do with gases. The properties of ideal gases are very different from the properties of real gases because there are no collisions in ideal gases and collisions are essential to the behaviour of gases. So while ideal gases may be an expedient in certain context, no conclusion about the dynamics of real gases should be drawn from them. Hence, the well-rehearsed examples do not establish that there is a gas-like system which behaves TD-like while failing to be

---

[13] The hypersurface of constant energy is usually compact, but the phase space is not, and the theorem cannot be rephrased as one about the energy hypersurface.

ergodic.[14] We now argue that this is not an artifact of the way the examples have been chosen; gases do seem to be epsilon-ergodic. We should point out that there are only few rigorous results about the dynamical properties of gases. Nevertheless, these, together with the results of some numerical studies, support the hypothesis that gases in SM are epsilon-ergodic.

The dynamics of a gas is specified by the potential which models the force between the particles. Two potentials are of particular importance: *the Lennard-Jones potential* and *the hard-sphere potential*. For two particles the Lennard-Jones potential has the form:

$$U(r) = 4\alpha \left(\left(\frac{\rho}{r} \right)^{12} - \left(\frac{\rho}{r} \right)^{6} \right), \tag{7.3}$$

where r is the distance between the particles, α corresponds to the depth of the potential well and ρ is the distance at which the inter-particle potential is 0. From this one obtains the potential of the entire system by summing over all two-particle interactions or by considering only the interactions between the nearest neighbours. The Lennard-Jones potential is among the most widely-used potentials because it agrees well with the data about inter-particle forces [31, pp. 236–237, 32, pp. 502–505].

The hard-sphere potential models the motion of impenetrable spheres of radius R that bounce off elastically. For two particles the hard-sphere potential is:

$$U(r) = \infty \quad \text{for} \quad r < R \quad \text{and 0 otherwise,} \tag{7.4}$$

where r is the distance between the particles. Again, one obtains the potential of the entire system by summing over all two-particles interactions. The hard-sphere potential simulates the steep repulsive part of realistic potentials [31, p. 234]. It is widely used in mathematical as well as numerical studies because it is the simplest potential.

Let us start by discussing the hard-sphere potential. Boltzmann [33] already studied this potential and conjectured that hard-sphere systems are ergodic when the number of balls is large. From a mathematical viewpoint it is easier to study the movement of particles on a *torus* rather than the movement of particles in box or in other containers with walls. For particles moving on a torus there are no walls; it is like if a ball reappears at the opposite side of the box instead of bouncing off the wall. Studying the motion of hard-spheres on a torus is important: if anything, the walls cause the motion to be more random than the motion on a torus (relative to the trivial invariants of motion; see Chernov [34]). Thus if the motion of hard-spheres on a torus is ergodic, this provides good evidence that the motion of hard

---

[14] However, the case of solids highlights an important issue. Namely that the approach to equilibrium in solids is an unsolved problem and that this problem deserves more attention than it has received so far.

spheres in a box (or other containers) is ergodic. Sinai [35] hypothesised that a system of N hard-spheres moving on T^2 and on T^3 is ergodic for all $N \geq 2$ where T^m is the m-torus [36]; this hypothesis became known later as the 'Boltzmann-Sinai ergodic hypothesis'. The first step towards proving this hypothesis was made by Sinai [37], who showed that the motion of two hard spheres on T^2 is ergodic.[15] Since then several important proofs have been accomplished; taken together, they add up to an almost complete proof of the Boltzmann-Sinai ergodic hypothesis (and mathematicians in this field expect that a full proof will be forthcoming soon). Three results are particularly important. First, Simányi [42] showed that a system of N hard-spheres moving on T^m is ergodic for all $m \geq N$, $N \geq 2$. Second, Simányi [43] proved that a system of N hard spheres moving on T^m is ergodic for all $N \geq 2$ and all $m \geq 2$ and for almost all values $(M_1, \ldots, M_N, r)$, where M_i is the mass of the i-th ball and r is the radius of the balls.[16] Third, Simányi [44] showed that a system of N hard spheres moving on T^m is ergodic for all N and all m provided that the Sinai-Chernov Ansatz is true (mathematicians who work in this field widely expect that the Sinai-Chernov Ansatz holds).[17]

Obtaining strict mathematical results about the more realistic case of hard-spheres moving in a box (rather than on a torus) is more difficult. Only few results have been obtained here. Most importantly, Simányi [45] proved that the system of two balls moving in an m-dimensional box is ergodic for all m. Numerical studies suggest that the same result holds true for an arbitrary number of balls. Zheng et al. [46] found evidence that systems of identical hard-spheres in a two-dimensional and a three-dimensional box are ergodic. Dellago and Posch [47] studied systems of a large number of identical hard-spheres in a three-dimensional box, and obtained numerical evidence that the motion is ergodic.

We now turn to the Lennard-Jones potential, which is much harder to treat mathematically. Donnay [48] showed that a system of two particles moving on T^2 where there is a generalised Lennard-Jones type potential is not ergodic for certain values of the energy of the system.[18] However, this result does not say anything

---

[15] All hard-sphere systems which are discussed in this section are not only ergodic but are also strongly chaotic – they are Bernoulli systems (for a discussion of the meaning of Bernoulli systems, see [38–41]).

[16] We are most interested in the case where the system has equal masses. Unfortunately, it is unknown whether the system is ergodic for equal masses because the proof does not provide an effective method of checking whether a given $(M_1, \ldots, M_N, r)$ is among the values where the system is ergodic [44, p. 383].

[17] Consider ∂M, the boundary of all possible states M of the hard-sphere system. Define SR^+ as the set of all states x in δM which correspond to singular reflections with the post-collision velocity v_0, for any arbitrary v_0. According to the Chernov-Sinai Ansatz, the forward solution originating from x is geometrically hyperbolic for almost every $x \in SR^+$ [44, p. 392].

[18] The set of generalised Lennard-Jones potentials consists of potentials of the same general shape as the Lennard-Jones potential and potentials which share some characteristics with the Lennard-Jones potential. More specifically, generalised Lennard-Jones potentials as considered by Donnay are smooth potentials where (a) for large r the potential is attracting, (b) as r goes to zero the

about the cases of interest in *SM*, namely systems with a large number of particles. And there is a general tendency that the larger the number of particles, the more often systems are ergodic. Important for us is that even if systems with Lennard-Jones potentials and with a large number of particles should turn out to be non-ergodic, they are likely to be epsilon-ergodic [49]. Donnay [48, p. 1024] expresses this as follows:

> Even if one could find such examples [generalised Lennard-Jones systems with a large number of particles that are non-ergodic], the measure of the set of solutions constrained to lie near the elliptic periodic orbits is likely to be very small. Thus from a practical point of view, these systems may appear to be ergodic.

Indeed, it is widely believed that Lennard-Jones type systems are epsilon-ergodic because similar systems are epsilon-ergodic and numerical studies provide evidence that they are epsilon-ergodic. More specifically, numerical studies of systems with Lennard-Jones potentials have found that there exists an energy threshold (a specific value of the energy) such that the system is epsilon-ergodic for values above the energy threshold and fails to be epsilon-ergodic for values below the threshold. Whether for energy values below the threshold the system is really not epsilon-ergodic, or is epsilon-ergodic but appears to be not so because it needs a very long time to approach equilibrium is still discussed [49–51]. Important for our purpose is that the energy values below the energy threshold are very low. This implies that the classical statistical mechanical description breaks down because quantum effects cannot be ignored any longer [17, 27, 32]. Consequently, the behaviour of these systems with very low energy values is irrelevant. In conclusion, there is evidence that gases with a Lennard-Jones potential are epsilon-ergodic for the relevant energy values.

After having discussed the hard-sphere potential and the Lennard-Jones potential, we want to briefly mention two important results about other potentials of relevance in SM. First, Donnay and Liverani [52] proved that the motion of two particles moving on T^2 is ergodic for three types of potentials, namely for a general class of repelling potentials, a general class of attracting potentials, and a class of potentials with attracting and repelling parts (the latter are called mixed potentials). Of particular importance here are the mixed potentials because they are everywhere smooth. Everywhere smooth potentials are regarded as more realistic than potentials with singularities, and Donnay and Liverani's [52] mixed potentials were the first smooth potentials which were proven to lead to ergodic motion. Second, among the systems with many degrees of freedom which have been most extensively investigated is the one-dimensional self-gravitating system consisting of N plane-parallel sheets with uniform density; this system models processes in plasma physics. Numerical investigations suggest that for $N \geq 11$ the system is

---

potential approaches infinity, and (c) the potential has finite range, i.e., there exists an $R>0$ such that $U(r) = 0$ for all $r \geq R$ ((c) is assumed because it considerably simplifies the mathematical treatment).

ergodic [53–55]. To conclude, the mathematical and numerical results provide evidence for the claim that all gases in SM are epsilon-ergodic.

7.7 Conclusion

This paper aimed to explain why gases exhibit thermodynamic-like behaviour. We have argued that there is thermodynamic-like behaviour when the system is epsilon-ergodic, i.e., ergodic on the entire accessible phase space except for a small region of measure epsilon. Then we have shown that the common objections against the ergodic approach are misguided and that there are good reasons to believe that the relevant systems in statistical mechanics are indeed epsilon-ergodic. Therefore, epsilon-ergodicity seems to be the sought-after explanation of why gases show thermodynamic-like behaviour. However, our approach presupposes that the equilibrium macro region is dominant, which can be shown only for gases. The situation might well be different in liquids and solids. Whether, and if so how, the current approach generalises to liquids and solids is an open question.

Acknowledgements A proto-version of this paper was first presented at a conference at the Van Leer Institute in Jerusalem in December 2008. We want to thank the audience for a helpful discussion. We also would like to thank Scott Dumas, David Lavis and Pierre Lochak for valuable comments.

References

1. Callender, C.: Taking thermodynamics too seriously. Stud. Hist. Philos. Mod. Phys. **32**, 539–553 (2001)
2. Dizadji-Bahmani, F., Frigg, R., Hartmann, S.: Who's afraid of Nagelian reduction? Erkenntnis **73**, 393–412 (2010)
3. Lavis, D.: Boltzmann and Gibbs: an attempted reconciliation. Stud. Hist. Philos. Mod. Phys. **36**, 245–273 (2005)
4. Callender, C.: Reducing thermodynamics to statistical mechanics: the case of entropy. J. Philos. **96**, 348–373 (1999)
5. Goldstein, S.: Boltzmann's approach to statistical mechanics. In: Bricmont, J., Dürr, D., Galavotti, M., Ghirardi, G., Pettrucione, F., Zanghi, N. (eds.) Chance in Physics: Foundations and Perspectives, pp. 39–54. Springer, Berlin/New York (2001)
6. Frigg, R.: Typicality and the approach to equilibrium in Boltzmannian statistical mechanics. Philos. Sci. (Suppl.) **76**, 997–1008 (2009)
7. Frigg, R.: Probability in Boltzmannian statistical mechanics. In: Ernst, G., Hüttemann, A. (eds.) Time, Chance and Reduction. Philosophical Aspects of Statistical Mechanics, pp. 92–118. Cambridge University Press, Cambridge (2010)
8. Frigg, R.: A field guide to recent work on the foundations of statistical mechanics. In: Rickles, D. (ed.) The Ashgate Companion to Contemporary Philosophy of Physics, pp. 99–196. Ashgate, London (2008)
9. Uffink, J.: Compendium to the foundations of classical statistical physics. In: Butterfield, J., Earman, J. (eds.) Philosophy of Physics, pp. 923–1074. North-Holland, Amsterdam (2007)

10. Emch, G.G., Liu, C.: The Logic of Thermostatistical Physics. Springer, Berlin/Heidelberg (2002)

11. Frigg, R., Werndl, C.: Entropy – a guide for the perplexed. Forthcoming in: Beisbart, C., Hartmann, S. (eds.) Probabilities in Physics. Oxford University Press, Oxford (2011)

12. Birkhoff, G.: Proof of the ergodic theorem. Proc. Natl. Acad. Sci. USA **17**, 656–660 (1931)

13. Frigg, R.: Why typicality does not explain the approach to equilibrium. In: Suárez, M. (ed.) Probabilities Causes and Propensities in Physics, pp. 77–93. Springer, Berlin (2010)

14. Lavis, D.: An objectivist account of probabilities in statistical physics. In: Beisbart, C., Hartmann, S. (eds.) Probabilities in Physics. Oxford University Press, Oxford (2011)

15. Werndl, C.: What are the new implications of chaos for unpredictability? Br. J. Philos. Sci. **60**, 195–220 (2009)

16. Sklar, L.: Physics and Chance: Philosophical Issues in the Foundations of Statistical Mechanics. Cambridge University Press, Cambridge (1993)

17. Vranas, P.B.M.: Epsilon-ergodicity and the success of equilibrium statistical mechanics. Philos. Sci. **65**, 688–708 (1998)

18. Frigg, R., Hoefer, C.: Determinism and chance from a Humean perspective. In: Dieks, D., Gonzalez, W., Stephan, H., Weber, M., Stadler, F., Uebel, T. (eds.) The Present Situation in the Philosophy of Science, pp. 351–372. Springer, Berlin/New York (2010)

19. Arnold, V.I.: Mathematical Methods of Classical Mechanics. Springer, New York/Heidelberg/Berlin (1980)

20. Arnold, V., Kozlov, V., Neishtat, A.: Dynamical Systems III. Springer, Heidelberg (1985)

21. Earman, J., Rédei, M.: Why ergodic theory does not explain the success of equilibrium statistical mechanics. Br. J. Philos. Sci. **47**, 63–78 (1996)

22. Pettini, M.: Geometry and Topology in Hamiltonian Dynamics and Statistical Mechanics. Springer, New York (2007)

23. Pettini, M., Cerruti-Sola, M.: Strong stochasticity thresholds in nonlinear large Hamiltonian systems: effect on mixing times. Phys. Rev. A **44**, 975–987 (1991)

24. Markus, L., Meyer, K.R.: Generic Hamiltonian dynamical systems are neither integrable nor ergodic. Mem. Am. Math. Soc. **144**, 1–52 (1974)

25. Bengtzelius, U.: Dynamics of a Lennard-Jones system close to the glass transition. Phys. Rev. A **34**, 5059–5069 (1986)

26. De Souza, V.K., Wales, D.J.: Diagnosing broken ergodicity using an energy fluctuation metric. J Chem Phys **123**, 134–504 (2005)

27. Penrose, O.: Foundations of statistical physics. Rep. Prog. Phys. **42**, 1937–2006 (1979)

28. Thirumalai, D., Mountain, R.: Activated dynamics, loss of ergodicity, and transport in supercooled liquids. Phys. Rev. E **47**, 479–489 (1993)

29. Bricmont, J.: Bayes, Boltzmann and Bohm: probabilities in physics. In: Bricmont, J., Dürr, D., Galavotti, M., Ghirardi, G., Pettrucione, F., Zanghi, N. (eds.) Chance in Physics: Foundations and Perspectives, pp. 3–21. Springer, Berlin/New York (2001)

30. Uffink, J.: Nought but molecules in motion (review essay of Lawrence Sklar: physics and chance). Stud. Hist. Philos. Mod. Phys. **27**, 373–387 (1996)

31. McQuarrie, D.A.: Statistical Mechanics. University Science, Sausalito/California (2000)

32. Reichl, L.: A Modern Course in Statistical Physics. Wiley, New York (1998)

33. Boltzmann, L.: Einige allgemeine Sätze über Wärmegleichgewicht. Wiener Berichte **53**, 670–711 (1871)

34. Chernov, N., Markarian, R.: Chaotic Billiards. American Mathematical Society, Providence (2006)

35. Sinai, Y.: On the foundations of the ergodic hypothesis for a dynamical system of statistical mechanics. Soviet Math. Dokl. **4**, 1818–1822 (1963)

36. Szász, D.: Boltzmann's ergodic hypothesis: a conjecture for centuries? Stud. Sci. Math. Hung. **31**, 299–322 (1996)

37. Sinai, Y.: Dynamical systems with elastic reflections: ergodic properties of dispersing billiards. Uspekhi Matematicheskikh Nauk **25**, 141–192 (1970)

38. Werndl, C.: Are deterministic descriptions and indeterministic descriptions observationally equivalent? Stud. Hist. Philos. Mod. Phys. **40**, 232–242 (2009)
39. Werndl, C.: Justifying definitions in mathematics – going beyond Lakatos. Philos. Math. **17**, 313–340 (2009)
40. Werndl, C.: On the observational equivalence of continuous-time deterministic and indeterministic descriptions. Eur. J. Philos. Sci. **1**(2), 193–225 (2011)
41. Werndl, C.: Observational equivalence of deterministic and indeterministic descriptions and the role of different observations, In: Hartmann, S., Okasha, S., De Regt, H. (eds.) Proceedings of the Second Conference of the European Philosophy of Science Association. Springer, Dordrecht (2011)
42. Simányi, N.: The K-property of N billiard balls. Invent. Math. **108**, 521–548 (1992)
43. Simányi, N.: Proof of the Boltzmann-Sinai ergodic hypothesis for typical hard disk systems. Invent. Math. **154**, 123–178 (2003)
44. Simányi, N.: Conditional proof of the Boltzmann-Sinai ergodic hypothesis. Invent. Math. **177**, 381–413 (2009)
45. Simányi, N.: Ergodicity of hard spheres in a box. Ergodic Theor. Dyn. Syst. **19**, 741–766 (1999)
46. Zheng, Z., Hu, G., Zhang, J.: Ergodicity in hard-ball systems and Boltzmann's entropy. Phys. Rev. E **53**, 3246–3253 (1996)
47. Dellago, C., Posch, H.: Mixing, Lyapunov instability, and the approach to equilibrium in a hard-sphere gas. Phys. Rev. E **55**, 9–12 (1997)
48. Donnay, V.J.: Non-ergodicity of two particles interacting via a smooth potential. J. Stat. Phys. **5**(6), 1021–1048 (1999)
49. Stoddard, S.D., Ford, J.: Numerical experiments on the stochastic behaviour of a Lennard-Jones gas system. Phys. Rev. A **8**, 1504–1512 (1973)
50. Bocchieri, P., Scotti, A., Bearzi, B., Loinger, A.: Anharmonic chain with Lennard-Jones interaction. Phys. Rev. A **2**, 213–219 (1970)
51. Diana, E., Galgani, L., Casartelli, G., Casati, G., Scotti, A.: Stochastic transition in a classical nonlinear dynamical system: a Lennard-Jones chain. Theor. Math. Phys. **29**, 1022–1027 (1976)
52. Donnay, V.J., Liverani, C.: Potentials on the two-torus for which the Hamiltonian flow is ergodic. Commun. Math. Phys. **135**, 267–302 (1991)
53. Fröschle, C., Schneidecker, J.-P.: Stochasticity of dynamical systems with increasing degrees of freedom. Phys. Rev. A **12**, 2137–2143 (1975)
54. Reidl, C.R., Miller, B.N.: Gravity in one dimension: the critical population. Phys. Rev. E **48**, 4250–4256 (1993)
55. Wright, H., Miller, B.N.: Gravity in one dimension: a dynamical and statistical study. Phys. Rev. A **29**, 1411–1418 (1984)

Chapter 8
Revising Statistical Mechanics: Probability, Typicality and Closure Time

Alon Drory

Abstract Standard statistical mechanics routinely assumes that the probable behavior of a system is determined by the phase-space volume of its present macrostate. In this context, typicality is merely the latest language in which one expresses this presumed relation between phase-space measure and probability. I argue that such a connection cannot hold in general, as we cannot, for example, reconstruct without further information the history of a system found to be in equilibrium now. Even if the system is not in equilibrium, we cannot in general know its history, unless it has been closed for an extremely long time, in which case, its present state most likely arose from equilibrium. As a consequence, there is no way in general to relate probabilities to phase-space measure. The standard exposition of statistical mechanics cannot be expected to adequately cover non-equilibrium behavior, therefore. I show that the past hypothesis requires an incredible degree of fine-tuning to explain this behavior, one that is as hard to explain as the observed behavior itself. Finally, an analysis of the diffusion equation suggests that the problem is independent of microscopic time-reversibility, and lies instead with the loss of microscopic information entailed in the very definition of macrostates.

8.1 Introduction

Several papers in the present volume are concerned with the relation between probability and the notion of typicality. In the context of statistical mechanics, which is the focus of the present work, the relation between probability and typicality can be said to be the relation between probability and measure, and as such can be traced back to the work of Maxwell and Boltzmann (even though the language of typicality is much more recent).

A. Drory (✉)
Department of Basic Sciences, Afeka College of Engineering, Tel-Aviv, Israel
e-mail: adrory@zahav.net.il

Y. Ben-Menahem and M. Hemmo (eds.), *Probability in Physics*, The Frontiers Collection, 115
DOI 10.1007/978-3-642-21329-8_8, © Springer-Verlag Berlin Heidelberg 2012

Much of the literature addresses the problem of justification, which is the question whether the typicality of a property justifies its being highly probable. Here, on the other hand, my claim will be that in statistical mechanics, when considering the type of systems that most interest us, the problem is more difficult than we think, because typicality and probability do *not* agree. In other words, I will argue that what we observe in practice, i.e., what we take to be the most probable observation, is not always the typical state, and that the question is not therefore of justifying *why* the typical states are the most probable but rather to find *when* (and *if*) the typical states are the most probable ones.

In statistical mechanics, the basic objects of interest are the microstates of a system, defined to be the complete set of positions and momenta of its N constituent particles, $(\mathbf{q},\mathbf{p})$. Here, $\mathbf{q}$ is the multi-dimensional (usually 3 N-dimensional) vector of the generalized positions of all the particles, and $\mathbf{p}$ the multi-dimensional vector of all their generalized momenta. These microstates may be grouped in similarity classes according to some criteria while still remaining distinct (i.e., there are other criteria that can differentiate between them). These similarity classes are said to be "properties" of the microstates. One particular set of property classes holds special significance. This is the set of macrostates of the system, which represent the macroscopically (which we take to mean experimentally[1]) distinguishable states of the system. Thus microstates belonging to the same class are said to represent the same macrostate. This means that such microstates are indistinguishable by macroscopic observation performed at the time when the microstates are said to represent the macrostate (the "present" time).[2]

Typicality requires a measure on the set of microstates, and a microstate is said to be typical with respect to a certain property if it belongs to the maximal similarity class defined by the property. Here we take maximal to mean the class that has the largest measure, assuming there is such a class.

Thus, according to the standard account of statistical mechanics, among the macrostates of the system there is one state of maximum size with respect to the Lebesgue measure. This maximal macrostate is overwhelmingly larger than all the other states put together. This is taken to be the equilibrium state of the system, and empirically, any closed system will attain this state at some point and remain in it henceforth. One of the main problems of statistical mechanics is to explain this behavior.

---

[1] The definition of what constitutes a macrostate is hardly trivial. See for example Pitowsky [1]. For the present purpose, however, I shall assume that some such states can be defined meaningfully, no matter how.

[2] This distinction is important because it may be the case that future macroscopic observations reveal distinctions between these microstates (for example, some may evolve to a non-equilibrium state while others do not). Such future observations are not acceptable for the purpose of defining the macroscopic equivalence of microstates. The determination of the macrostate corresponding to a given microstate must be performed at the time to which the correspondence refers.

Through the work of Maxwell and Boltzmann it has become clear that although the empirical (macroscopic) observation is of a unidirectional evolution towards the equilibrium state, in fact this behavior cannot be expected to be absolute and is only extremely "likely".

It is in order to clarify the meaning of this likelihood that the notion of probability is introduced into statistical mechanics. But it turns out that there are two apparently different ways to formalize this.

8.2   Two Notions of Probability

The first formalization, developed mainly by Boltzmann, centers on the probability that a system observed over an interval of time T will be found to be in the equilibrium state. In this view, if we observe the system at random times, the most probable observation is that the system is in equilibrium. If we observe it in a non-equilibrium state, the most probable outcome of the next observations is that the system will be found to evolve towards equilibrium. Evolution towards the equilibrium state is taken to mean that the next observation is most likely to be in a state that has a larger Lebesgue measure than the original state. Thus, if the next observed macrostate is still not the maximal state, it is nevertheless highly probable that its Lebesgue measure is larger than that of the original state, and so on from one observation to the next until we reach the maximal state of equilibrium. At this point, we expect that the most probable outcome of any further observation is that the system remains in equilibrium.

The second way to formalize the likelihood of a system reaching equilibrium originates with Maxwell but was mainly developed by Gibbs. In this view, we imagine a large (tending to infinity) collection of systems observed to be in the same macrostate, but in possibly different microstates. From this ensemble, we pick at random a system and perform an observation on it, asking specifically whether this system evolves towards equilibrium within a certain time frame T. Not *all* systems in the ensemble will evolve towards equilibrium, but according to this view, there is an overwhelming probability that the system we have picked will do so.

These Boltzmannian and Gibbsian formalizations are sometimes presented as mutually exclusive, but that is not the case. Nor is there any need to choose one over the other. Indeed *both* types of probability are required to obtain a full picture of the system's behavior.

Consider Boltzmann's view first. It is hardly obvious that *every* microstate will evolve so that the system spends most of its time in the equilibrium state. This does hold if the system is ergodic, but this appears to be a rare property, and many realistic systems are probably non-ergodic. Fortunately, we do not need such a strong property. All we require is that when we observe a system, there should be a *high probability* that it will spend most of its time in equilibrium in the future.

In the last sentence, however, the type of probability we mean is not Boltzmannian, but rather Gibbsian. What we are actually claiming is that out of many observations of the system (basically, we are considering an ensemble of systems), the most probable future is that the system will spend most of its time in equilibrium. In other words, Boltzmann's formalization should more properly be viewed along such lines: Suppose that we repeatedly observe a system to be in some macrostate. In each observation, the system is in one of the microstates that correspond to the observed macrostate. Then, in overwhelmingly most of these observations, the system will be in such a microstate that its future phase-space path remains overwhelmingly most of the time in the equilibrium state. Equivalently, we could say that, in overwhelmingly most of the observations, the system will be in such a microstate that, in future observations, the system is overwhelmingly likely to be found in the equilibrium state. In this (and the previous) sentence, the various occurrences of "overwhelmingly" refer to two different types of probability. The first is Gibbsian, the second Boltzmannian. Another (slightly cumbersome) way to say this is the following: Empirically, a statistical mechanical system has a very high (Gibbsian) probability of being in such a microstate that whenever we perform an observation, it will be found to be in equilibrium with a high (Boltzmannian) probability.

Conversely, suppose we start with the Gibbsian formalization, i.e., from the requirement that the microstate in which the system is found when we observe it is highly likely to evolve towards the equilibrium state. This is insufficient for what we seek to explain, because it could happen, in principle, that the system reaches equilibrium but does not *remain* there. Clearly, part of what we mean by a state of equilibrium is that the system should also *stay* there. This means that we must also require that the temporal behavior of the system be such that it spends most of its time in the equilibrium state. Just as not all microstates are likely to evolve to the equilibrium state, so not all microstates are likely to remain forever in the equilibrium state once they reach it. We know that fluctuations do occur, so that often enough, the system evolves out of equilibrium at least for a short while. Hence, we must not only require that the system should evolve towards the equilibrium state, but also that it should return to it swiftly whenever it fluctuates out of equilibrium. Now the description of the system's behavior in time calls for the Boltzmann formalization. Indeed, the Gibbsian formalization could be expressed in the following way: In a series of experimental observations of a statistical mechanical system, there is an overwhelmingly high (Gibbsian) probability that the system will be found in a microstate that evolves towards equilibrium and to a very high (Boltzmannian) probability, remains there.

Thus, Gibbsian and Boltzmannian views complete each other, and we need both types of probabilities to account for the statistical mechanical behavior of systems.

It is immaterial what view we take of probability itself, whether it is a frequentist view, or a subjective view or any other, but we ought to be able to justify the claims put forward. On what grounds can we assess whether a system is most probably to be found in the equilibrium state or to belong to the class that evolves towards this state? Classical mechanics has no built-in concept of probability. Hence, to extract from it some notion of probability requires a linkage to something that can be based on dynamical properties alone. Typicality represents an attempt to make just such a linkage by relating probability to measures and identifying "high probability" with "majority".

8.3   Measure and Typicality

The idea is that the dynamics can be used to establish (or at least to suggest) that in any macrostate, the majority of microstates will evolve towards the equilibrium state. Here, "majority" means this: consider the set of microstates that correspond to a given macrostate, and define the subset of microstates that will evolve towards equilibrium. Then the measure of this subset is overwhelmingly larger than the measure of the set of microstates that do not evolve towards equilibrium. This claim is unrelated to probabilities. It is a matter of counting phase-space paths, and is entirely determined by the underlying microdynamics. Thus, according to the definition given above, microstates that evolve towards equilibrium are typical. Advocates of typicality as an explanation of the behavior of macroscopic systems claim that this suffices to establish that with high probability, we will observe systems to evolve towards equilibrium, because we select microstates "at random" and thus in the overwhelming majority of cases, we will select a microstate that evolves towards equilibrium.

As Pitowsky [2], Hemmo and Shenker [3] argue, however, this argument lacks justification because "at random" requires a probability distribution of its own on the microstates. This serves as a measure on the set of microstates, and nothing requires using the same measure for the phase-space size of macrostates. Of course, one can always claim that this is simply one of the axioms of statistical mechanics. As Hemmo and Shenker put it, this means relying on pure induction. And while there is nothing wrong with such a step, it does rob us of a justification in terms of the microdynamics.

It is instructive to compare this situation with what happens in the Boltzmannian view. In Boltzmann's program, the microdynamics is used to justify or at least make plausible the claim that the phase-space path of the system's microstate corresponds to the equilibrium state during overwhelmingly most of the time. This again is not a probabilistic claim, but one that relies on the microdynamics (and the definition of the macrostates). Although rarely expressed in this manner, this too is in fact a claim of typicality. Indeed, what we mean by "most of the time" must be something along these lines: consider the set of moments when the microstate corresponds to the equilibrium state, and compare it to the set of moments when it corresponds to some other macrostate. Then, we expect that the measure of the set of "equilibrium moments" will be overwhelmingly larger than the measure of the set of remaining moments. Here the transition to probability is made by assuming that we observe the system at random times, and therefore, so the standard claim goes, there is an overwhelmingly high probability of finding it in the equilibrium state.

What happens if we try to raise the same objections as before? After all, a random selection of the observation times requires a probability distribution on the set of moments, according to which these random moments are selected. And nothing *logically* requires that we should select, e.g., a uniform probability distribution on the observation times.

There is a fundamental difference between this and the previous case, however. We are free to perform observations in any way we please, so the selection of a probability distribution over the observation times is truly a matter of choice. Thus, while there is no logical necessity to choose the moments from a uniform distribution, we are nevertheless *free* to stipulate that we should do so. By contrast, whenever we select a microstate from an ensemble of systems in the same macrostate, we have no control over the distribution from which the selection is performed. Empirical conditions do the selection for us.

But this does raise another problem. If we are truly free to determine the set of observation times as we please, why *should* we use a uniform distribution? Indeed, it would appear that *any* experimental result could be established by merely adopting whatever distribution over observation times we wish so that it would yield the desired result. Thus, if these distributions are truly free, they are in danger of becoming meaningless. Unless, that is, we could establish that the precise form of the distribution is of little importance.

To some extent, we can actually hope to achieve this, provided we recall that any observation requires some time to perform, so that any series of observations necessitates a minimal extent. We cannot, for example, take a realistic series of experiments and treat the series of times in which they are performed as mere points on the number line. Were this the case, we could imagine an infinite series of experiments performed during a well chosen couple of seconds that would yield whatever result we wish. But real experiments take a minimum time to perform. As the number of observations increases, the minimal extent of the series must also increase, therefore. Thus, if we make the series infinite, it must extend over an infinite time. If, indeed, the system spends most of its time in the equilibrium state, we can hope that in such a series, the minimal time-separation between the various observations will force the results to mimic a uniform distribution, at least in the sense that most observations will show the system to be in equilibrium. It may still be the case, of course, that a particular series of observations hits a fluctuation every time so that the system appears to be constantly out of equilibrium. In fact, that might happen even if the observations times are uniformly distributed. But there is no way to ensure such a result beforehand. There is no particular choice of a time series that will likely yield such a result *a-priori*, even if it might happen accidentally. Thus, it seems that the form of the distribution over observation times is not crucial and that the typicality argument has at least a reasonable chance of working here.

But this is only true if the series of observations times is infinite. Clearly, if the series is finite, not only are there no grounds for preferring one distribution over another, but more seriously, even a uniform distribution can yield systematically skewed results. Indeed, if the series of observations is short enough, the system may not have enough time to reach equilibrium, so that this apparently typical state is not observed at all. Hence, if we seek to connect Boltzmannian probability to measures in the manner of the typicality argument, it seems that we must have a system that remains closed for a very long time. In fact it must remain closed forever.

This relation between probability, measure and closure time is not limited to this case, however. It turns out to be absolutely crucial to statistical mechanical arguments. The way to see this is to consider the past of systems rather than their future.

8.4 The Reversibility Paradox

Let us accept for a moment the typicality argument in either of its two forms. This implies the following. Assume we observe a system and find it in a non-equilibrium state. What can we say about this state? If probability and measure are directly related, i.e., if typicality determines probability, then there is an overwhelming probability that we are observing a microstate that will evolve into the equilibrium state. This is the ensemble view. Alternatively, we can say that future observations are overwhelmingly likely to yield a state closer to equilibrium. This is the time evolution view.[3]

Now for both these conclusions to hold, the system must remain closed in the future. We must assume, in other words, that the system is completely described by a Hamiltonian of the form:

$$H = H_{int}(\mathbf{q}, \mathbf{p}), \tag{8.1}$$

where H_{int} represents the Hamiltonian of the internal interactions, which depends entirely on the generalized positions and momenta of the particles in the system, $(\mathbf{q}, \mathbf{p})$, and contains no parameter describing external interventions.

We know from experience that such external interactions, if they existed, could allow us to change drastically the behavior of the system we observe (for example, we can connect it to a refrigerator and cause its entropy to drop). Thus, an open system need not even reach equilibrium (which is the supposedly typical state in which the system should be observed), much less stay in it. Hence, the requirement that the system remains closed in the future is a necessary one.

Now what about the past of the system? We know that a Hamiltonian of the type expressed in Eq. 8.1 is time-reversible. This means that the phase-space paths of the system exhibit a certain symmetry, expressible through an operation that transforms every microstate $\mathbf{r} = (\mathbf{q}, \mathbf{p})$ into a time-reversed state, denoted $R\mathbf{r}$, which is a state in

---

[3] As argued above, both views must be combined to yield a proper understanding of the system's behavior. However, for the sake of this analysis, it is sometimes simpler to treat each aspect separately. I will therefore continue to speak of these as alternative views, though they are actually complementary.

Fig. 8.1 A microstate
$\mathbf{r}_A$ evolves in time into a
microstate $\mathbf{r}_B$. The time
reversed microstate $R\mathbf{r}_B$ will
evolve in the same time into
the reversed microstate $R\mathbf{r}_A$

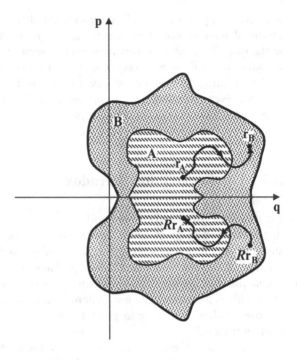

which all momenta have been reversed without altering the positions; in other
words $R\mathbf{r} = (\mathbf{q},-\mathbf{p})$. In thermodynamic macrostates, the direction of the velocities
will generally not be of importance, so that if $\mathbf{r}$ corresponds to a macrostate A, so
will $R\mathbf{r}$. Figure 8.1 shows the symmetry associated with the time-reversal operation.
A microstate $\mathbf{r}_A = (\mathbf{q}_A,\mathbf{p}_A)$, corresponding to a macrostate A, evolves over a time
t into a microstate $\mathbf{r}_B = (\mathbf{q}_B,\mathbf{p}_B)$ corresponding to a different macrostate B. To each
of these microstates corresponds a time-reversed state, $R\mathbf{r}_A$ and $R\mathbf{r}_B$, respectively.
Because the Hamiltonian equations of motions are invariant under time-reversal,
the evolution in time of $R\mathbf{r}$ looks like the time evolution of $\mathbf{r}$ run backwards. Thus, if
$\mathbf{r}_A$ evolves into $\mathbf{r}_B$, $R\mathbf{r}_B$ will evolve into $R\mathbf{r}_A$ over the same time period. The paths
described by these processes are mirror-images of each other, with respect to the
position or $\mathbf{q}$-hyperplane. This result has nothing to do with probabilities. It is
merely a geometric fact of phase-space, directly derivable from the microdynamics.

This result further implies the following geometric fact. Consider the times prior
to the moment at which the system is in the macrostate A, i.e., the past history of the
microstate $\mathbf{r}_A$. Suppose then that over a time period τ this microstate has evolved
from a past microstate $\mathbf{r}_P$, which corresponds to a macrostate P. Figure 8.2 shows
the implication of time-reversal symmetry, which is merely the reversal of Fig. 8.1.
To every microstate $\mathbf{r}_A$ there corresponds a reversed microstate $R\mathbf{r}_A$, which *will*
evolve over a time τ into the microstate $R\mathbf{r}_P$. Thus to every past history of the
system, there corresponds a future unfolding, which is its mirror-image through the
$\mathbf{q}$-hyperplane.

Fig. 8.2 A microstate $\mathbf{r}_A$ has evolved in time from a microstate $\mathbf{r}_P$. The time reversed microstate $R\mathbf{r}_A$ will evolve in the same time into the reversed microstate $R\mathbf{r}_P$

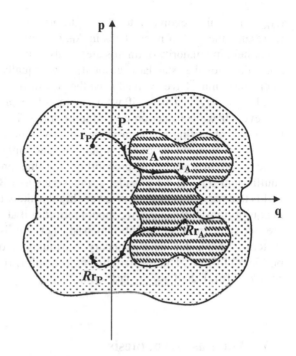

It bears stressing that this one-to-one correspondence is unrelated to probabilities, and represents a mere dynamical fact. For this very reason, it relates to typicality, however. Because both $\mathbf{r}_A$ and $R\mathbf{r}_A$ belong to the same macrostate, any claim that holds of the majority of microstates holds of the majority of the reversed microstates as well. This is because every macrostate can be divided into two symmetrical parts, the first being a set of microstates $A_F = \{\mathbf{r}\}$ and the second being the times reversed set $A_R = \{R\mathbf{r}\}$. Obviously, we have that:

$$A = A_F + A_R, \tag{8.2a}$$

$$A_F \cap A_R = \varnothing,^4 \tag{8.2b}$$

$$\mu(A_F) = \mu(A_R), \tag{8.2c}$$

where $\mu(A)$ represents the Lebesgue measure of the set A. Thus, if we claim that the *overwhelming* majority of microstates in a macrostate A evolves into the equilibrium macrostate E, this cannot hold of A_F alone or of A_R alone, because the measure of each of these sets is half the measure of the set A, so that neither can represent on its own an overwhelming majority. The overwhelming majority of reversed states

---

[4] Unless A contains the anomalous microstates $(\mathbf{q}, 0)$, but these have measure zero in phase space; hence property (8.2c) still holds, and this is the main point.

(A_R) must also evolve into the equilibrium state, therefore. This evolution represents the past of microstates in A_F, however. Thus, in any macrostate, if the overwhelming majority of microstates evolves into the equilibrium state (as it must according to the standard account), an equally overwhelming majority of microstates must have evolved *from* the equilibrium state.

This is very bad news if we wish to relate probability directly to measure, however. If we consider an ensemble of states, à la Gibbs, and wish to maintain that when observing a system we are overwhelmingly likely to observe a typical microstate, then the preceding argument leads inexorably to the conclusion that there is an overwhelming probability that the present state evolved from the equilibrium state. If we consider a time series à la Boltzmann, the same argument yields that past observations were very highly likely to have shown the system to be in equilibrium. Either way, we should retrodict that the system has evolved into its present state from an equilibrium past. Yet this is surely not the case for real systems, so we are in a quandary, variously known as "Boltzmann's second problem" [4], or "The reversibility objection" ([5], p. 128), or "the reversibility paradox" [6]. I shall use the last term.

8.5   The Past Hypothesis

There seems to be a relative consensus that the reversibility paradox is solved by invoking cosmology, namely some special initial conditions for the beginning of the universe (see, e.g., [7]). This "past hypothesis" posits that the universe began in a "low entropy" state (i.e., one very dissimilar to equilibrium),[5] and that this initial state is *sufficient* to explain the observed behavior of actual systems. I have no objection to the idea that the universe may have begun in a special state, but this would be merely a *necessary* condition. The idea that it could be sufficient to explain the behavior of every cup of coffee seems to be highly unlikely, given what we know of cosmology today.

To begin with, the observed background cosmic radiation is highly uniform, which implies that the early universe was also highly uniform. In fact, since this radiation fits the black-body curve to a very high precision, the early universe must have been in thermal equilibrium [9]. This is already bad news for the past hypothesis, which would require it to be very far from equilibrium.

But perhaps this notion of thermal equilibrium, involving as it does a mixture of radiation and matter is not quite what proponents of the past hypothesis have in

---

[5] I have purposely avoided up to now the use of entropy, as I believe the problem is of a more fundamental nature. In particular, I have no clear idea of how to define the entropy of the universe. John Earman [8] has analyzed this concept thoroughly and concludes it is merely an "intuition pump". Nevertheless, even given some such concept, there is a growing body of opinion that the past hypothesis will not solve the reversibility paradox.

mind when discussing the "initial" state of the universe. One possibility, for example, is that by "equilibrium", they really mean "uniformity". To say that the universe began in some special condition would then mean that it possessed some internal structure that led it to evolve into what we observe today. And to be sure, the early universe was not completely uniform, as this would not have led to the evolution of stars and planets.[6] Variations in density were shown to relate directly to variations in the temperature of the background cosmic radiation, so we should be able to estimate just how much "structure" the early universe possessed. Observational data shows inhomogeneities in the cosmic radiation of the order of 10^{-5} [9]. These are presumed to be the seeds of super-clusters of galaxies, so that the seed of a single galaxy should be a fluctuation of the order of about 10^{-7}–10^{-8}. The mass of a galaxy such as our Milky Way can be estimated at around 10^{41}–10^{42} kg. Now consider the notion that the initial state of the universe determines what happens to a cup of coffee. Since such a cup has a mass of about 0.1 kg, the initial state of the universe must be determined down to sizes that are at least 10^{-42} smaller than that of a galaxy, which corresponds to a fluctuation of relative size 10^{-50} at the time from which we obtain the background cosmic radiation. Therefore, to rely on the "initial" conditions of the universe to determine the behavior of local systems implies that these conditions must be determined to at least 50 digits. Such a precision is nothing short of miraculous, and one will be hard put to explain why at least some local systems do not exhibit a non-thermodynamic behavior.

This quandary becomes even greater when considering that the past hypothesis requires merely "low entropy" beginnings. But at best this could only explain why the present state of the universe as a whole is not one of equilibrium. There is no reason why some – or even many – closed subsystems should not evolve from equilibrium states into non-equilibrium states (as suggested by the reversibility paradox) provided the overall entropy of the universe is still growing. It is in this sense that the past hypothesis seems to fail as a *sufficient* condition to guarantee that every single closed subsystem should evolve from what we consider to be the correct past – a state of *individual* low entropy, unrelated to the state of its surroundings. To obtain this stronger result seems to require not merely low entropic initial conditions, but much more highly specific conditions that determine the future behavior of every single subsystem to which statistical mechanics applies. This brings us back to a kind of fantastic conspiracy on the cosmic scale, in which the state of the universe is determined to dozens of significant figures, in such a way to give the impression of a law-like behavior (thermodynamics) whilst being in fact a huge coincidence. It may be, of course, that this is the correct explanation and that the universe does behave in this ridiculous manner. Before we

---

[6] Whether this relates to some concept of universal entropy or not is not essential. Clearly, the early universe could not have been in a state of complete uniformity, and as much as I can make out, this is the property usually meant when one talks of "equilibrium" and "high entropy" in the cosmological context.

embrace such a stringent constraint on the universe, however, we should be certain that we have understood the problem correctly.

8.6   Closure Time and Probability

Actually, what the reversibility paradox leads to is the realization that probability and measure are not as simply related as we thought up to now. To see why, consider a system closed from an initial time $t_0 = 0$ up to a certain moment 2T. For illustration, assume that this is a thermos bottle. Suppose that we observe the system at the mid-point T, and find that it is not in equilibrium; the thermos bottle could contain some ice and lukewarm water, for example. When asked what the state of the system is at t_0 and at 2T, experience leads us to expect that at 2T the system will be at equilibrium or closer to it (smaller ice chunks and cooler water) while at t_0 the system was further away from equilibrium (larger ice chunks and warmer water). But as the previous argument shows, this contradicts the notion that the system is in a typical microstate. While the overwhelming majority of states will evolve towards equilibrium, in accordance with our experience, it is also true that the overwhelming majority *has* evolved from equilibrium, contrary to what we expect. Thus the typical microstate is one that represented, at t_0, a system in equilibrium or close to it (closer than it is at T). The horns of the dilemma are quite pointed, and we cannot have it both ways. One cannot claim that observed microstates must be typical and that this explains their future behavior, without necessarily accepting that the same typicality leads to incorrect explanations of their origins. One might posit the typicality of the behavior in the future but reject it for the past, but this would void the typicality argument of any explanatory power, and introduce time asymmetry into the argument by fiat.

As mentioned, the time reversal argument that leads to this retrodiction only works if the system is closed, however. The Hamiltonian of an open system contains terms that describe the interaction of the particles of the system with outside sources, so that it can be written as:

$$H = H_{int}(\mathbf{q}, \mathbf{p}) + H_{ext}(\mathbf{q}, \mathbf{p}, \alpha_1, \dots, \alpha_R), \tag{8.3}$$

where the term H_{ext} , which describes the interaction with the external world, depends on some set of parameters $\alpha_1 , \dots , \alpha_R$ that may be functions of time. Because of this, reversing the momenta will cause the external interaction term to behave differently in the time reversed and the original processes, unless all the parameters $\alpha_1 , \dots , \alpha_R$ are also time reversed. However, to do so would mean to extend, in effect, the definition of the system so that it includes the sources of these external interactions, thereby creating a larger system that can be considered closed. Only then will time-reversing all the parameters of the system cause it to retrace its history and end up in its time-reversed original state.

Let us concentrate on closed systems, therefore. As long as we consider the system's microstate, it is mathematically obvious that the previous argument only requires the system to be closed between the times $t_0 = 0$ and 2T. If that is the case, the equations of motion allow us to determine uniquely the initial microstate of the system if we know its microstate at the mid-point T (or indeed at any time between 0 and 2T). The system's history prior to t_0 and posterior to 2T is of no import, so that the system may cease to be closed after 2T or may have been open right up to the moment t_0 without it influencing the argument.

What holds at the microscopic scale does not hold at the macroscopic scale, however. In particular, the amount of closure time prior to the moment t_0 does influence considerably our macroscopic retrodictions.

To see why, imagine two copies of the thermos bottle, which we open at a time T. In both cases we find some ice chunks and lukewarm water. In both cases we are told the bottles have been closed and isolated from the time t_0, and we are asked to retrodict what the state of the system was at t_0. There seems to be no reason to retrodict anything other than the standard result, namely, that at t_0, the system was further away from equilibrium than it is at time T.

But now imagine that without changing in any way what happened between the time t_0 and the time T, we are given additional information. One bottle, we are told, was closed at t_0 or shortly before. But the other had been closed and isolated for 5 years (assume for the sake of the argument that this is possible and that we are convinced that the isolation was complete). For this second bottle, the retrodiction we are likely to make will change drastically. For if the bottle had indeed been isolated for such a long time, then, however improbable it may be, the most likely explanation for what we observe at the time T is that we are witnessing a spontaneous fluctuation from equilibrium. No amount of ice and more or less warm water that we would have placed in the bottle 5 years ago could explain a present state of non-equilibrium unless it involved some form of thermodynamic fluctuation, uncaused by external interventions.

But such fluctuations are rare, and the larger they are the rarer. Which is more unlikely, therefore: a very recent fluctuation that produced just about the state we observe at T, or a larger fluctuation at t_0, that would then have relaxed to the extent we observe at T? Clearly, the smaller fluctuation is more probable. But this means that at t_0 the system was most likely to be in equilibrium. We should therefore retrodict a completely different state than in the case of the first bottle. Now, both systems were under identical conditions from t_0 up to the time T, both were observed to be in the same macrostate at T, and yet we retrodict different past states for each. To be sure, they are in different microstates at the time T, but this is a type of information to which we can have no access. Empirically, there is no observable difference in the conditions of the two systems between t_0 and T, and no difference between their observable states at T. Yet these observations are insufficient to determine the most likely *macrostate* of the system at t_0.

This should come as no surprise, actually, and the problem can be made far more acute. Consider a slightly different case, in which a thermos bottle that has been closed for 10 min is opened and found to contain only water at a uniform temperature, but one slightly lower than the ambient room temperature. What can we say of the system's state 10 min ago? Clearly nothing, for it is impossible to make any retrodiction at all. The water may have been placed 10 min ago in the container in the same condition in which it is observed to be now, uniformly cool. Or it may have contained slightly warmer water and small chunks of ice that have melted in the meantime. Even the quantity of such ice chunks may not be determined. Small chunks and cooler water could lead to the same observed equilibrium state as slightly larger chunks and warmer water. A 10 min closure time would not permit very large chunks of ice (barring an improbable fluctuation), but there are still infinitely many possibilities of choosing the size of initial ice chunks and of the water temperature that would all lead to uniformly cool water after 10 min. There is no way to distinguish between them or to assign probabilities to these initial macrostates. This is clearly not a probabilistic question at all, but rather a lack of sufficient information.

What is important to notice is that the additional necessary information regards what happened *before* the time t_0. There is no way to establish *any* probability regarding the initial state on the basis of the present observed state and of the conditions that prevailed from t_0 to T. In other words, the present macrostate and the Hamiltonian of the system between t_0 and T are insufficient to provide any information, *even a probabilistic one*, on the state of the system at t_0 (or indeed even at times later than t_0 but prior to T).

This is a drastic conclusion, but it is important to realize that it only holds if the system could have been open at or around t_0. The situation is very different if we know that the bottle has been closed several years. If that is the case, then we will consider it highly likely that 10 min ago, the contents of the bottle were in the same state in which we find them now, namely in equilibrium. Any other state could only represent a highly unlikely fluctuation.

What can we conclude from all this regarding the validity of the connection between probability and macrostate measure, or probability and typicality? If there is to be any chance of determining probabilities purely on the basis of the macrostate of the system, then we should at least be able to assign probabilities to different histories based on the measures of the possible initial states. But if the system was recently open, there is no way of making such an assignment at all, neither on the basis of macrostate volume nor on any other macrostate parameter. The history of a recently open system cannot be determined, either probabilistically or otherwise, *solely* on the basis of its present macrostate. It requires some knowledge of the interactions between the system and its environment *prior* to its becoming closed.

The situation is dramatically different if the system has been closed for a very long time, however. In that case, the effect of such prior interactions has had time to dissipate before the observation or the time we are requested to retrodict about. In other words, the far history of the system has no longer any influence on our predictions/retrodictions, and on our assignments of probabilities. This is a necessary precondition for any probabilistic analysis of the system. Thus, only long-closed systems can possibly be described probabilistically in the standard statistical-mechanical approach. Because an assignment of probabilities on the basis of macrostate measure implies that the history of the system is irrelevant, standard statistical mechanics can only apply to systems that have no "memory" of such histories, i.e., systems that have been closed long enough to have such histories "erased".

As we have seen, such systems force us to make a drastically different retrodiction than the usual one. No matter the macrostate of the contents of the thermos bottle at the time T, if we know that the bottle has been closed and isolated for years, we shall retrodict that 10 min ago, the system was in equilibrium. This is precisely the retrodiction that we should have made on the basis of the typicality/measure argument. Typical system do indeed originate from equilibrium as well as evolve towards it.

These arguments do not prove the relation of probability to measure or typicality, nor can they serve as any justification of it. Instead, they put in question the very existence of this relation for the most common systems we encounter. Only in systems that have been closed for an extremely long time (long with respect to the thermalization time of the system) can we hope to find a direct relation between the measure of a macrostate and its probability. For systems that were open recently we have no basis for such a connection, because although it does seem to give the right answer regarding future behavior, it gives the wrong answer regarding the system's history. Such a track record would be considered a rather poor one in any other theory.

I have suggested elsewhere [6] that for the purposes of constructing a logically consistent theory, we could add an explicitly time-asymmetrical assumption to the usual axioms of statistical mechanics, one that recognizes this state of affairs. I will not defend this proposal here, as it is not the main subject of this paper. I merely note that that proposal was not meant to be an explanation of this state of affairs, merely an attempt to clarify the nature of the question.

8.7 Macrostates and Information Loss

The previous argument stresses the problem of time reversal and its consequences. But the question of time reversal only serves to heighten the problem. The origin of the problem lies elsewhere and this is what I would like to turn to now.

I shall argue that the origin of the problem lies in the very fact that for macrostates to be defined at all, there must be a certain loss of information about the system. Indeed, the problem we have identified, namely the influence of histories prior to the time t_0 at which we know the system to be already closed, does not exist on the microscopic scale. Knowledge of the present microstate $(\mathbf{q}, \mathbf{p})$ of the system at a time T completely determines its future and past states, and in particular the microstate $(\mathbf{q_0}, \mathbf{p_0})$ at t_0. For example, one can imagine several copies of the system in question, each influenced by a different Hamiltonian up to the moment $t_0 = 0$. It is perfectly possible that each of these systems is brought to an identical state $(\mathbf{q_0},\mathbf{p_0})$ by a different Hamiltonian, but this has no influence on its subsequent history. Thus, the initial state $(\mathbf{q_0},\mathbf{p_0})$ stands in a one-to-one correspondence with any state $(\mathbf{q},\mathbf{p})$ observed at a later time t, via the Hamiltonian of the system between t_0 and t. It makes no difference if ones goes forwards or backwards in time, and it is a well-known mathematical property of the equations of motion that they can be expressed just as easily on the basis of final conditions as on the basis of initial ones.

But as the above argument shows, it is precisely this property that fails in the case of macroscopic systems. When the states of the system are only defined macroscopically, the final state does not in general determine the initial state of the system. Identical macrostates can have very different origins, and there is no simple way to attach probabilities to them, certainly not on the basis of their measure alone. The system's present macrostates do not determine its past states. Only the microstates do.

Now, that problem has nothing to do with time reversal. To see this, consider a phenomenological macroscopic dynamical equation like the diffusion equation

$$\frac{\partial \rho}{\partial t} = D\nabla^2 \rho \tag{8.4}$$

This equation describes the macroscopic behavior of the system, and it is explicitly non time-reversible. Thus, it offers us the opportunity to check the source of the under-determination of the system's past separately from the reversibility argument.

The diffusion equation can be solved by Fourier transforming the spatial components, thus yielding the solution in the form of an integral

$$\rho(\mathbf{r}, t) = \int \hat{\rho}(\mathbf{k}, t) e^{-i\mathbf{k}\cdot\mathbf{r}} d\mathbf{r} \tag{8.5}$$

The coefficients $\hat{\rho}(\mathbf{k}, t)$ are easily found from the diffusion equation, and they are of the form

$$\hat{\rho}(\mathbf{k}, t) = \hat{\rho}(\mathbf{k}, 0) e^{-(Dk^2)t} \tag{8.6}$$

The coefficients $\hat{\rho}(\mathbf{k}, 0)$ are the spatial Fourier transforms of the initial state of the function $\rho(\mathbf{r}, t)$, i.e., the transforms of $\rho(\mathbf{r}, 0)$.

As noted above, the diffusion equation is time-irreversible because of the first order time derivative. Yet mathematically speaking, the value of the function $\rho(\mathbf{r}, t)$ determines its initial value (and thus its past history) completely, because given the present Fourier coefficients $\hat{\rho}(\mathbf{k}, t)$, one can always use Eq. 8.6 to calculate the initial coefficients $\hat{\rho}(\mathbf{k}, 0)$, and from them the initial condition $\rho(\mathbf{r}, 0)$. Conversely, one needs nothing more than the initial function to completely determine the state of the system at any future time. Thus, both past and future can be determined from the present state of the system, even though the system is not time-reversible. This property is therefore independent of time reversibility.

But if the diffusion equation does determine completely the history of the system, does this not contradict the previous argument that macroscopic systems have determined pasts only if they are long-closed? Indeed it would if the diffusion equation (and other phenomenological equations like it) were correct and complete descriptions of the system. But a look at Eq. 8.6 shows where the problem lies. As time goes by, every coefficient diminishes exponentially, except $\hat{\rho}(0, t)$ which generates a uniform function. This means that the spatial variations of $\rho(\mathbf{r}, t)$ decrease exponentially, but never truly vanish. It is for this reason that the system exhibits – formally – an infinitely long memory. This is a mathematical fiction, however, and this is precisely where we reach the root of the problem.

The very notion of macroscopic states requires the *in*ability to distinguish between microscopic states. Thus, at some time, the variations in $\rho(\mathbf{r}, t)$ *must* become macroscopically unnoticeable.. This means that *observationally*, all the Fourier components except $\hat{\rho}(0, t)$ are zero. Such a solution corresponds to a uniform density, which means, for the processes described by the diffusion equation, that the system has reached equilibrium.

But if all the other Fourier components must be considered observationally zero, then any microscopic information is truly lost and the initial state of the system can no longer be reconstructed.

For the equilibrium state this is not very surprising, however, and one may be tempted to think that we are being excessive. After all, what matters is reconstructing a system's history from it present *non*-equilibrium state. But a moment's thought shows that this is an equally vain hope. The equilibrium state may be the most dramatic occurrence of information loss, but it is by no means the only one. Any macrostate contains many microstates that will differ from one another by some small variations in the Fourier components, variations that are macroscopically indistinguishable. Thus, at any point, a given function $\rho(\mathbf{r}, t)$ is macroscopically indistinguishable from myriad other such functions that differ from it by small variations. The past history of the macrostate cannot be determined, since one can no longer be sure of the exact values of the Fourier components at present. Yet these past histories may correspond to macroscopically different states.

This is seen most clearly by rewriting Eq. 8.6 in the following form:

$$\hat{\rho}(\mathbf{k}, -T) = \hat{\rho}(\mathbf{k}, 0)e^{Dk^2T}, \tag{8.7}$$

which represents a retrodiction of the system's state at $-T$. As we trace back the system's history from $t_0 = 0$ to the time $-T$, every Fourier components is exponentially magnified. Thus, if two states differ at a certain time by minute amounts, their histories will diverge exponentially as we look back in time. States that are macroscopically indistinguishable now can have evolved from very different pasts, and there is no way to determine the correct history on the basis of the present macrostate alone (though there would be, of course, if we had access to the actual microstate).

Now, the diffusion equation is phenomenological, and can make no pretense to being a fundamental description of the system, as statistical mechanics aspires to be. I do not mean to imply that the details of the foregoing analysis are directly applicable to statistical mechanics. But the general situation is similar. Any macroscopic description must, by its very definition, be insensitive to microscopic differences in the system's state. Yet these now immeasurable differences arise from vastly different histories. In the past, they represented macroscopic differences. Thus, there is no way in principle of reconstructing the system's past – either probabilistically or otherwise – solely on the basis of its observed present macrostate.

It is an empirical fact that while minute microscopic differences in the present state can build up to vast macroscopic differences in the past, they (most probably) do not do so in the future. To an overwhelming probability, microscopic states that are very similar now (in the sense that they belong to the same macrostate) will evolve similarly in the future, i.e., into the same macrostate. This fact cannot be explained by mere typicality or by the measure of the present macrostate, however. For any such claim would hold of the system's history as well, independently of whether the system is time-reversible or not. The question is not whether a process can be time-reversed, but whether its history can be determined at all. Typical states do have a common typical history, but in most cases that interest us, this is not the actual history of the system. Indeed, only if the system has been closed for a very long time is the typical history also the actual history.

8.8 Conclusions

It is well known that statistical mechanics is a highly successful theory for describing a system's equilibrium state. It is equally well-known that there is no single complete theory of the dynamics of systems far from equilibrium (though some specific systems are better understood than others).

In view of the present analysis, this is no accident. Standard statistical mechanics relates probabilities to measure in phase-space, whether in Boltzmann's formulation or in Gibbs's. But in order for such a relation to hold, the system must be closed for a very long time, both in the future and in the past.

As noted in Sect. 8.3, if the system does not remain closed for a very long future time, we cannot apply Boltzmannian probabilities, and as noted in Sect. 8.2, this type of analysis is required even if we take the Gibbsian point of view. Thus the requirement of long closure time in the future is necessary for the standard probabilistic analysis to hold.

The same holds for the past. If the system was not closed for a very long time in the past, we cannot reconstruct its recent history on the basis of its present condition alone. If we observe the system to be in a non-equilibrium state now, it only appears that we can reconstruct its history because we implicitly *assume* that the system was open recently, and what type of interaction with the external world operated then. But if we make no such assumptions, we can truly make no retrodiction at all. We would only retrodict a non-equilibrium past state if we explicitly added the information that the system was recently open, and that we have some knowledge of the type of intervention it could have been subjected to.

This last point is important. Suppose for example that the thermos bottle of Sect. 8.4 was open recently, but that we are told the only external intervention was that a light was shone on it. This type of external intervention would not help explain how we happen to find ice chunks in it now. If this is the only recent interaction with the external world, we would still have to assume that we are witnessing a fluctuation and that the system was recently in equilibrium. Our retrodictions are different only if we specifically assume that someone could have put ice chunks in the bottle, namely if we assume a *specific* kind of recent external intervention in order to be able to retrodict the system's state. We implicitly make such assumptions continually, but we must recognize them as additional information on the system. Adding a different information, namely that the system had been closed for a long time, would lead us to retrodict an equilibrium past state.

As noted in Sect. 8.6, the situation can be pushed to extreme by imagining that the system is observed to be in equilibrium. Then adding the information that the system was recently open is no longer sufficient. Even assuming this, along with any kind of intervention we wish, there is still no way to retrodict the system's recent past, because there are infinitely many histories that fit the present observation. Only if the added information is that the system had been closed for a very long time can we reach a decisive retrodiction, namely that the system was recently in the equilibrium state. Hence, retrodictions about recently open systems require more information than merely the system's present state and the conditions between t_0 and T.

Thus, the only systems in which we might expect probabilities to be directly related to phase-space measure are those that have been long-closed and will remain so in the future. In such systems, the typical state is equilibrium, and any non-equilibrium state must be the result of fluctuations. Such fluctuations are time-symmetrical, and long-closed systems do not exhibit any time arrow, therefore. If standard statistical mechanics only applies to such systems, as I claim here, then we cannot expect it to explain the thermodynamic time arrow.

This arrow of time is only present in systems that have interacted with the external world recently. But for these systems, we cannot hope for a complete probabilistic description based solely on phase-space measures of macrostates. Observing the specific macrostate of a system that was recently open does not suffice to determine its history in any way, probabilistic or otherwise, without further information or assumptions. It bears repeating that it *does* suffice if the system has been closed for a very long time, and that no additional assumptions or details are required.

That the present macrostate of a recently opened system does suffice to probabilistically determine its future is an empirically surprising property, and it should be treated thus. Our inability to derive a probable *history* from the present macrostate alone is not a problem to be explained away. Rather, it is our success in establishing from the present macrostate a probable *future* that is surprising.

This asymmetry is deeply related to the fact that every macrostate contains many microstates that cannot be distinguished experimentally. This loss of detail is essential for understanding the behavior of the system, and the temporal asymmetry of its behavior must be somehow related to it, as I tried to argue in Sect. 8.7. The nature of this relation is yet unclear. We must better understand the notion of macrostates, and how the temporal evolution of the system combines with empirical limitations on our knowledge to generate the observed thermodynamic time arrow. This element is still underdeveloped in statistical mechanics. I believe that this is the direction in which we must seek further developments.

References

1. Pitowsky, I.: On the definition of equilibrium. Stud. Hist. Philos. Mod. Phys. **37**, 431–438 (2006)
2. Pitowski, I.: Typicality and the role of the Lebesgue measure in statistical mechanics, in: Y. Ben-Menahem and M. Hemmo (eds.), Probability in Physics, The Frontiers Collection, p. 41, Springer-Verlag Berlin Heidelberg (2011) (this volume)
3. Hemmo, M., Shenker, O.: Measures over initial conditions, in: Y. Ben-Menahem and M. Hemmo (eds.), Probability in Physics, The Frontiers Collection, p. 87, Springer-Verlag Berlin Heidelberg (2011) (this volume)
4. Brown, H.R., Uffink, J.: The origins of Time-asymmetry in thermodynamics: the minus first law. Stud. Hist. Philos. Mod. Phys. **32**, 525–538 (2001)
5. Leeds, S.: Foundations of statistical mechanics – two approaches. Philos. Sci. **70**, 126–144 (2003)
6. Drory, A.: Is there a reversibility paradox? Recentering the debate on the thermodynamic time arrow. Stud. Hist. Philos. Mod. Phys. **39**, 889–913 (2008)
7. Albert, D.: Time and Chance. University of Harvard Press, Cambridge (2000)
8. Earman, J.: The "past hypothesis": not even false. Stud. Hist. Philos. Mod. Phys. **37**, 399–430 (2006)
9. Scott, D., Smoot, G.F.: "Cosmic microwave background", *in* The Review of Particle Physics, C. Amsler et al. (Particle Data Group). Phys. Lett. **B667**:1 (2008)

Chapter 9
How Many Maxwell's Demons, and Where?

Eric Fanchon, Klil Ha-Horesh Neori, and Avshalom C. Elitzur

Abstract Maxwell's demon has been conceived as a tool for challenging the law of entropy increase. Several resolutions of the paradox have been proposed, making it clear that the demon does not violate the second law of thermodynamics. Nevertheless, since recent experiments come close to realizing some variants of Maxwell's demon, it is interesting to revisit it. In this article we first address two questions, left unnoticed despite many years of intensive study: (1) on which side of the door should the demon be located when the door is shut? and (2) how is kinetic energy exchanged between the two compartments due to the demon's sorting? We propose a simple setting which is more realistic than the current versions, in which the demon monitors and accesses both sides of the partition, so as to enable the sorting task. Next we study the impact of this sorting on the molecular kinetic energy exchanges. We show that the temperature difference between compartments grows till the cold part of the gas approaches 0 K. We then emphasize that this setting yields to the familiar resolution of the paradox. In the last part we derive the expression of the average rate of energy flow between the two compartments of the system, based on the new setting proposed.

In loving memory of Itamar Pitowsky
"...forasmuch as thou knowest how we are to encamp in the wilderness, and thou mayest be to us instead of eyes" (*Num.* 10, 31).

A.E.

E. Fanchon (✉)
UJF-Grenoble 1, CNRS, Laboratoire TIMC-IMAG UMR 5525, F-38041 Grenoble, France
e-mail: Eric.FanchonRM@imagRM.fr

K.H.-H. Neori
Department of Physics, University at Albany (SUNY), 1400 Washington Ave, Albany, NY 12222, USA
e-mail: kn431498@albany.edu

A.C. Elitzur
UJF-Grenoble 1, CNRS, Laboratoire TIMC-IMAG UMR 5525, F-38041 Grenoble, France

Y. Ben-Menahem and M. Hemmo (eds.), *Probability in Physics*, The Frontiers Collection, 135
DOI 10.1007/978-3-642-21329-8_9, © Springer-Verlag Berlin Heidelberg 2012

Maxwell's [1] demon is a famous paradox, the various resolutions of which have provided new insights into the nature of entropy and information. Yet some basic details of the demon's alleged operation remain unclear. Below we address these still-open issues.

9.1 The Paradox

For the sake of clarity we begin with a simple and precise protocol for the demon's operation. Let a closed box be divided into two halves by a compartment that has a small hole in it. The box is full with a gas at temperature T_o. As the gas's molecules freely move between the box's two halves through the hole, its entropy is maximal. Now consider a tiny demon positioned next to the hole. It measures the molecules' velocities. It then determines the initial root mean square velocity, $s_0 = (<v^2>)^{1/2}$, as the threshold. Every time a fast molecule, $v^2 > s_0^2$, approaches the hole from the left, the demon lets it pass to the right side, whereas every slow molecule, $v^2 < s_0^2$, is denied passage by closing the door. Conversely, the demon allows slow molecules to pass from the right to the left side and denies such passage to fast molecules. In time, fast molecules will accumulate on the right side and slow molecules on the left, with the resulting temperatures $T_r > T_o > T_l$.

The apparent paradox is clear. Entropy has been lowered in defiance of the Second Law: The energy required to open and close the slit can be made negligible, therefore operating the slit does not disperse energy and hence does not increase entropy outside of the box [2, 3].

Two species of the demon are known. The above demon, which Maxwell first had in mind, is a "temperature demon". Later he considered also a "less intelligent" one, named "pressure demon," whose task is simpler, namely, to concentrate all the gas in one side of the box [4]. Its protocol in the latter case is therefore simpler: Let all molecules pass from one side to the other, never *vice versa* (see [1] for Maxwell's own references to both species). In what follows we discuss the commoner "temperature demon."

The paradox importance stems from its being deeply rooted in some of thermodynamics' basic principles. It also bears on some fundamental issues in biology and nanotechnology [5].

9.2 The Resolutions

Equally famous is the paradox's resolution, due to Szilard [6] and Brillouin [7]: The acquisition of the information needed for the sorting operation has its cost in energy (e.g., warranting an additional light source), which increases entropy more than that decreased by the sorting.

Landauer [8] and Bennett [9] have made the resolution more precise, proving that the principal energetic cost goes to the erasure stage: In order for the demon to perform continuously, it must erase one bit of information before registering a new one. It is this erasure that takes the basic energetic toll.

For the discussion's completion we also mention a dissenting view that the above resolutions to the paradox are not satisfactory, hence the demon may violate the second law nonetheless [10].

9.3   How Many Demons, and Where?

Surprisingly, in all the discussions of Maxwell's demon a seemingly trivial question has been left unaddressed: *On which side of the door should the demon wait during the time interval it is shut?*

For the pressure demon the answer is simple: He must always be positioned within the half of the box that has to be emptied, waiting for *any* molecule that approaches the hole at whatever velocity, in order to open the door and let it pass to the other side. But for the commoner, temperature demon, the question is trickier (Fig. 9.1): In order to maintain an equal number of molecules on both sides of the partition, the demon must have access to both of them. Yet the door must be shut every time a molecule is about to make an inappropriate passage. From which side, then, should the demon shut it?

We note in passing that this question has been indirectly dealt with by Leff and Rex's [2] review, which shows that some authors depicted the demon as located on one side of the door, while others depicted it as operating from outside the box, on which the reviewers themselves comment: "placing a temperature-demon outside the gas is questionable because of the need for thermal isolation" (p. 7). Yet, surprisingly, the authors did not indicate their own opinion concerning this essential point.

Two possibilities, then, come to mind:

1. *One demon moving back and forth between the two compartments.* As the demon is supposed to be material, made of at least one atom and possessing at least one

Fig. 9.1   On which side of the door should the demon be placed?

degree of freedom. This switching between compartments would therefore give rise to an additional thermal energy transfer. In other words, it will thermalize the gas, hence this procedure will fail.

2. *Two demons, each located at another side of the door.* The inefficiency of this method is even more obvious. Two demons must work in accordance with one another, so as to avoid, for example, the case in which one opens the door at the same moment that the other has to shut it. For this propose, they need to communicate between them, necessitating again the energy investment for the information acquisition invoked in Szilard's [6] solution.

In what follows we present still another mode of operation involving a single demon, and show that it leads to formulae different from those of Leff [11] which are based on the selective effusion process (see Sect. 9.7 below). Our proposal is based on the following reasoning. Rather than performing bulk measurements, it is more practical to measure the velocity of a single isolated molecule within a small protected volume accessible to the demon.

Having indicated a single demon, we now deal with the "which side" problem. Let the two compartments be not in direct communication, but rather connected through a small intermediate chamber with two doors, one opening to the cold side and the other to the hot side (see Fig. 9.2).

Calling $n = N/V$ the density of molecules in the compartments, the volume of the connecting chamber should be $1/n$, so that it contains on average one molecule. The working cycle of the demon, starting from a state in which the door on the hot (right) side is open and the other door is closed, is as follows:

1. Close the door on the hot side.
2. Make sure a single molecule is trapped.
3. Measure its velocity.
4. If the velocity is above the threshold, open the door on the cold side, wait for some preset time Δt_0 (to let the molecule go back to where it came from), and return to step 1. Otherwise open the other door to realize a transfer and wait some preset time Δt_0.

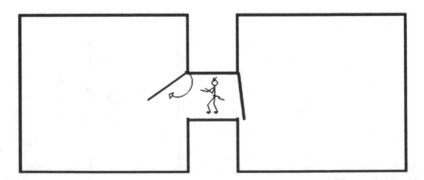

Fig. 9.2 The proposed answer to the "which side" question

5. Close the door on the cold side.
6. Make sure a single molecule is trapped.
7. Measure its velocity.
8. If the velocity is below the threshold, open the door on the hot side, wait for some preset time Δt_0 (to let the molecule go back to where it comes from), and return to step 5. Otherwise open the other door to realize a transfer (from the hot to the cold compartment) and wait some preset time Δt_0.

A few remarks are in order:

1. At any time, only one door is in the "open" position.
2. The protocol ensures that transfers occur in pairs: a transfer in one direction is necessarily followed by a transfer in the reverse direction.
3. There is a non-zero probability that zero, two or more molecules would get trapped simultaneously during steps 1 and 5. In that case, the door should be reopened to the same side for another try.
4. Having verified that exactly one molecule is present (steps 2 and 6), and after the velocity measurement, the door must remain open for a sufficiently long time Δt_0 to allow for the trapped molecule to go out and for another to get in. This interval Δt_0 is a parameter of the protocol and its value depends on the velocity of the slowest molecules to be processed.
5. Neglecting the measurement times, the minimum time for the total cycle is thus equal to $2 \cdot \Delta t_0$. In any case the total cycle time is $k \cdot \Delta t_0$, with $k > 1$, when at least one of the transfers fails on the first trial.
6. Once a molecule has been trapped, there is no need to measure its position, nor to perform complicated computations of individual trajectories to compute the time at which the trapdoor should be opened.

This defines the framework of the derivation in Sect. 9.7. The demon picks at random *one* molecule at each cycle, and its velocity is thus drawn from Maxwell-Boltzmann (MB) distribution (divided by 2 because half the velocity space is involved) at the current temperature.

Von Neumann [12] suggested a superficially similar model, in terms of "windows" in a membrane separating two compartments, but he posits the success of his model (and its dissimilarity from Maxwell's demon, of which he explicitly states that it is not an example) precisely on the fact that it ignores which side of the membrane the molecules originate from, a point which, as we have shown above, is essential to the operation of Maxwell's demon.

9.4   Intermolecular Kinetic Energy Exchanges

Our next question concerns the details of the heat extraction performed by the demon. All texts conclude Maxwell's *gedankenexperiment* with demon's apparent completion of the molecular sorting task, proceeding to the paradox's resolution.

It thus appears that the demon's only operation on the molecules is their placement in the appropriate compartment. O. Penrose [13] (who incidentally preceded Landauer [8] and Bennett [9] in pointing out the role of erasure in resolving Maxwell's paradox) gives what seems to be the common view of the demon's task, namely, "collecting all the fast molecules on one side of a diaphragm and the slow ones on the other" (p. 1994).

But is it the case? A pure sorting based on velocities would result in truncated Maxwell-Boltzmann distributions, the above-threshold temperature part of the gas now being in the hot compartment, and the below-threshold temperature part in the cold one.

Let us analyze the process in detail. In what follows we are interested only in energy transfers and do not need to worry about the exact working cycle of the demon. Consider again the above demon's box divided into two compartments, each containing N molecules of mass m, at the same initial temperature T_0. For a (monoatomic) ideal gas, the internal energy is $U = N.(1/2).m.s_0^2$ (N times the average kinetic energy). As a function of temperature T_0: $U = (3/2) N kT_0$. For convenience, we will henceforth refer to the right compartment as the "hot" and to the left one as the "cold" compartment.

The demon executes the following protocol: It first lets one molecule with velocity v, such that $v^2 < s_0^2$, to go from the hot to the cold compartment. The removal of this slower-than-average molecule results in a small increase of temperature of the hot side. The demon's next step is to let a molecule go from the cold to the hot side, in order to recover an equal number of molecules on both sides. This time it chooses a molecule that satisfies $v^2 > s_0^2$. This contributes to a slight decrease of the temperature of the cold side. The time between the two transfers is long enough for the system to relax toward statistical equilibrium, so that the demon always draws molecules from Maxwell-Boltzmann distributions (the one associated to the compartment considered). The variables T_{Hi} and T_{Ci} represent the temperature of these compartments, respectively, at step number i. By convention $i = 0$ represents the initial step where $T_{H0} = T_{C0} = T_0$. An odd value of i is associated with a state with $N + 1$ molecules in the cold compartment, and $N-1$ in the hot one. An even value is associated to a state with N molecules in each compartment.

Let us now show that, following the above simple protocol, the temperature difference $T_{Hi}-T_{Ci}$ grows monotonically. Let us assume that we are at step $i = 2p$. We want to express $T_{H,2p+2}$ and $T_{C,2p+2}$ as functions of $T_{H,2p}$ and $T_{C,2p}$. In the transition from $i = 2p$ to $i = 2p + 1$, the demon lets one molecule with $v_{2p+1}^2 < s_0^2$ from the hot compartment go to the cold one. Denoting by e_{2p+1} the kinetic energy of this molecule, we can write: $e_{2p+1} = (3/2).kT_0-\varepsilon_{2p+1}$, where $\varepsilon_{2p+1} > 0$ and $(3/2).kT_0$ represents the average kinetic energy of the molecules at T_0. The internal energies of the compartments after the transfer of this molecule are:

$$U_{H,2p+1} = U_{H,2p} - e_{2p+1}$$

$$U_{C,2p+1} = U_{C,2p} + e_{2p+1}.$$

In the next step, $i = 2p + 2 = 2(p + 1)$, the demon transfers a molecule with kinetic energy e_{2p+2} higher than $(3/2).kT_0$ from the cold to the hot side: $e_{2p+2} = (3/2).kT_0 + \varepsilon_{2p+2}$, where $\varepsilon_{2p+2} > 0$. Conservation of energy gives:

$$U_{H,2p+2} = U_{H,2p+1} + e_{2p+2} = U_{H,2p} - e_{2p+1} + e_{2p+2}$$

$$U_{C,2p+2} = U_{C,2p+1} - e_{2p+2}. = U_{C,2p} + e_{2p+1} - e_{2p+2}$$

Using the expressions for the e_i's, and for the U's in terms of the T's, we get:

$$T_{H,2(p+1)} = T_{H,2p} + 2(\varepsilon_{2p+1} + \varepsilon_{2(p+1)})/(3Nk)$$

$$T_{C,2(p+1)} = T_{C,2p} - 2(\varepsilon_{2p+1} + \varepsilon_{2(p+1)})/(3Nk).$$

From these expressions one deduces that indeed, when one considers only pairs of consecutive transfers (i.e., even steps), T_H monotonically increases and T_C monotonically decreases (since ε_{i+1} and ε_{i+2} are both strictly positive). This was not obvious because a molecule with $v_{2p+1}^2 < s_0^2$ leaving the hot side contributes to increase $T_{H,2p+1}$, but may well also increase the *cold* side, $T_{C,2p+1}$, when v_{2p+1}^2 is larger than the average squared velocity in the cold compartment at the current temperature $T_{C,2p+1}$. It turns out, however, that the increase in $T_{H,2p+1}$ is larger than the increase in $T_{C,2p+1}$, hence the difference $T_{H,2p+1} - T_{C,2p+1}$ is monotonically increasing. The same is true for a transfer in the reverse direction, from cold to hot: The criterion $v^2 > s_0^2$ insures that T_C will decrease, but the temperature T_H might decrease too, the important point being that the temperature difference is growing nevertheless.

Two remarks can be made at this point. First, the arithmetic average $<T>_{2(p+1)}$ of $T_{H,2(p+1)}$ and $T_{C,2(p+1)}$ is equal to the average $<T>_{2p}$. From this we deduce, by recurrence: $<T>_{2p} = T_0$. Second, the monotonicity property of T_C and T_H holds true if the protocol compares the squared velocities to $(3k_BT_s)/m$, where T_s is any temperature within the interval $[T_{C,2p}, T_{H,2p}]$. The only requirement is that the *same* temperature T_s should be chosen (in $[T_{C,2p}, T_{H,2p}]$) for each pair of consecutive transfers (hot to cold, and cold to hot, the reference states being those in which there are exactly N molecules in each compartment).

The entropy in the transition from step $2p$ to step $2(p + 1)$ can be calculated from the well-known formula for the change in entropy of an ideal gas, when the volume and number of molecules remain the same, and only the temperature changes:

$$S_{H,2(p+1)} - S_{H,2p} = (3/2)Nk \, ln\left(T_{H,2(p+1)}/T_{H,2p}\right)$$

$$S_{C,2(p+1)} - S_{C,2p} = (3/2)Nk \, ln\left(T_{C,2(p+1)}/T_{C,2p}\right)$$

Using the fact that $ln(y) = y - 1$ when y is close to 1, we get:

$$S_{H,2(p+1)} - S_{H,2p} = (3/2) Nk \left(T_{H,2(p+1)} - T_{H,2p}\right) / T_{H,2p}$$

$$S_{C,2(p+1)} - S_{C,2p} = (3/2) Nk \left(T_{C,2(p+1)} - T_{C,2p}\right) / T_{C,2p}$$

Then, using the expressions we obtained above for the temperature variations:

$$S_{H,2(p+1)} - S_{H,2p} = \left(\varepsilon_{2p+1} + \varepsilon_{2(p+1)}\right)/T_{H,2p}$$

$$S_{C,2(p+1)} - S_{C,2p} = -\left(\varepsilon_{2p+1} + \varepsilon_{2(p+1)}\right)/T_{C,2p}$$

Finally, the total variation of entropy for both compartments is:

$$S_{Total,2(p+1)} - S_{Total,2p} = -\left(\varepsilon_{2p+1} + \varepsilon_{2(p+1)}\right)\left(T_{H,2p} - T_{C,2p}\right)/\left(T_{H,2p}T_{C,2p}\right)$$

It can thus be verified that the entropy variation is negative, as expected (since ε_{2p+1} and $\varepsilon_{2(p+1)}$ are strictly positive quantities).

In conclusion of this part, with the protocol considered, the temperatures of the two compartments are growing apart monotonically, and the average of the two temperatures stays equal to T_0 (only at even steps $i = 2p$, strictly speaking).

How far can the demon go? It is clear that, when the internal energy $U_{C,i} = N.(m/2).<v^2>_i$ becomes smaller than the threshold value $(m/2).s_0^2$, the process stops: Even if a fluctuation concentrates the total kinetic energy of the N molecules into one of them, this molecule would not be selected by the demon. The lower bound for $T_{C,i}$ is $T_{C,min} = T_0/N$, and is thus very close to 0 K (N being of the order of the Avogadro constant).

However, with reference to our above remark on monotonicity, it is still conceivable to go beyond that limit by decreasing the temperature T_s used in the selection criterion, instead of keeping it constant at T_0 during the process. Of course the durations involved now become unphysically long well before reaching T_0/N. In other words, given infinite time, and neglecting the fact that the gas would eventually experience a phase transition, the demon could get very close to 0 K.

9.5   The Paradox Made More Precise

Let us summarize the above analysis. The demon monitors not only velocity differences between one molecule and another, but also variations in the velocity of the same molecule over time. Thus, *a molecule that was denied passage at one instant may be allowed to pass at the next, and a molecule that has passed from one side to the other may make the passage back and forth, in accordance with the changing velocities it acquires due to collisions.* What the demon actually accumulates in the hot compartment, therefore, is not merely fast molecules, but kinetic energy itself.

Gradually, the cold and hot portions of the gas would deviate further and further from the initial temperature T_0, decreasing the likelihood of velocity fluctuations large enough to enable molecules crossing sides. The demon will now have to expend another resource, namely, time [11].

One could think of optimizing the durations by devising a meta-protocol for T_s decrease (requiring an external control device), as explained at the end of Sect. 9.4, but the time between two successive transfers would nevertheless remain extremely long. The non trivial result is that, in theory, with this protocol Maxwell's demon could extract almost all the energy from one compartment, and the temperature would approach 0 K.

9.6   Reaffirming the Thermodynamic Resolution

We next show how our refined analysis of the demon's feat is subject to the same restrictions that hold for the familiar one, namely, that the second law of thermodynamics is not violated.

In the standard experiment, these restrictions say that in order to transfer molecules between the hot and cold compartments, the demon must measure their velocities [6, 7], and further erase each measurement's result so as to enable further measurements [8, 9], for which it must pay with energy.

How do these restrictions apply to our version? A possible source of confusion lies in the use of the word *sorting* itself, which implies that the sorted objects have permanent attributes. We have stressed, however, that "fast" and "slow" are not permanent but temporary attributes of molecules. The demon's task is therefore different from sorting, say, ^{16}O and ^{17}O isotopes, whose properties remains unchanged. Therefore, as the demon is likely to measure the changing velocities of the same molecule several times, that means that it has to carry out a larger number of measurements than that envisaged in the common version described in Sect. 9.4. In view of the fact that the present protocol allows the demon to achieve much higher energy difference, it perfectly makes sense that it has to carry out a much larger number of measurements, for which the thermodynamic price in energy is naturally higher.

Notice also that, as the probability for molecules to deviate from their ensemble average velocity decreases with the increase of temperature difference between compartments, the demon must wait for longer and longer periods, eventually for an infinite time.

9.7 Analysis of the Rate of Energy Transfer for a Variant of the Improved Protocol

Leff [11] presented a derivation of the rate of energy transfer between the two compartments, which is based on the kinetic theory of the effusion process. In Leff's picture the demon measures the velocity of the molecules by emitting a photon in one of the box's two compartments and measuring its scattering upon collision with a gas molecule. This scheme may give rise to some errors, because a photon does not necessarily interact with the nearest molecule. In addition, the molecule's *position* is not measured by this method, making it difficult to open the trapdoor during just the time needed to let the right molecule go through. In other words, bulk measurements are not practical. Moreover, Leff does not mention the question addressed in Sect. 9.3 concerning the demon's position on whither side of the door.

These problems are avoided by our setting, where the demon first *isolates* a molecule and then measures its velocity. In what follows this setting enables us also to compute the limit temperatures reached in infinite time and present a new derivation of the rate of energy transfer using the setting presented in Sect. 9.3 with a single demon residing within an intermediate chamber.

We now consider a variant of the above setting: The demon uses the current temperatures T_C and T_H instead of T_0. It lets a molecule go from the hot to the cold compartment if $e_{i+1} < (3/2)kT_H$, and in the next step lets a molecule from the cold to the hot compartment if $e_{i+2} > (3/2)kT_C$. In other words, there are now two different threshold values, and the selection criteria are now less stringent. This results in a larger number of transfers realized per time unit, and in the loss of the monotonicity property. At a given temperature T the Maxwell-Boltzmann (MB) velocity distribution is:

$$f(v) = (m/2\pi kT)^{3/2} exp(-mv^2/2kT).$$

The distribution of the norm v of the velocity is: $F(v) = 4\pi v^2 f(v)$ where the factor 4π results from the integration of the surface element $sin\theta d\theta d\varphi$ (spherical coordinates) over a sphere. In our case the demon selects molecules in a half space and this factor should be replaced by 2π. The root mean square velocity at temperature T is denoted ρ_T (equal to $(3kT/m)^{1/2}$).

Starting from a state where the temperature is T_0 in both compartments, the temperatures will grow apart steadily during the first steps. The system will thus reach a state where the two F distributions are significantly separated. Consequently, a molecule from the hot side which passes the test might well heat up the cold side (if its kinetic energy is between $(3/2)kT_C$ and $(3/2)kT_H$). In the long run, the system will reach a point where the balance of the energy transfers is 0 when averaged over time. To make this more precise, let us consider a series of measurement cycles. We assume the measurement time itself is negligible, and only take into account the time Δt_0 during which the door remains open (see Sect. 9.3).

Every Δt_0 a molecule is trapped in the intermediate chamber and its velocity measured. If the molecule does not meet the above criterion, it is sent back to where it came from, and the cycle is repeated until a molecule passes the test. Thus each transfer is characterized by the number of trials n which have been made, and by the kinetic energy e of the selected molecule, these two quantities being both stochastic variables. Starting from step $i = 2p$, a pair of consecutive H-C and C-H transfers is described by $((n_{2p+1}, e_{2p+1}), (n_{2(p+1)}, e_{2(p+1)}))$, meaning that the transfer to $2p + 1$ took $n_{2p+1} \cdot \Delta t_0$ unit of time and the transfer to $2(p + 1)$, in the reverse direction, took $n_{2(p+1)} \cdot \Delta t_0$. Each pair of transfers changes the total energy of the compartments, and consequently their temperature, by an infinitesimal amount, so that a very large number of transfers is required to obtain a significant displacement of the F_{TC} and F_{TH} distributions. Now let us assume that the demon has made a large number M of pairs of consecutive transfers over a period of time. M is chosen to be large enough for the averages of n_{2p} and e_{2p} to make sense, and small enough to consider that the F_{TC} and F_{TH} distributions are constant. The variation of temperature δT resulting from a molecule transfer is of the order T/N, where T is the temperature of the compartment considered and N the number of molecules. If N is taken as being the Avogadro number, then M can be fairly large before having a significant impact on the MB distributions. The average over M transfer pairs of the kinetic energy flowing from the cold to the hot compartment is $\Delta e_{CH} = \langle e_{2(p+1)} - e_{2p+1} \rangle$, which can be written $\langle e_{2(p+1)} \rangle - \langle e_{2p+1} \rangle$ because the draws are independent. It should be understood that $<e>_{HC}$ is the arithmetic average over the odd indices, corresponding to transitions from hot to cold, and similarly for $<e>_{CH}$. The average time taken by the M transfer pairs is equal to $(<n>_{CH} + <n>_{HC})\Delta t_0$. Finally the rate κ of energy transfer from C to H is

$$\kappa = \frac{1}{2} \cdot (m/\Delta t_0) \cdot (<v^2>_{CH} - <v^2>_{HC})/(<n>_{CH} + <n>_{HC}).$$

We assume next that the above arithmetic averages over M independent draws, M being large, can be approximated by ensemble averages. To express these, we define $I_C(\rho, \alpha)$ and $I_H(\rho, \alpha)$ as the integrals of $v^4 exp(-\alpha v^2)$ from ρ to $+ \infty$, and from 0 to ρ, respectively. Similarly $J_C(\rho, \alpha)$ and $J_H(\rho, \alpha)$ are the integrals of $v^2 exp(-\alpha v^2)$ from ρ to $+ \infty$, and from 0 to ρ, respectively. Thus $<v^2>_{HC} = I_H(\rho_{TH}, \alpha_{TH})/J_H(\rho_{TH}, \alpha_{TH})$, as it samples from the cold portion of the hot distribution, and similarly $<v^2>_{CH} = I_C(\rho_{TC}, \alpha_{TC})/J_C(\rho_{TC}, \alpha_{TC})$, as it samples from the hot portion of the cold side's distribution, with $\alpha_{TH} = m/(2kT_H)$ and $\alpha_{TC} = m/(2kT_C)$.

The stationary regime is reached when the average $<v^2>_{HC}$ of molecules drawn from the hot compartment is equal to the average $<v^2>_{CH}$ of the ones drawn from the cold compartment, which can be rewritten as: $<v^2>_{HC} = <v^2>_{CH}$.

The integrals to compute are well-known and the results are expressed in term of the Gaussian error function $erf(x)$ and $erfc(x) = 1 - erf(x)$:

$$J_H(\rho, \alpha) = \sqrt{\pi}/4\alpha^{-3/2} erf(\rho\sqrt{\alpha}) + (1/2)(\rho/\alpha)exp(-\alpha\rho^2)$$

$$J_C(\rho\alpha) = \sqrt{\pi}/4\alpha^{-3/2}erfc(\rho\sqrt{\alpha}) - (1/2)(\rho/\alpha)exp(-\alpha\rho^2)$$

$$I_H(\rho,\alpha) = (3/8)\sqrt{\pi}\alpha^{-5/2}erf(\rho\sqrt{\alpha}) - (1/2)(3/2 + \alpha\rho^2)(\rho/\alpha^2)exp(-\alpha\rho^2)$$

$$I_C(\rho,\alpha) = (3/8)\sqrt{\pi}\alpha^{-5/2}erfc(\rho\sqrt{\alpha}) + (1/2)(3/2 + \alpha\rho^2)(\rho/\alpha^2)exp(-\alpha\rho^2)$$

Considering that $\alpha_T = m/2kT$ and $\rho_T = (3kT/m)^{1/2}$, a simplification occurs: $\alpha\rho^2 = 3/2$.

$$<v^2>_{HC} = \left(kT_H/m\right).\left[\left(3\sqrt{\pi}erf\left(\left(3/2\right)^{1/2}\right)\right.\right.$$
$$\left.\left. -6\sqrt{6}exp\left(-3/2\right)\right)/\left(\sqrt{\pi}erf\left(\left(3/2\right)^{1/2}\right) - \sqrt{6}exp\left(-3/2\right)\right)\right]$$

$$<v^2>_{CH} = (kT_C/m).\left[\left(3\sqrt{\pi}erfc\left((3/2)^{1/2}\right)\right.\right.$$
$$\left.\left. +6\sqrt{6}exp(-3/2))/\left(\sqrt{\pi}erfc\left((3/2)^{1/2}\right) + \sqrt{6}exp(-3/2)\right)\right]$$

A remarkable result ensues: The condition for a stationary state reduces to a linear relationship between T_C and T_H. Numerically: $T_H = 3.63\ T_C$. Conservation of energy entails $T_C + T_H = 2\ T_0$, so that:

$$T_H = 1.568T_0, T_C = 0.432T_0$$

With $T_0 = 298K$ (*room temperature*), for instance, we get $T_H = 467.3K$ (*close to the melting point of Lithium*) and $T_C = 128.7\ K$ (*close to the condensation point of NO*)!

In order to obtain the full expression of the rate κ, we still need to compute $<n>\_HC$ and $<n>CH$. We denote by q_C the probability that $v^2 > \rho_C^2$ when a molecule is picked at random in the distribution F_{TC}. It is easy to show that the probability $pr(n_{2(p+1)} = a)$ for the number of trials to be equal to a is $(1-q_C)^{a-1}q_C$. Consequently:

$$<n>_{CH} = \sum_{a=1}^{\infty} a(1 - q_c)^{a-1}q_C = -q_c\,d/dq_C\left(\sum_{a=1}^{\infty}(1 - q_c)^a\right)$$

From this we deduce $<n>_{CH} = 1/q_C$. Since by definition $q_C = 2\pi(m/2\pi kT_C)^{3/}$ $^2 J_C(\rho_{TC}, \alpha_{TC})$, this gives the expression of $<n>_{CH}$ in term of an integral we have already computed. The same derivation would give for odd steps: $<n>_{HC} = 1/q_H$, with $q_H = 2\pi\ (m/2\pi kT_H)^{3/2} J_H(\rho_{TH}, \alpha_{TH})$. Since the rate κ is expressed in term of

$<v^2>_{HC}, <v^2>_{CH}, <n>_{HC}$ and $<n>_{CH}$, we now have a full expression of K as a function of the temperatures (T_H and T_C), Δt_0 and basic constants:

$$\kappa = \frac{1}{2\Delta t_0} \sqrt{\frac{m^5}{2\pi k^3}} \cdot \frac{J_H I_c - J_c I_H}{J_H T_C^{3/2} + J_c T_H^{3/2}}$$

Remember that Δt_0, the time interval during the doors remain open, is a parameter of the protocol, as explained in Sect. 9.3.

We now revisit Leff's [11] derivation of the flux generated by the demon and the rate of energy exchange between the compartments. The setting is different (selective effusion process), and the protocol is also slightly different from the one considered here: A molecule is simply allowed to enter into H, or C, if its kinetic energy is greater than $(3/2)kT_H$, or lower than $(3/2)kT_C$, respectively. The second difference is minor as it suffices to adapt the integration bounds in the above computation. The expressions obtained are different, and a comparison of the two approaches is interesting.

In Leff's derivation the action of the demon is viewed as a selective effusion process. Two comments can be made:

1. The validity of Leff's derivation is questionable. The demon manipulates one molecule at a time. Consequently, the energy transferred, and the time between two arrivals, should both be considered as stochastic variables. The rate of energy transfer should then be computed from the probability distribution of these variables. It is nevertheless possible that Leff's results are qualitatively correct if the fluxes and energy rates are interpreted as averages over a large number of transfers (a method we used explicitly in our derivation).
2. In Leff's picture, the demon measures the velocity of the molecules by emitting a photon in one of the box's two compartments and measuring the scattered photon. This scheme would generate many errors because the photon would not necessarily interact with the closest molecule. In addition the molecule's position is not measured this way, and it is thus difficult to open the trapdoor during just the time needed to select the right molecule. In other words, bulk measurements are not practical.

In conclusion, two categories of protocols have been considered in this paper for Maxwell's demon operation. In the first category the demon uses a single threshold to select "hot" and "cold" molecules. Then, this single threshold can either stay fixed at the initial temperature, or be adjusted as the demon proceeds. We showed that the evolution of T_C and T_H is monotonous as long as the threshold temperature used at each step belongs to the interval $[T_C, T_H]$. This allows, in theory, to reach a temperature very close to 0 K. Next we studied a different kind of protocol in which the demon uses two different thresholds, one for each compartment. We showed that with this protocol the system reaches a stationary regime, and that is a linear relationship holds between the temperatures, involving only mathematical constants.

9.8 Summary

In this paper we presented a new, refined scheme for the operation mode of Maxwell's demon, in which the demon resides in a little intermediate chamber between the box's two compartments, and makes measurements on isolated molecules trapped within the chamber. In addition we have made the Maxwell demon paradox more precise, stressing the fact the demon does not actually *sort* molecules on the basis of a fixed attribute, but transfers kinetic energy from one compartment to the other. In the last section we analyzed a variant of the demon protocol in which the molecule velocity is not compared to a fixed temperature. Using the scheme mentioned above, we proposed a new derivation of the rate of energy transfer, and of the stationary temperature reached. We then commented on a similar study proposed by Leff [12] and argued that our scheme is more realistic.

References

1. Maxwell, J.C.: Letter to P. G. Tait, 11 December 1867. In: Knott, C.G. (ed.) Life and Scientific Work of Peter Guthrie Tait. Cambridge University Press, London (1911). Reproduced in [2]
2. Leff, H.S., Rex, A. (eds.): Maxwell's Demon 2: Entropy, Classical and Quantum Information, Computing. Institute of Physics Publishing, Bristol (2003)
3. Maruyama, K., Nori, F., Vedral, V.: Physics of Maxwell's demon and information. Rev. Mod. Phys. **81**, 1–23 (2009)
4. Leff, H.S., Rex, A.: Overview. In: Leff, H.S., Rex, A. (eds.) Maxwell's Demon 2: Entropy, Classical and Quantum Information, Computing, pp. 1–32. Institute of Physics Publishing, Bristol (2003)
5. Serreli, V., Lee, C.-F., Kay, E.R., Leigh, D.A.: A molecular information ratchet. Nature **445**, 523–527 (2007)
6. Szilard, L.: Uber die Entropieverminderung in einem thermodynamischen System bei Eingriffen intelligenter Wesen. Z. Phys. **53**, 840–856 (1929)
7. Brillouin, L.: Maxwell's Demon cannot operate: information and entropy. I. J. Appl. Phys. **22**, 334–337 (1951)
8. Landauer, R.: Irreversibility and heat generation in the computing process. IBM J. Res. Dev. **3**, 183–191 (1961)
9. Bennett, C.: Logical reversibility of computation. IBM J. Res. Dev. **17**, 525–532 (1973)
10. Hemmo, M., Shenker, O.: Maxwell's Demon. J. Philos. **107**(8), 389–411 (2010)
11. Leff, H.S.: Maxwell's Demon, power, and time. Am. J. Phys. **58**, 135–142 (1990)
12. von Neumann, J.: Mathematical Foundations of Quantum Mechanics. Princeton University Press, Princeton (1932/1955)
13. Penrose, O.: Foundations of statistical mechanics. Rep. Prog. Phys. **42**, 1937–2006 (1979)

Chapter 10
Locality and Determinism: The Odd Couple

Yemima Ben-Menahem

Abstract This paper examines the conceptual relations between the notions of determinism and locality. From a purely conceptual point of view, determinism does not appear to imply locality, nor (contrapositively) does nonlocality appear to imply indeterminism. The example of Newtonian mechanics strengthens this impression. It turns out, however, that in the context of quantum mechanics, a more complex connection between determinism and locality emerges. The connection becomes crucial when nonlocality is distinguished from no signaling. I argue that it is indeterminism that allows nonlocal theories such as quantum mechanics to comply with the no signaling constraint. I examine a number of interpretations of quantum mechanics, among them that of Schrödinger, Pitowsky and Popescu and Rohrlich, to support this claim.[1]

In the physics literature, the term 'causality' usually refers to either determinism or locality. This ambiguity could be taken to pertain to the very meaning of the notion of cause, or, less dramatically, to the different constraints that physicists believe to be relevant to its application. Either way, the conceptual relationship between these two pivotal notions, and in particular, the question of whether they are interdependent, merits consideration. Despite the importance of this question for a better understanding of causal relations in general, the philosophical literature on causation has all but ignored it. Here I will not attempt a general analysis of causation[2];

---

[1] I would like to thank Meir Hemmo, Hilary Putnam, Daniel Rohrlich, Orly Shenker and Mark Steiner for their critique of earlier versions of this paper.

[2] Such a general analysis involves, in my view, additional notions and constraints such as stability, conservation laws and extremum principles, which will not be considered in this paper.

Y. Ben-Menahem (✉)
Department of Philosophy, The Hebrew University of Jerusalem, Jerusalem, Israel
e-mail: msbenhy@mscc.huji.ac.il

Y. Ben-Menahem and M. Hemmo (eds.), *Probability in Physics*, The Frontiers Collection, 149
DOI 10.1007/978-3-642-21329-8_10, © Springer-Verlag Berlin Heidelberg 2012

I focus exclusively on the relationship between determinism and locality (as well as the contrary notions of indeterminism and nonlocality). Let me stress at the outset that although I begin by considering strict determinism and its relation to locality, the analysis extends to probabilistic correlations of the kind found in quantum mechanics (QM). In fact, both determinism and locality need not be understood as binary notions: It is helpful to replace strict determinism with quantitative assessments of degrees of correlation and conceive of locality as well as coming in degrees, so that theories can be more or less deterministic as well as more or less local.

In what follows, I consider the relation between determinism and locality first in abstract terms (Sect. 10.1), and then in the context of QM, where traditional conceptions of determinism and locality have been most radically challenged (Sect. 10.2). Limiting myself to the standard interpretation of QM, I compare three interpretative approaches, due to Schrödinger, Pitowsky, and Popescu and Rohrlich.[3] The common denominator of these approaches (and the characteristics guiding my choice) is that all of them concentrate on conceptual characteristics of QM rather than dynamical or mechanical ones. Though these authors do not address the question of the relation between determinism and locality explicitly, their work provides important clues for answering it. I will argue that determinism and locality are independent concepts. Nonetheless, and this can be taken to be the thrust of this paper, under certain conditions (to be discussed below), determinism and locality counterbalance one another so that the violation of one makes room for the satisfaction of the other.

10.1 Determinism and Locality

Determinism requires that two copies of a closed system that agree in their fundamental physical parameters at some time t, agree on the values of these parameters at all other times (or at least future times).[4] Although in many contexts, for example, that of the free will problem, the concern is over the truth of determinism 'in reality', it is more accurate to think of determinism (indeterminism) as a characteristic of theories. A theory is deterministic if it implies the above condition, that is, if, given the equality of the states of a closed system at one point,

---

[3] Clearly there is no single 'standard' interpretation, but the term is used here, as is common, to refer to descendents of the Copenhagen interpretation. Pitowsky's interpretation, for instance, is based on the Birkhoff von Neumann axiomatization, the corner stone of the standard interpretation. Rival interpretations such as Bohm's, GRW, modal interpretations and the many world interpretation deserve separate analyses.

[4] The concept of a closed system is of course an idealization that should be weakened to meet more realistic conditions. Further, the notion of the value of a parameter at a specific time also needs refinement, for some physical magnitudes such as velocity involve change over time, convergence to a limit, and so on.

the theory implies the equality of states at other (or future) times. Ascribing determinism (indeterminism) to the world is then just a short way of ascribing this property to our best theory of the world.[5] Also, in principle, it could be the case that a theory is not deterministic with regard to all of its parameters but only with regard to some of them.

Determinism resembles the traditional 'same-cause-same-effect' postulate, and thus, when this postulate is considered constitutive of causality, determinism and causality are identified. Alternatively, determinism has also been identified with the idea that every event has a cause so that spontaneous events are excluded. Note that the two postulates, the universality postulate, asserting that every event has a cause, and the regularity requirement, according to which the same causes have the same effects, are quite distinct principles. On the one hand, it could happen that whenever the same (type of) cause recurs, the same (type of) effect(s) recur but that uncaused, random, events may also happen. On the other hand, it could be the case that every event has a cause but the same (type of) cause does not invariably lead to the same (type of) effect. If, however, we *define* causation by means of the same-cause-same-effect requirement, then, a world in which every event has a cause is, *ipso facto*, a world in which the same (type of) cause leads to the same (type of) effect. The reverse does not follow; causation can be defined via the regularity requirement but a world obeying regularity may still contain chance events. Whereas the two traditional requirements of universality and regularity, as well as the ensuing characterizations of determinism, are thus distinct, the contemporary definition of determinism given above (although it is not construed in terms of an ontology of events or causes), seeks to cover both of them: If any two systems occupying the same physical state at one instance continue to be identical in their physical parameters at all other times, then deviation from both universality and regularity is excluded. A number of writers construe determinism (usually as opposed to causation) in epistemic terms: a theory is deterministic if knowledge of a set of initial conditions of a system enables the prediction of any other state. This condition is stronger than the one given above, for a state may be predetermined but nonetheless unpredictable due to difficulties in ascertaining the initial conditions, or due to complexity and computability considerations. As noted, it may become necessary to introduce quantitative measures—degrees of correlation—rather than make do with the binary division between determinism and indeterminism. Depending on the context, therefore, conclusions about the links between determinism and locality may apply to a whole spectrum of situations ranging from perfect to weak correlation. (Analogously, the notion of indeterminism could be applied to a variety of situations from less-than-perfect correlation to genuine randomness).

---

[5] Naturally, by 'state' what is meant here is a state-*type* (rather than a token), which involves the question of the description-sensitivity of the state. This question will not affect what follows and is not discussed in this paper. See Davidson [24]; Ben-Menahem [1].

Locality has two components: it asserts the continuity of causal influence and constrains its speed to forbid instantaneous propagation of impact and information.[6] Like determinism, locality has classical origins, e.g. in the classical notion of contiguity and the idea that there are no 'jumps' in Nature. In contemporary physics, locality is generally taken to mean Lorentz invariance, and the upper bound on the speed of propagation is the speed of light. (The requirement of asymmetry in time, namely, that the cause precedes the effect, is often added.[7]) Comparing the notion of locality with that of determinism, the two notions appear completely independent[8]: Locality requires that *if* there is a cause it must act locally, that is, continuously and at a finite speed, but it asserts neither that every event has a cause, nor that the same cause must have the same effect. In the same vein, continuous and finite-speed interactions can be deterministic or indeterministic. The latter possibility describes a case in which despite the continuity and finite speed of an interaction, there are no laws guaranteeing that a recurrence of the same initial conditions dictates the repetition of the trajectory in its entirety. Conversely, deterministic interactions can in principle be continuous or discontinuous, instantaneous or of finite speed. Naturally, there are circumstances that lead us to conjoin determinism and locality. The deterministic laws that make up classical physics, for example, have specific mathematical characteristics, such as analyticity, which guarantee the continuity of a physical interaction even if not its finite speed. We are thus accustomed to the picture of laws that are both continuous and deterministic. In this picture, there is no room for spatial 'jumps', but the legitimacy of infinite speed still allows for temporal 'jumps'. In other words, in classical physics, we are accustomed to think in terms of a combination of determinism and continuity, even if not full-blown locality. Still, this combination is not forced on us by the concepts of determinism or locality *per se*. From the logical point of view, all four combinations therefore seem conceivable:

locality and determinism
nonlocality and determinism
locality and indeterminism
nonlocality and indeterminism

Note, however, that the fourth combination, nonlocality and indeterminism, although logically possible, poses a serious epistemological difficulty. Nonlocality, i.e., an instantaneous interaction between distant events, or a transmission of signals

---

[6] Note that I here refer to locality, not Bell-locality; see below.

[7] See Frisch [2].

[8] They were not conceived as independent in Antiquity and the Middle Ages; see Glasner [3] Chap. 3. One of the reasons for the difference between ancients and moderns on this point is that the former were inclined to understand determinism in terms of the universality requirement— every event has a cause –rather than in terms of the same-cause-same effect requirement. On this construal, it is easier to appreciate why the contiguity of interaction appeared to exclude spontaneous occurrences.

between them that exceeds the speed limit, can only be demonstrated via the existence of recurring *correlations* between such distant events. Individual nonlocal interactions would not be identified by us as *interactions*, but seen, instead, as the occurrence of independent and causally unrelated events. But recurring correlations, even merely probabilistic ones, introduce at least some degree of regularity, or determinism, into the picture. A *local* indeterministic influence could perhaps still be identified as such when the actual trajectory of the action is visible: For example, if we witnessed that identical pushes of a ball result in its moving haphazardly in different directions, we could perhaps still think of the push as a cause, albeit one that does not act deterministically. By contrast, a *nonlocal* indeterministic interaction could not be perceived in this way. Hence, the fourth possibility, combining nonlocality and indeterminism, can only appear in our theories in a tempered form; such theories will not be totally indeterministic with regard to all parameters. Surprisingly, then, a grain of determinism turns out to be *de facto*, even if not *de jure*, necessary for nonlocality; necessary, that is, for the formulation of a nonlocal theory.[9] In our actual theories, therefore, we do not have absolute independence between determinism and locality after all.

The history of physics provides concrete examples of possible combinations of locality and determinism. While the special theory of relativity (STR) is local and deterministic, exemplifying the first possibility above, Newtonian mechanics, which is nonlocal and deterministic, exemplifies the second.[10] Determinism is evidently not a sufficient condition for locality. But is it a necessary condition? Or, putting the same question counterpositively, is indeterminism sufficient for nonlocality? Here again, from the purely conceptual point of view, the answer seems to be negative. Nonetheless, in actual theories, QM in particular, we get a more complicated picture. Let me briefly review the situation.

Famously, the threat to locality appears in the context of QM with the phenomenon of quantum entanglement—the existence of states exhibiting long-distance correlations that are maintained even when the systems occupying these states are separated by spacelike intervals. Entanglement was identified by Schrödinger in 1935 and is amply demonstrated by experiment. It is a well-known fact that correlations challenge our causal intuitions more than singular events. While most people can envisage random singular events, systematic correlations seem to demand a causal explanation. Systematic correlations are thus generally explained either as the result of direct causal influences or as the result of 'common

---

[9] Again, by 'a grain of determinism' I do not mean the necessity of strict universal laws; probabilistic dependence would be sufficient to indicate nonlocality. Indeed this is what happens in some of the quantum mechanical cases.

[10] See, however Earman [4] and Norton [5] for counter examples to determinism in Newtonian mechanics. Despite these examples, Newtonian mechanics countenances numerous processes that are deterministic but nonlocal, attesting to the insufficiency of determinism for locality. It is also questionable as to whether STR is necessarily deterministic, but it is certainly compatible with determinism, which is all we need in order to demonstrate the feasibility of the first combination.

causes' acting on the systems in question at an earlier time and producing the correlated states.[11] Such common causes are said to 'screen off' the dependence between correlated states, meaning that, conditional on the common cause, the states no longer appear inter-dependent.[12] If the explanatory options of direct influence and common causes are exhaustive, entangled quantum states should likewise be seen as indicating either that the systems in question exert direct, albeit nonlocal, influences on one another, or that there are common causes responsible for the linkage between entangled states. Accepting the former option liberates causation from the locality constraint, generating overt conflict between QM and STR, a theory centered on locality. The alternative of explaining entanglement by means of common causes (local hidden variables, as they are usually referred to) that predetermine each of the correlated states independently is, therefore, much more attractive.[13]

There is a battery of arguments, however, beginning with the violation of Bell (and Bell-type) inequalities, that are generally thought to preclude this alternative.[14] In response to this conundrum, the following distinction between types of locality (nonlocality) has been introduced: The nonlocal correlations exhibited by entangled states are tolerated and considered consistent with STR as long as they do not allow super-luminal 'signaling'—transmission of information, between the remote states. Causality cum locality (as defined above) is thus narrowed down to a constraint on signaling rather than correlation in general.[15] Based on this understanding of locality, entangled states escape the horns of the dilemma: they result neither

---

[11] Reichenbach [6] Chap. 19.

[12] To test the existence of a common cause, one therefore compares the conditional probability of the joint event (on the common cause) with the product of the conditional probabilities of the individual events (on the common cause). When these probabilities are equal (unequal), one talks of factorizability (non-factorizability). See Chang and Cartwright [7] for an analysis of the relationship between factorizability and the existence of a common cause. They argue that since, in the probabilistic (indeterministic) case, factorizability is in general not a necessary condition for the existence of a common cause, the non-factorizability of quantum distributions does not exclude the possibility of common causes of the EPR correlations. They go on to propose such a common cause model, but their model requires discontinuous causal influences and is manifestly nonlocal.

[13] In the context of discussions of Bell's inequalities, the assumption that a common cause (whether deterministic or stochastic) exists is sometimes referred to as locality, or Bell-locality. Note, however, that the Bell-locality is not identical to the requirement of locality characterized above, for it is committed not only to the continuity and finite speed of any causal interaction *if it exists*, but to the very existence of a cause—a 'screening-off' event.

[14] Bohmians reject this conclusion. See note no. 17 below.

[15] The notion of signaling has an anthropomorphic flavor, but I will not attempt to refine it. It should be noted that no signaling is not identical with Lorentz invariance; a theory can prohibit signaling while failing to be Lorents invariant. Note, further, that in the traditional understanding of the concept of locality the only possibilities were locality plus no signaling or nonlocality plus signaling. The distinction seeks to make room for a new possibility—nonlocality *and* no signaling which was previously seen as incoherent. (The fourth possibility, locality plus signaling, remains incoherent).

from nonlocal causal interaction, nor from preexisting common causes. In the new terminology, QM can be said to observe relativistic locality, for, despite entanglement, nonlocal signaling is prohibited. The merit of this route is obvious—no conflict with STR. Its drawback is that the strange correlations are not only left unexplained, but are deemed inexplicable, or rather, not even in need of explanation.[16]

Clearly, the distinction between (legitimate) nonlocal correlations and (illegitimate) nonlocal signaling would not make a genuine difference if the correlations could have been used for signaling. But why can't they? The very idea that correlation and signaling could be separated seems paradoxical: correlation is certainly necessary for signaling and at a first glance it also seems sufficient. How then can we conceive of correlations between systems that nonetheless observe the no signaling constraint? It turns out that entangled states can be prevented from becoming a means of signaling precisely because they are not predetermined! Had the results of measurement been predetermined, the experimenter at one end of the entangled system, by looking at her results, could in fact immediately know whether the experimenter at the other end had made a measurement that interfered with the predicted outcome.[17] In the absence of such determination, even though her results are correlated with those at the other end, they do not disclose information about them. In other words, it is the *indeterminism* of QM (on the standard interpretation) which saves it from signaling and thereby saves its consistency with STR. What an ironic twist of Einstein's vision!

I have argued above that nonlocality cannot be identified as such if the nonlocal correlations are completely random. Nonlocal theories must therefore accommodate *correlations*, i.e. at least some measure of determinism. In the case of QM, it is indeed the correlations exhibited by entangled states (*to wit*, not only strictly deterministic correlations) that suggest nonlocality. In contrast, we now see that abiding by the no signaling constraint is made possible by the fact that QM also accommodates (at least a certain measure of) *in*determinism. The upshot of these dual considerations is that theories such as QM (there could be a family of such theories, see below), which sanction entanglement, but prohibit signaling, strike a very delicate balance between their deterministic and indeterministic features.

To better understand how this balance is maintained in QM, I would like to link both determinism and locality to a central tenet of QM—the uncertainty relations. While the connection between the uncertainty relations and indeterminism is conspicuous, their connection with nonlocality and entanglement is far less

---

[16] See Redhead [8] and Maudlin [9] for finer distinctions between locality notions. Although the distinction between correlation and signaling is widely accepted, many physicists and philosophers feel that non-signaling correlations still violate the spirit, if not the letter of STR.

[17] Bohmian QM seems to provide a counter-example, for despite being deterministic, it does not allow signaling. Recall, however, that in Bohmian QM the equilibrium state excludes knowledge of the predetermined states. In the absence of this information, the experimenter cannot use the correlations for signaling.

obvious. Nonetheless, there are arguments to the effect that not only indeterminism, but entanglement, and nonlocality as well, are implicit in the uncertainty relations and mediated by them. I will mention three approaches that link uncertainty and nonlocality: The first two, due to Schrödinger and Pitowsky, move from the uncertainty relations to entanglement, the second, following Popescu and Rohrlich, moves from entanglement to the uncertainty relations.

10.2 Three Approaches to Locality and Determinism in QM

10.2.1 Schrödinger's Approach

For Schrödinger, the crux of the uncertainty relations is their restriction on the determinacy of the basic physical parameters. Accordingly, "The classical notion of *state* becomes lost in that at most a well-chosen half of a complete set of variables can be assigned definite numerical values" [10, p. 153]. Remarkably, Schrödinger does not point to an epistemic problem. He does not see the uncertainty relations as referring to our knowledge or the possibility of measurement, but to the *assignment of values*—the very assignment of definite values to all variables is excluded. Schrödinger therefore goes on to rule out the possibility that quantum probabilities and uncertainties are analogous to probabilities in statistical mechanics, reflecting our ignorance rather than genuine indeterminacy in physical reality. Further, Schrödinger, who was in general a friend of continuity and very critical of the idea of 'quantum jumps' [11], acknowledges candidly that the basic characteristic that distinguishes quantum from classical mechanics is not merely the admission of discrete processes, but the different structure of the event space. Rather than stressing (along with many of his colleagues) the difference between the deterministic character of classical mechanics and the probabilistic nature of QM, Schrödinger stresses the *non-classical nature of quantum probability*, an insight that was driven home decades later by the work of Bell, Gleason, Kochen and Specker, and others. (This is the core of Pitowsky's approach discussed in the next section). Schrödinger makes the following pertinent observation:

> One should note that there was no question of any time-dependent changes. It would be of no help to permit the model to vary quite 'unclassically,' perhaps to 'jump.' Already for the single instant things go wrong. At no moment does there exist an ensemble of classical states of the model that squares with the totality of quantum mechanical statements of this moment. The same can also be said as follows: if I wish to ascribe to the model at each moment a definite (merely not exactly known to me) state, or...to *all* determining parts definite (merely not exactly known to me) numerical values, then there is no supposition as to these numerical values *to be imagined* that would not conflict with some portion of quantum theoretical assertions. [10, p. 156]

The move to indeterminism is straightforward: "If even at any given moment not all the variables are determined by some of them, then of course neither are they all determined for a later moment by data obtainable earlier" (p. 154). And further, "if a classical state does not exist at any moment, it can hardly change *causally*. What do change are... the probabilities; *these*, moreover, causally".

Schrödinger takes the Ψ function to represent a maximal catalog of possible measurements. It embodies "the momentarily-attained sum of theoretically based future expectations, somewhat as laid down in a *catalog*. ... It is the determinacy bridge between measurements and measurements" (p. 158). As such, upon a new measurement the Ψ function undergoes a change that "*depends on the measurement result obtained*, and so *cannot be foreseen*" (ibid. italics in original.) The maximality or completeness of the catalog—a consequence of the uncertainty relations—entails that we cannot have a more complete catalog, that is, cannot have two Ψ functions of the same system one of which is included in the other. "Therefore, if a system changes, whether by itself or because of measurement, there must always be statements missing from the new function that were contained in the earlier one" (p. 159). In other words, any additional information, arrived at by measurement, must change the previous catalog by *deleting* information from it. This is the basis of the 'disturbance' that the measurement brings about. True statements, that had been part of the catalog prior to the measurement, now become false. This means that at least some of the previous values have been destroyed.

So far Schrödinger has derived from the uncertainty relations three features of OM, indeterminism, 'disturbance' by measurement and the bold claim that no classical ensemble will recover the quantum probabilities. As I have noted, only the first two were fully acknowledged at the time. But now comes entanglement. This new feature also follows from the maximality or completeness of the Ψ function, that is, from the uncertainty relations. He argues as follows: A complete catalog for two separate systems is, *ipso facto*, also a complete catalog of the combined system, but the reverse does not follow. "*Maximal knowledge of a total system does not necessarily include total knowledge of all its parts, not even when these are fully separated from each other and at the moment are not influencing each other at all*" (p. 160, italics in original). The reason we cannot infer such total information is that the maximal catalog of the combined system may contain conditional statements of the form: *if* a measurement on the first system yields the value x, a measurement on the second will yield the value y, and so on. He sums up: "Best possible knowledge of a whole does not necessarily include the same for its parts...The whole is in a definite state, the parts taken individually are not" (p. 161). In other words, separated systems can be correlated or entangled via the Ψ function of the combined system, but this does not mean that their individual states are already determined! Schrödinger's argument clarifies the conclusion reached above regarding the merit of a combination of determinism—to detect nonlocality—and indeterminism—to prevent signaling. The grain of determinism is supplied by the conditional statements governing the correlations and derived from conservation laws. Indeterminism pertains to the individual outcomes.

Schrödinger's argument is purely conceptual. As we have seen, he does not examine entanglement as a physical process in space and time (such processes would be described as intuitive or *anschaulich* in the idiom of the time), but rather as a conceptual possibility emerging from the uncertainty relations and the notion of a maximal catalog. Similarly, he does not see the collapse of the wave function as a physical process. It too is a formal property of the Ψ function, a function which is anyway in configuration space rather than real space. The only causal, or intuitive, consideration that figures in Schrödinger's argument is that, to be entangled, the two systems must have interacted in the past. We cannot create entanglement between the systems at a distance.

Schrödinger tells us (in f.n. no. 7) that his 1935 paper was written in response to the EPR paper published earlier that year. One tends to think of Schrödinger as Einstein's ally in opposing the Copenhagen interpretation, and, indeed, Schrodinger's paper leaves no doubt as to the fact that he is unhappy with "the present situation in QM". One can therefore easily miss the point that without saying so explicitly, Schrödinger also assumes here the role of Einstein's critic, launching a much more lucid and effective critique of the EPR argument than that of Bohr. The EPR argument purports to show that the correlations between the remote parts of the system—the conditional statements—entail that each individual state has a determinate value *prior* to measurement. By contrast, Schrödinger argues, first, that such determinacy is precluded by the uncertainty relations properly understood, and second, that, given his reading of the Ψ function as a maximal catalog of possible measurements, the indeterminacy of individual outcomes makes perfect sense. In other words, whereas the EPR argument seeks to understand the correlations in terms of predetermined values, which amounts to understanding them in terms of common causes, Shrödinger sees that this solution does not work. He therefore suspects that QM might be incompatible with STR. Schrödinger's pioneering argument is of course still programmatic, but in later years there are more rigorous derivations of several of his claims.

10.2.2 Pitowsky's Approach

In his "Quantum Mechanics as a Theory of Probability" [12], Pitowsky elaborates the axiomatic approach originating with Birkhoff and von Neumann [23]. Building upon their classic axiomatization in terms of the Hilbert space structure of quantum events and its relation to projective geometry, Pitowsky seeks to incorporate later developments, such as Gleason's [13] theorem and the Bell (and Bell-type) inequalities, and identify their roots in the axiom system. In this work, Pitowsky wraps up much of his earlier work on the foundations of QM, highlighting, in particular, the non-classical nature of quantum probability. The ramifications of the non-classical structure of the quantum probability space, he argues, include indeterminism, the loss of information upon measurement, entanglement and the Bell-type inequalities. They are also closely related to Gleason's [13] and Kochen and Specker's [14] theorems, and (as he argues in a later work with Jeffrey Bub) to the information-theoretic principle of no-cloning (or no broadcasting). I will mention those aspects of Pitowsky's system and interpretation that enhance our understanding of locality and determinism

The non-classical nature of quantum probability manifests itself in the violation of basic classical constraints on the probabilities of interrelated events and is already reflected in simple paradigm cases such as the two-slit experiment. For example, in the classical theory of probability, it is obvious that if we have two events E_1 and E_2 with probabilities p_1 and p_2, and their intersection whose probability is $p_1.p_2$, the probability of the union $(E_1 U E_2)$ is $p_1 + p_2 - p_{12}$, and cannot exceed the sum of the probabilities $(p_1 + p_2)$.

$$0 \leq p_1 + p_2 - p_{12} \leq p_1 + p_2 \leq 1$$

In the two-slit experiments, however, the predictions of quantum mechanics violate this classical condition, for there are areas on the screen that get more hits when the two slits are open together for a certain time Δt, than when each slit is open separately for the same time interval Δt. In other words, contrary to the classical principle, we get a *higher* probability for the union than the sum of the probabilities of the individual events. (Since we get this violation in different experiments—different samples—it does not constitute an outright logical contradiction) This phenomenon is usually described in terms of interference, superposition, the wave particle duality, the nonlocal influence of one open slit on a particle passing through the other, and so on. Pitowsky's point is that before we venture an *explanation* of the pattern predicted by QM (and confirmed by experiment), we must acknowledge the bizarre phenomenon that it displays—nothing less than a violation of the highly intuitive principles of the classical theory of probability.

QM predicts analogous violations of almost any other classical condition of probability. The most famous of these violations is that of the Bell inequalities which, like the above rule, can be derived from classical combinatorial considerations. In fact, the analog of the above classical condition for three events says that the probability of the union event $(E_1 U E_2 U E_3)$ cannot exceed $p_1 + p_2 + p_3 - p_{12} - p_{13} - p_{23}$. From here, Pitowsky showed ([12] and references therein), that it is just a short step to the Bell inequalities, which are violated by QM (and experiment). Bell's famous inequalities are thus directly linked by Pitowsky to Boole's classical "conditions of possible experience". What makes the violated conditions 'classical' is their underlying assumption of stable properties obtained by the entities in question independently of measurement: Just as balls in an urn are red or wooden, to derive Bell's inequalities it is assumed that particles have a definite polarization, or a definite spin in a certain direction, and so on. The violation of the classical principles of probability compels us to discard this picture, replacing it with a new understanding of quantum states and quantum properties. What does it mean for a particle to *be* in a certain state, say, spin-1 in the x direction, and what is the role of measurement in revealing this state? More generally, what is the meaning of the quantum state function? Pitowsky's answer is similar to that of Schrödinger's: Rather than an analogue of the classical state, which represents physical entities and their properties prior to measurement, the quantum state function only keeps track of the probabilities of measurement results, "a book-keeping device" in Pitowsky's terms or "a catalog of possible experiments" in Schrödinger's.

This understanding of the state function, in turn, led Pitowsky to two further observations: First (and unlike Schrödinger), he interpreted the book-keeping picture subjectively, i.e., quantum probabilities are understood as degrees of partial belief. Second (in agreement with Schrödinger), the notorious collapse problem is not as formidable as when the state function is construed realistically, for if what collapses is not a real entity in physical space, then there is no reason why the collapse should be construed as a real physical process abiding by the constraints of locality and Lorentz invariance. There is thus a direct link between Pitowsky's seeing QM primarily as a theory of non-classical probability and his renouncing of what he and Bub, in their joint paper [15], dub "two dogmas" of the received view, namely the reality of the state function and the need for a dynamic account of the measurement process.

Like Schrödinger, Birkhoff and von Neumann, Pitowsky takes the uncertainty relations to be "the centerpiece that demarcates between the classical and quantum domain" [12, p. 214]. The only non-classical axiom in the Birkoff von Neumann axiomatization, and thus the logical anchor of the uncertainty relations, is the axiom of *irreducibility*.[18] While a classical probability space is a Boolean algebra where for all events x and z

$$x = (x \cap z) \cup (x \cap z^{\perp}) \text{ (reducibility)}$$

in QM, we get irreducibility, i.e., (with 0 as the null event and 1 the certain event): If for some z and for all $x, x = (x \cap z) \cup (x \cap z^{\perp})$ then $z = 0$ or $z = 1$

Irreducibility signifies the non-Boolean nature of the algebra of possible events, for the only irreducible Boolean algebra is the trivial one $\{0,1\}$. As Birkhoff and von Neumann explain, irreducibility means that there are no 'neutral' elements z, $z \neq 0 \ z \neq 1$ such that for all x, $x = (x \cap z) \ U \ (x \cap z^{\perp})$. (If there would be such 'neutral' events, we would have non-trivial projection operators commuting with all other projection operators). Intuitively, irreducibility embodies the uncertainty relations for when x cannot be presented as the union of its intersection with z and its intersection with $z^{\perp}$, the complement of z, then x and z cannot be assigned definite values at the same time. Thus, whenever $x \neq (x \cap z) \cap (x \cap z^{\perp})$ x and z are incompatible and, consequently, a measurement of one yields no information about the other. The axiom further implies genuine uncertainty, or indeterminism— probabilities strictly between (unequal to) 0 and 1. This result follows from a

---

[18] Pitowsky's formulation is slightly different from that of Birkhoff and von Neumann, but the difference is immaterial. Pitowsky makes significant progress, however, in his treatment of the representation theorem for the axiom system, in particular in his discussion of Solér's theorem. The theorem, and the representation problem in general, is crucial for the application of Gleason's theorem, but will not concern us here.

theorem Pitowsky calls the *logical indeterminacy principle*, and which proves that
for incompatible events x and y

$$p(x) + p(y) < 2$$

The loss of information upon measurement—the phenomenon called 'disturbance' by the founders of QM—also emerges as a formal consequence of the
probabilistic picture.

Having shown that the axiom system entails genuine uncertainty, Pitowsky
moves on to demonstrate the violation of the Bell-inequalities, i.e. the phenomenon
of entanglement and nonlocality. These violations already appear in finite-
dimensional cases and follow from the calculation of the probabilities of the intersection of the subspaces of the Hilbert space representing the (compatible) measurement
results at the two ends of the entangled system. He shows, in both logical and
geometrical terms, that the quantum range of possibilities is indeed larger than the
classical range so that we get *more* correlation than is allowed by the classical
conditions.[19] Whereas the usual response to this phenomenon consists in attempts
to discover the dynamics that makes it possible, Pitowsky emphasizes that this is a
logical-conceptual argument, independent of specific physical considerations over
and above those that follow from the non-Boolean nature of the event structure. He
says

> Altogether, in our approach there is no problem with locality and the analysis remains intact
> no matter what the kinematic or the dynamic situation is; the violation of the inequality is a
> purely probabilistic effect. The derivation of Clauser-Horne inequalities. . . is blocked since
> it is based on the Boolean view of probabilities as weighted averages of truth values. This,
> in turn, involves the metaphysical assumption that there is, simultaneously, a matter of fact
> concerning the truth-values of incompatible propositions. . . . [F]rom our perspective the
> commotion about locality can only come from one who sincerely believes that Boole's
> conditions are really conditions of possible experience. . . . . But if one accepts that one is
> simply dealing with a different notion of probability, then all space-time considerations
> become irrelevant. [12, pp. 231–232]

Recall that in order to countenance nonlocality without breaking with STR, the
no signaling constraint must be observed. As nonlocality is construed by Pitowsky
in formal terms—a manifestation of the quantum mechanical probability calculus,
uncommitted to a particular dynamics—it stands to reason that the no signaling
principle will likewise be derived from probabilistic considerations. Indeed, it turns
out that no signaling can be construed as an instance of a more general principle
known as the non-contextuality of measurement [17]. In the spirit of the probabilistic approach to QM, Pitowsky and Bub therefore maintain that "'no signaling' is
not specifically a relativistic constraint on superluminal signaling. It is simply a
condition imposed on the marginal probabilities of events for separated systems

---

[19] In his book [16], Pitowsky studied the geometrical meaning of Boole's classical conditions on
probability. More details can be found in the introduction to this volume.

requiring that the marginal probability of a B-event is independent of the particular set of mutually exclusive and collectively exhaustive events selected at A, and conversely" (2010, p. 443).[20]

I have used Pitowsky's formal approach to trace both indeterminism and nonlocality to the event structure of QM. As made clear in the introduction to this volume, it is a basic tenet of this approach that there is no deeper foundation to QM than this formal structure. Once we accept QM as a new and non-classical theory of probability (or information), the argument goes, the intriguing problems of how nonlocal correlations are brought about, why measurement generates disturbance, and so on, can be set aside. Pitowsky and Bub draw an analogy between their formal approach to QM and the widely held understanding of STR, according to which relativistic effects such as the contraction of rods in the direction of motion and time dilation are kinematical effects that need no further dynamical explanation. These issues need not be addressed here. Pitowsky's approach, like Schrödinger's, was used here only to tackle the conceptual question I raised regarding the relation between determinism, locality and the uncertainly relations. The further claim regarding the explanatory value of this approach, or the redundancy of alternative dynamical ones, is not part of my argument.

10.2.3   Popescu and Rohrlich's Approach

So far we have seen that according to both Schrödinger and Pitowsky, a formalism that incorporates the uncertainly relations (or, equivalently, incorporates the axiom of irreducibility and countenances incompatible events), also gives rise to both indeterminism and nonlocality. The question now arises of whether we can also move in the opposite direction, that is, does quantum entanglement yield the uncertainty relations. A series of papers by Rohrlich and Popescu [26] shed light on this intriguing question. The original question addressed in these papers was whether the nonlocal correlations of QM could be tempered with and destroyed by a third party. The idea was that if the nonlocal correlations reflect superluminal communication between distant systems, it might be possible to interfere with this mysterious communication channel. To test this idea Popescu and Rohrlich consider Jim the Jammer, situated in a position that enables him to jam the EPR correlations between Bob and Alice. We have noted repeatedly that nonlocality in itself could be sanctioned as long as it does not entail super-luminal signaling.

---

[20] In the literature, following in particular Jarrett [30], it is customary to distinguish outcome independence, violated in QM, from parameter independence, which is observed, a combination that makes possible the peaceful coexistence with STR. The non-contextuality of measurement amounts to parameter independence. See, however Redhead [8] and Maudlin [9], among others, for a detailed exposition and critical discussion of the distinction between outcome and parameter independence and its implications for the compatibility with STR.

Accordingly, the envisaged jamming must be such that Bob and Alice will not notice it.

Popescu and Rohrlich seek a quantitative assessment of the relation between nonlocality and no signaling: they look for the *maximal* amount of nonlocality that does not lead to signaling, i.e. the maximal nonlocality that avoids a clash with STR. Their initial conjecture was that the constraints of maximum nonlocality and no signaling suffice to precisely recover QM, no more, no less. Intuitive support for this conjecture came from the observation made above that an indeterministic theory could be both nonlocal and consistent with STR. Hence the feasibility of the idea that QM strikes exactly the right balance between nonlocality, no signaling and indeterminism. Are nonlocality and no signaling then sufficient to generate QM? Surprisingly, the answer reached by Rohrlich and Popescu was negative: nonlocality plus no signaling give us a family of theories that includes, in addition to QM, a range of theories that are *more nonlocal* than QM.[21] Yet in all members of this family we get uncertainty relations analogous to those of QM, but possibly differing from them in the value of the numerical limit they set. Here both (non) locality and (in)determinism have become quantitative, rather than binary, notions. Moreover, they have become mutually interdependent. The combination of nonlocality and no signaling is linked to, and made possible by, the existence of uncertainty relations and the indeterminism that follows from them.

The Popescu-Rohrlich argument suggests that indeterminism is at least (part of) a sufficient condition for the peaceful coexistence of nonlocality and STR. On the face of it, the stronger claim that indeterminism is also a necessary condition for this coexistence clashes with Bohmian QM, which is deterministic but does not allow signaling.[22] However, given the fact that in Bohmian QM, due to the equilibrium conjecture, the pre-determined states are unknown to the experimenter, and are thus useless for signaling, this clash may turn out to be illusory. If so, a kind of epistemic indeterminism (such as is found even in Bohmian QM) is not only a sufficient condition for the peaceful coexistence of nonlocality and no signaling, but also a necessary one. The interconnection between nonlocality and indeterminism is further supported by two recent papers. Oppenheim and Wehner [19] argue that the two basic features of QM, nonlocality and the uncertainty relations, "are inextricably and quantitatively linked" (p. 1072), so that QM cannot be more nonlocal than it is without violating the uncertainty principle. From a different perspective, Goldstein et al. [20], in their critique of Conway and Kochen's [21] "free will theorem", also stress the difference between deterministic and stochastic

---

[21] In the Clauser-Horn-Shimony-Holt inequality the classical limit reached by local realist considerations is $-2 \leq S \leq 2$. In QM this inequality can be violated, but as Boris Tsirelson has shown there is an upper bound to this violation: $-2\sqrt{2} \leq S \leq 2\sqrt{2}$. Rohrlich and Popescu show that the Tsirelson bound can be violated without violation of STR, that is, without violation of the no signaling requirement.

[22] Although it does not allow signaling, Bohmian QM is not Lorentz invariant; see Albert [18] Chap. 7.

theories in so far as their compliance with the no signaling constraint is concerned. Their argument is particularly significant in view of Tumulka's relativistic version of the GRW theory [22]. Were Tumulka's argument to apply to a deterministic analogue of the GRW theory, we would have a deterministic Lorentz invariant version of quantum mechanics and thus a counterexample to the argument of this paper. Goldstein et al. show, however, that indeterminism plays a crucial role in Tumulka's argument.

To conclude, we have seen that, conceptually, determinism is neither sufficient nor necessary for locality, but epistemically, a degree of determinism is necessary for the detection of nonlocalty. We have seen, further, that once we distinguish between nonlocality and no signaling, the lesson of QM is that a combination of nonlocality and no signaling is made possible by indeterminism. In theories that, like QM, sanction nonlocal correlations, nonlocality and indeterminism cooperate to prevent signaling and protect compatibility with STR. The uncertainty relations play a major role in maintaining this cooperation.

References

1. Ben-Menahem, Y.: Direction and description. Stud. Hist. Philos. Mod. Phys. **32**, 621–635 (2001)
2. Frisch, M.: The most sacred tenet'? causal reasoning in physics. Br. J. Philos. Sci. **60**, 459–474 (2009)
3. Glasner, R.: Averroes' Physics. Oxford University Press, Oxford (2009)
4. Earman, J.: A primer on Determinism. Reidel, Dordrecht (1986)
5. Norton, J. D.: Causation as folk science. In: Price and Corry, pp. 11–44 (2007)
6. Reichenbach, H.: The Direction of Time. University of California Press, Los Angeles (1956)
7. Chang, H., Cartwright, N.: Causality and realism in the EPR experiment. Erkenntnis **38**, 169–190 (1993)
8. Redhead, M.: Incompleteness, Nonlocality and Realism. Clarendon, Oxford (1987)
9. Maudlin, T.: Quantum Non-Locality and Relativity. Blackwell, Oxford (1994)
10. Schrödinger, E.: The present situation in quantum mechanics. In: Trimmer, J.D. (trans.) Wheeler and Zurek, pp. 152–167 (1983) [1935]
11. Ben-Menahem, Y.: Struggling with causality: schrödinger's case. Stud. Hist. Philos. Sci. **20**, 307–334 (1989)
12. Pitowsky, I.: Quantum mechanics as a theory of probability. In: Demopoulos and Pitowsky, pp. 213–239 (2006)
13. Gleason, A.M.: Measurement on closed subspaces of a Hilbert space. J. Math. Mech. **6**, 885–893 (1957)
14. Kochen, S., Specker, E.P.: The problem of hidden variables in quantum mechanics. J. Math. Mech **17**, 59–89 (1967)
15. Bub, J., Pitowsky, I.: Two dogmas about quantum mechanics. In: Saunders, S., Barrett, J., Kent, A., Wallace, D. (eds.) Many Worlds? Everett, Quantum Theory and Reality, pp. 433–459. Oxford University Press, Oxford (2010)
16. Pitowsky, I.: Quantum Probability, Quantum Logic Lecture Notes in Physics 321. Springer, Heidelberg (1989)
17. Barnum, H., Caves, C.M., Finkelstein, J., Fuchs, C.A., Schack, R.: Quantum probability from decision theory? Proc. R. Soc. Lond. A **456**, 1175–1190 (2000)

18. Albert, D.Z.: Quantum Mechanics and Experience. Harvards University Press, Cambridge (1992)
19. Oppenheim, J., Wehner, S.: The uncertainty principle determines the nonlocality of quantum mechanics. Science **330**, 1072–1074 (2010)
20. Goldstein, S.,Tausk, D.V., Tumulka, R., Zanghi, N.: What does the free will theorem actually prove? arXiv: 0905.4641v1 [quant-ph] (2009)
21. Conway, J.H., Kochen, S.: The free will theorem. Found. Phys. **36**, 1441–1473 (2006)
22. Tumulka, R.: A relativistic version of the Ghirardi-Rimini-Weber model. J. Stat. Phys. **125**, 821–840 (2006)
23. Birkhoff, G., von Neumann, J.: The logic of quantum mechanics. Ann. Math. **37**, 823–845 (1936)
24. Davidson, D.: Causal relations In: Essays on Actions and Events. Clarendon Press, Oxford, pp. 149–162, [1967] (1980)
25. Demopoulos, W., Pitowsky, I. (eds.): Physical Theory and its Interpretation. Springer, Dordrecht (2006)
26. Popescu, S., Rohrlich, D.: The joy of entanglement. In: Popescu and Spiller, pp. 29–48 (1998)
27. Popescu, S., Spiller, T.P. (eds.): Introduction to Quantum Computation. World Scientific, Singapore (1998)
28. Price, H., Corry, R. (eds.): Causation, Physics and the Constitution of Reality. Oxford University Press, Oxford (2007)
29. Wheeler, J.A., Zurek, W.H.: Quantum Theory and Measurement. Princeton University Press, Princeton (1983)
30. Jarrett, J. P.: On the physical significance of the locality condition in the Bell arguments. Nous. **18**, 569–589 (1984)

Chapter 11
Why the Tsirelson Bound?

Jeffrey Bub

Abstract Wheeler's question 'why the quantum' has two aspects: why is the world quantum and not classical, and why is it quantum rather than superquantum, i.e., why the Tsirelson bound for quantum correlations? I discuss a remarkable answer to this question proposed by Pawłowski et al. [1], who provide an information-theoretic derivation of the Tsirelson bound from a principle they call 'information causality.'

11.1 Introduction

In a remarkable information-theoretic derivation of the Tsirelson bound for quantum correlations by Pawłowski et al. [1], the authors derive the bound from a principle they call 'information causality.' Here I review the original derivation and the information-theoretic principle involved, and consider the significance of the result.

Einstein's special theory of relativity follows from just two principles: the light postulate and the principle of relativity. In a seminal paper [2], Popescu and Rohrlich asked whether quantum mechanics follows from relativistic causality, the principle that causal processes or signals cannot propagate outside the light cone, and nonlocality in the sense of Bell's theorem [3]. They showed that it does not: quantum mechanics is only one of a class of theories consistent with these two principles.

To see this, consider a 'nonlocal box,' a hypothetical device proposed by Popescu and Rohrlich, now called a 'Popescu-Rohrlich box' or PR-box. A PR-box has two inputs, $a \in \{0, 1\}$ and $b \in \{0, 1\}$, and two outputs, $A \in \{0, 1\}$ and

J. Bub (✉)
Philosophy Department and Institute for Physical Science and Technology, University of Maryland, College Park, MD, USA
e-mail: jbub@umd.edu

Y. Ben-Menahem and M. Hemmo (eds.), *Probability in Physics*, The Frontiers Collection,   167
DOI 10.1007/978-3-642-21329-8_11, © Springer-Verlag Berlin Heidelberg 2012

$B \in \{0, 1\}$,[1] and is defined by the following correlations between inputs and outputs:

$$A \oplus B = a \cdot b \tag{11.1}$$

where $\oplus$ is addition mod 2, i.e.,

1. Same outputs (i.e., 00 or 11) if the inputs are 00 or 01 or 10
2. Different outputs (i.e., 01 or 10) if the inputs are 11

together with a 'no signaling' constraint.

A PR-box is bipartite and nonlocal in the sense that the a-input and A-ouput can be separated from the b-input and B-output by any distance without altering the correlations. For convenience, we can think of the a-input as controlled by Alice, who monitors the A-output, and the b-input as controlled by Bob, who monitors the B-output. If we want the correlations of a PR-box to be consistent with relativistic causality, they should satisfy a 'no signaling' constraint: no information should be available in the marginal probabilities of Alice's outputs about alternative input choices made by Bob, and conversely, i.e.,

$$\sum_{b \in \{0,1\}} p(A,B|a,b) = p(A|a), \ A, a, b \in \{0,1\} \tag{11.2}$$

$$\sum_{a \in \{0,1\}} p(A,B|a,b) = p(B|b), \ B, a, b \in \{0,1\} \tag{11.3}$$

Note that 'no signaling' is not a relativistic constraint *per se* – it is simply a constraint on the marginal probabilities. But if this constraint is not satisfied, instantaneous (hence superluminal) signaling is possible, i.e., 'no signaling' is a necessary condition for relativistic causality.

It follows from (11.1) and 'no signaling' that the correlations are as in Table 11.1:

The probability $p(00|00)$ is to be read as $p(A = 0, B = 0|a = 0, b = 0)$, and the probability $p(01|10)$ is to be read as $p(A = 0, B = 1|a = 1, b = 0)$, etc. (I drop the

Table 11.1 PR-box correlations

b	a	0		1	
0		$p(00|00) = 1/2$	$p(10|00) = 0$	$p(00|10) = 1/2$	$p(10|10) = 0$
		$p(01|00) = 0$	$P(11|00) = 1/2$	$p(01|10) = 0$	$p(11|10) = 1/2$
1		$p(00|01) = 1/2$	$p(10|01) = 0$	$p(00|11) = 0$	$p(10|11) = 1/2$
		$p(01|01) = 0$	$P(11|01) = 1/2$	$p(01|11) = 1/2$	$p(11|11) = 0$

---

[1] In a simulation of PR-box correlations by classical or quantum correlations, inputs correspond to observables measured and outputs to measurement outcomes represented by real numbers, so it might seem more appropriate to use A, B for inputs and a, b for outputs. I follow the notation of Pawłowski et al. [1] here, since this is the result I discuss in detail below.

commas for ease of reading; the first two slots in $p(- - | - -)$ before the conditio-
nalization sign 'l'represent the two possible outputs for Alice and Bob, respectively,
and the second two slots after the conditionalization sign represent the two possible
inputs for Alice and Bob, respectively.) Note that the sum of the probabilities in each
square cell of the array in Table 11.1 is 1, and that the marginal probability of 0 for
Alice or for Bob is obtained by adding the probabilities in the left column of each
cell or the top row of each cell, respectively, and the marginal probability of 1 is
obtained for Alice or for Bob by adding the probabilities in the right column of each
cell or the bottom row of each cell, respectively. One could define a PR-box as
exhibiting the correlations in Table 11.1, which are 'no signaling,' rather than in
terms of the condition $A \oplus B = a. b$ and the 'no signaling' constraint.

Note that a PR box functions in such a way that if Alice inputs a 0 or a 1, her
output is 0 or 1 with probability 1/2, irrespective of Bob's input, and irrespective of
whether Bob inputs anything at all. Similarly for Bob. The requirement is simply
that whenever there are in fact two inputs, the inputs and outputs are correlated
according to (11.1). A PR-box can function only once, so to get the statistics for
many pairs of inputs one has to use many PR-boxes. This avoids the problem of
selecting the 'corresponding' input pairs for different inputs at various times, which
would depend on the reference frame. In this respect, a PR-box is like a quantum
system: after a system has responded to a measurement (produced an output for an
input), the system is no longer in the same quantum state, and one has to use many
systems prepared in the same quantum state to exhibit the probabilities associated
with a given quantum state.

What is the optimal probability that Alice and Bob can simulate a PR-box,
supposing they are allowed certain resources?

In units where $A = \pm 1, B = \pm 1,$[2]

$$\langle 00 \rangle = p(\text{same output}|00) - p(\text{different output}|00) \tag{11.4}$$

so:

$$p(\text{same output}|00) = \frac{1 + \langle 00 \rangle}{2} \tag{11.5}$$

$$p(\text{different output}|00) = \frac{1 - \langle 00 \rangle}{2} \tag{11.6}$$

and similarly for input pairs 01, 10, 11.

---

[2] It is convenient to change units here to relate the probability to the usual expression for the
Clauser-Horne-Shimony-Holt correlation, where the expectation values are expressed in terms of
± 1 values for A and B (the relevant observables). Note that 'same output' or 'different output'
mean the same thing whatever the units, so the probabilities $p(\text{same output} | AB)$ and $p(\text{different}$
$\text{output} | AB)$ take the same values whatever the units, but the expectation value $\langle AB \rangle$ depends on
the units for A and B.

It follows that the probability of successfully simulating a PR-box is given by:

$$p(\text{successful sim}) = \frac{1}{4}(p(\text{same output}|00) + p(\text{same output}|01) \tag{11.7}$$
$$+ \ p(\text{same output}|10) + p(\text{different output}|11))$$

$$= \frac{1}{2}\left(1 + \frac{K}{4}\right) = \frac{1}{2}(1 + E) \tag{11.8}$$

where $K = \langle 00 \rangle + \langle 01 \rangle + \langle 10 \rangle - \langle 11 \rangle$ is the Clauser-Horne-Shimony-Holt (CHSH) correlation.

Bell's locality argument in the Clauser-Horne-Shimony-Holt version [4] shows that if Alice and Bob are limited to classical resources, i.e., if they are required to reproduce the correlations on the basis of shared randomness or common causes established before they separate (after which no communication is allowed), then $|K_C| \leq 2$, i.e., $|E| \leq \frac{1}{2}$, so the optimal probability of successfully simulating a PR-box is $\frac{1}{2}(1 + \frac{1}{2}) = \frac{3}{4}$.

If Alice and Bob are allowed to base their strategy on shared entangled states prepared before they separate, then the Tsirelson bound for quantum correlations requires that $|K_Q| \leq 2\sqrt{2}$, i.e., $|E| \leq \frac{1}{\sqrt{2}}$, so the optimal probability of successful simulation limited by quantum resources is $\frac{1}{2}(1 + \frac{1}{\sqrt{2}}) \approx .85$.

Clearly, the 'no signaling' constraint (or relativistic causality) does not rule out simulating a PR-box with a probability greater than $\frac{1}{2}(1 + \frac{1}{\sqrt{2}})$. As Popescu and Rohrlich observe, there are possible worlds described by 'superquantum' theories that allow nonlocal boxes with 'no signaling' correlations stronger than quantum correlations, in the sense that $\frac{1}{\sqrt{2}} \leq E \leq 1$. The correlations of a PR-box saturate the CHSH inequality ($E = 1$), and so represent a limiting case of 'no signaling' correlations.

We see now that Wheeler's question 'why the quantum' has two aspects: why is the world quantum and not classical, and why is it quantum rather than superquantum, i.e., *why the Tsirelson bound?* In the following section, I discuss a remarkable answer to this question proposed by Pawłowski et al. [1].

11.2 Information Causality

Pawłowski et al. [1] consider a condition they call 'information causality,' that the information gain for Bob about an unknown data set of Alice, given all his local resources and m classical bits communicated by Alice, is at most m bits.[3] They

---

[3] The restriction to the communication of classical bits is essential here. Recall that entanglement correlations can be exploited to allow Alice to send Bob two classical bits by communicating just one quantum bit.

remark that the 'no-signaling' condition is just information causality for $m = 0$: if Alice communicates nothing to Bob, then there is no information in the statistics of Bob's outputs about Alice's data set. Pawłowski et al. show that the Tsirelson bound, $|E| \leq \frac{1}{\sqrt{2}}$, follows from this condition.

To see how they arrive at this startling result, it is convenient to consider the following game (related to oblivious transfer and communication complexity problems; see [5–7] and Sect. 11.4): At each round of the game, Alice receives N random and independent bits $\vec{a} = (a_0, a_1, \ldots, a_{N-1})$. Bob, separated from Alice, receives a value of a random uniformly distributed variable $b \in \{0, 2, \ldots, N-1\}$. Alice can send one classical bit to Bob with the help of which Bob is required to guess the value of the b-th bit in Alice's list, a_b, for some value of $b \in \{0, \ldots, N-1\}$. We assume that Alice and Bob are allowed to communicate and plan a mutual strategy before the game starts, but once the game starts the only communication between them is the one classical bit that Alice is allowed to send to Bob at each round of the game. They win a round if Bob correctly guesses the b-th bit for the round. They win the game if Bob always guesses correctly over any succession of rounds. Note that Alice must decide on the bit she sends to Bob at each round of the game independently of the value of b, which is given to Bob at each round and is unknown to Alice.

Clearly, Bob will be able to correctly guess the value of one of Alice's bits, assuming they agree in advance about the index k of the bit Alice sends at each round, but Bob's guess will be at chance when the value of $b \neq k$.

Now, suppose Alice and Bob are equipped with a supply of shared PR-boxes. Pawłowski et al. show that there is a strategy that will allow Alice and Bob to win the game, i.e., for any round, and for any $b \in \{0, 2, \ldots, N-1\}$, Bob will be able to correctly guess the value of any designated bit a_k in Alice's list $a_0, a_1, \ldots, a_{N-1}$.

Consider first the simplest case $N = 2$, where Alice receives two bits, a_0, a_1. The strategy in this case involves a single shared PR-box. Alice inputs $a_0 \oplus a_1$ into her part of the box (i.e., $a = a_0 \oplus a_1$) and obtains the output A. She sends the bit $x = a_0 \oplus A$ to Bob. Bob inputs the value of b, i.e., 0 or 1, into his part of the box and obtains the output B. He guesses $a_b = x \oplus B = a_0 \oplus A \oplus B$.

Now, the box functions in such a way that $A \oplus B = a \cdot b = (a_0 \oplus a_1) \cdot b$. So Bob's guess is $x \oplus B = a_0 \oplus A \oplus B = a_0 \oplus ((a_0 \oplus a_1) \cdot b)$. It follows that if $b = 0$, Bob correctly guesses a_0, and if $b = 1$, Bob correctly guesses $a_0 \oplus a_0 \oplus a_1 = a_1$.

Suppose Alice receives four bits, a_0, a_1, a_2, a_3. ($N = 4$). Bob's random variable labeling the bit he has to guess takes four values, $b = 0, 1, 2, 3$, and can be specified by two bits, b_0, b_1:

$$b = b_0 2^0 + b_1 2^1 = b_0 + 2b_1$$

The strategy in this case involves an inverted pyramid of PR-boxes: two shared PR-boxes, L and R, at the first stage, and one shared PR-box at the final second stage. Alice inputs $a_0 \oplus a_1$ into the L box, and $a_2 \oplus a_3$ into the R box. Bob inputs b_0 into both the L and R boxes and obtains the output B_0 (the input to one of these boxes will be irrelevant, depending on what bit Bob is required to guess; see below). At the second stage, Alice inputs $(a_0 \oplus A_L) \oplus (a_2 \oplus A_R)$ into the shared PR-box, where A_L is the Alice-output of the L box and A_R is the Alice-output of the R box, and obtains the output A. Bob inputs b_1 into this box and obtains the output B_1. Alice then sends Bob the bit $x = a_0 \oplus A_L \oplus A$.

Now, Bob could correctly guess either $a_0 \oplus A_L$ or $a_2 \oplus A_R$, using the elementary $N = 1$ strategy, as $x \oplus B_1 = a_0 \oplus A_L \oplus A \oplus B_1$. Here $A \oplus B_1 = (a_0 \oplus A_L \oplus a_2 \oplus A_R) \cdot b_1$. If $b_1 = 0$, Bob would guess $a_0 \oplus A_L$. If $b_1 = 1$, Bob would guess $a_2 \oplus A_R$.

So if Bob is required to guess the value of a_0 (i.e., $b_0 = 0$, $b_1 = 0$) or a_1 (i.e., $b_0 = 1, b_1 = 0$)— the input to the PR-box L—he guesses $a_0 \oplus A_L \oplus A \oplus B_1 \oplus B_0$, where B_0 is the Bob-output *of the L box*. Then:

$$a_0 \oplus A_L \oplus A \oplus B_1 \oplus B_0 = a_0 \oplus A_L \oplus B_0 \\ = a_0 \oplus (a_0 \oplus a_1) \cdot b_0 \tag{11.9}$$

If $b_0 = 0$, Bob correctly guesses a_0; if $b_0 = 1$, Bob correctly guesses a_1.

If Bob is required to guess the value of a_2 (i.e., $b_0 = 0$, $b_1 = 1$) or a_3 (i.e., $b_0 = 1, b_1 = 1$)— the input to the PR-box R—he guesses $a_0 \oplus A_L \oplus A \oplus B_1 \oplus B_0$, where B_0 is the Bob-output *of the R box*. Then:

$$a_0 \oplus A_L \oplus A \oplus B_1 \oplus B_0 = a_2 \oplus A_R \oplus B_0 \\ = a_2 \oplus (a_2 \oplus a_3) \cdot b_0 \tag{11.10}$$

If $b_0 = 0$, Bob correctly guesses a_2; if $b_0 = 1$, Bob correctly guesses a_3.

These strategies are winning strategies for $N = 2$, and $N = 4$ (the game for $N = 1$ is trivial). Clearly, the strategy for $N = 4$ is also a strategy for $N = 3$ (there is just one less value of b that Bob has to worry about). By adding more stages (levels) to the inverted pyramid, one obtains a strategy for $N = 8$ (four shared PR-boxes at the first stage, two shared PR-boxes at the next stage, and one shared PR-box at the third and final stage), and so on. This is also a strategy for $4 \leq N < 8$, so there is a strategy for any N.

The game can be modified to allow Alice to send m classical bits of information to Bob at each round, in which case Bob is required to guess the values of any set of m bits in Alice's list of N bits. In this case, Alice and Bob simply apply the above strategy for any N with m inverted pyramids of PR-boxes, one for each bit in the set of bits Bob is required to guess.

We have seen that Alice and Bob can win this game if they share PR-boxes ($E = 1$). What if they share non-signaling (NS) boxes with *any* 'no signaling' correlations corresponding to $|E| < 1$, such as classical correlations($|E| \leq \frac{1}{2}$), or the correlations of entangled quantum states ($|E| \leq \frac{1}{\sqrt{2}}$), or superquantum 'no signaling' correlations ($\frac{1}{\sqrt{2}} < E < 1$)?

The probability of simulating a PR-box with a NS-box is $\frac{1}{2}(1 + E)$, where E depends on the NS-box (the nature of the correlations). Consider the $N = 4$ game where Alice and Bob share NS-boxes, and Alice is allowed to communicate one bit to Bob. Bob's guess $x \oplus B_1 \oplus B_0$ will be correct if B_1 and B_0 are both correct or both incorrect (since $B_1 \oplus B_0$ will be the same in either case).

The probability of being correct at both stages is:

$$\frac{1}{2}(1+E) \cdot \frac{1}{2}(1+E) = \frac{1}{4}(1+E)^2 \tag{11.11}$$

The probability of being incorrect at both stages is:

$$\left(1 - \frac{1}{2}(1+E)\right) \cdot \left(1 - \frac{1}{2}(1+E)\right) = \frac{1}{2}(1-E) \cdot \frac{1}{2}(1-E) = \frac{1}{4}(1-E)^2 \tag{11.12}$$

So the probability P_k that Bob guesses correctly, i.e., the probability that $\beta = a_k$ when $b = k$, is:

$$P_k = \frac{1}{4}(1+E)^2 + \frac{1}{4}(1-E)^2 = \frac{1}{2}(1+E^2) \tag{11.13}$$

In the general case $N = 2^n$, Bob guesses correctly if he makes an even number of errors over the n stages $(B_0, B_1, B_2, \ldots)$ and the probability is:

$$P_k = \frac{1}{2^n}(1+E)^n + \frac{1}{2^n} \sum_{j=1}^{\lfloor \frac{n}{2} \rfloor} \binom{n}{2j}(1-E)^{2j}(1+E)^{n-2j} = \frac{1}{2}(1+E^n) \tag{11.14}$$

where $\lfloor \frac{n}{2} \rfloor$ denotes the integer value of $\frac{n}{2}$. For example, if $n = 3$, the probability of being correct at each stage is:

$$\frac{1}{2}(1+E) \cdot \frac{1}{2}(1+E) \cdot \frac{1}{2}(1+E) \tag{11.15}$$

and the probability of being incorrect at two out of the three stages (i.e., at B_0, B_1 or B_0, B_2 or B_1, B_2) is:

$$3 \cdot \frac{1}{2}(1-E) \cdot \frac{1}{2}(1-E) \cdot \frac{1}{2}(1+E) \tag{11.16}$$

so the probability that Bob guesses correctly is :

$$P_k = \frac{1}{8}(1+E)^3 + \frac{3}{8}(1-E)^2(1+E) = \frac{1}{2}(1+E^3) \tag{11.17}$$

11.3 The Tsirelson bound

In the game considered above, Alice has a list of N bits and Bob has to guess an arbitrarily selected one of these bits, $b = k$. If Bob knows the value of the bit he has to guess, $P_k = 1$. The binary entropy of P_k is defined as $h(P_k) = -P_k \log P_k - (1 - P_k) \log (1 - P_k)$, so $h(P_k) = 0$. If Bob has no information about the bit he has to guess, $P_k = 1/2$, i.e., his guess is at chance, and $h(P_k) = 1$.

If Alice sends Bob one classical bit of information, information causality requires that Bob's information about the N unknown bits increases by at most one bit. So if the bits in Alice's list are unbiased and independently distributed, Bob's information about an arbitrary bit $b = k$ in the list cannot increase by more than $1/N$ bits, i.e., for Bob's guess about an arbitrary bit in Alice's list, the binary entropy $h(P_k)$ is at most $1/N$ closer to 0 from the chance value 1, i.e., $h(P_k) \geq 1 - 1/N$.

It follows that the condition for a violation of information causality in this case can be expressed as:

$$h(P_k) < 1 - 1/N \tag{11.18}$$

or, taking $N = 2^n$, the condition is:

$$h(P_k) < 1 - \frac{1}{2^n} \tag{11.19}$$

Since $P_k = \frac{1}{2}(1 + E^n)$, we have a violation of information causality when:

$$h\left(\frac{1}{2}(1 + E^n)\right) < 1 - \frac{1}{2^n} \tag{11.20}$$

Pawłowski et al. [1] make use of the following inequality:

$$h\left(\frac{1}{2}(1 + y)\right) \leq 1 - \frac{y^2}{2\ln 2} \tag{11.21}$$

where $\ln 2 \approx .693$ is the natural log of 2 (base e). So information causality is violated if

$$1 - \frac{E^{2n}}{2\ln 2} < 1 - \frac{1}{2^n} \tag{11.22}$$

i.e., if

$$(2E^2)^n > 2\ln 2 \approx 1.386 \tag{11.23}$$

If $2E^2 = 1$, i.e., if $E = E_T = \frac{1}{\sqrt{2}}$ (the Tsirelson bound), the inequality (11.23) is satisfied. This is a sufficient condition for a violation of information causality, but it is not necessary: even if $(2E_T^2)^n /2\ln 2$, we could still have a violation of information causality for some n if $h(\frac{1}{2}(1 + E_T^n)) < 1 - \frac{1}{2^n}$. See the Appendix for a proof that information causality is satisfied for $E = E_T$, i.e., $h(\frac{1}{2}(1 + E_T^n)) \geq 1 - \frac{1}{2^n}$ for any n.

If $E > E_T$, i.e., if $2E^2 = 1 + a$, for some a, no matter how small, there is a violation: $(2E^2)^n > 1 + na,$[4] but $1 + na > 2\ln 2 \approx 1.386$ for some n. That is, for any a, however small, there is a value of n such that $n > \frac{.386}{a}$, hence a value of n for which information causality is violated.

To appreciate the significance of this result, consider some numbers for E and n. The condition for a violation of information causality is $h(P_k) < 1 - \frac{1}{2^n}$. Recall that
$$\log_2 x = \frac{\log_{10} x}{\log_{10} 2} \approx \frac{\log_{10} x}{.301}.$$
Consider first the case where $E = E_T = \frac{1}{\sqrt{2}} \approx .707$, the Tsirelson bound. When $n = 1$, Alice has $2^1 = 2$ bits:

$$
\begin{aligned}
h(P_k) = - & \left(\frac{1}{2}\left(1 + \frac{1}{\sqrt{2}}\right) \frac{\log_{10}\frac{1}{2}(1 + \frac{1}{\sqrt{2}})}{.301} \right. \\
& \left. + \frac{1}{2}\left(1 - \frac{1}{\sqrt{2}}\right) \frac{\log_{10}\frac{1}{2}(1 - \frac{1}{\sqrt{2}})}{.301} \right) \\
& \approx .600
\end{aligned}
\tag{11.24}
$$

There is no violation of information causality because $.600 > 1 - \frac{1}{2^1} = \frac{1}{2}$. When $n = 10$, Alice has $2^{10} = 1,024$ bits:

$$
\begin{aligned}
h(P_k) = - & \left(\frac{1}{2}\left(1 + \frac{1}{\sqrt{2}^{10}}\right) \frac{\log_{10}\frac{1}{2}(1 + \frac{1}{\sqrt{2}^{10}})}{.301} \right. \\
& \left. + \frac{1}{2}\left(1 - \frac{1}{\sqrt{2}^{10}}\right) \frac{\log_{10}\frac{1}{2}(1 - \frac{1}{\sqrt{2}^{10}})}{.301} \right) \approx .99939
\end{aligned}
\tag{11.25}
$$

There is still no violation of information causality because $.99939 > 1 - \frac{1}{2^{10}} = 1 - \frac{1}{1024} \approx .9990$.

Now consider the case where $E > E_T$. Take $E = .725$ and $n = 7$. In this case, there is a violation of information causality:

$$
\begin{aligned}
h(P_k) = - & \left(\frac{1}{2}(1 + .725^7) \frac{\log_{10}\frac{1}{2}(1 + .725^7)}{.301} \right. \\
& \left. + \frac{1}{2}(1 - .725^7) \frac{\log_{10}\frac{1}{2}(1 - .725^7)}{.301} \right) \\
& \approx .99208
\end{aligned}
\tag{11.26}
$$

---

[4] Recall that $(1 + a)^n$ can be expanded as $(1 + a)^n = 1 + na + \frac{n(n-1)}{2!} + \frac{n(n-1)(n-2)}{3!} + \cdots$

There is a violation of information causality because $.99208 < 1 - \frac{1}{128} \approx .99218$. There is no violation for $n = 6$ because $.9848 > 1 - \frac{1}{64} \approx .9844$.

Note that the inequality (11.21) has not been used in the above calculations. The only role of the inequality is to allow one to easily see that information causality is violated for *some* value of n if $E > E_T$, i.e., if $2E^2 > 1 + a$ for any a. In fact, information causality could be violated for a lower value of n. In the case above, $E = .725$, $a \approx .05125$. Using the inequality, we find that information causality is violated when $n > \frac{.386}{a}$, i.e., when $n \geq 8$.

If E is very close to the Tsirelson bound, then n must be very large for a violation of information causality. For $n = 10$ and $E = .708$:

$$h(P_k) = -\left(\frac{1}{2}(1 + .708^{10}) \frac{\log_{10}\frac{1}{2}(1 + .708^{10})}{.301} \right.$$
$$\left. + \frac{1}{2}(1 - .708^{10}) \frac{\log_{10}\frac{1}{2}(1 - .708^{10})}{.301} \right) \qquad (11.27)$$
$$\approx .99938$$

There is no violation of information causality because $.99938 > 1 - \frac{1}{1024} \approx .9990$. Using the inequality, with $a = .708 - \frac{1}{\sqrt{2}}$, we find that $n \geq 432$ for a violation of information causality.

Another way to look at this: If $E = E_T = \frac{1}{\sqrt{2}}$, $P_k = \frac{1}{2}(1 + E^n) \to \frac{1}{2}$ and $h(P_k) \to 1$ as $n \to \infty$. So, if Alice has a very long list and sends Bob one bit of information, Bob's ability to correctly guess an arbitrary bit in Alice's list is essentially at chance if the correlations are bounded by the Tsirelson bound. For a PR-box, $E = 1$, $P_k = 1$, $h(P_k) = 0$, so Bob can correctly guess any arbitrary bit in Alice's list.

11.4   Comments

The analysis in Sect. 11.3 related information causality directly to a condition on the binary entropy. In Pawłowski et al. [1], the authors relate information causality directly to a condition on the mutual information between Alice and Bob, and only indirectly to the binary entropy:

> Ideally, we wish to define that information causality holds if, after transfer of the m-bit message, the mutual information between Alice's data $\vec{a}$ and everything that Bob has—that is, the message $\vec{x}$ and his part B of the previously shared correlation—is bounded by m. Intuitively appealing though such a definition is, it has the severe issue that it is not theory-independent. Specifically, a mutual information expression '$I(\vec{a} : \vec{x}, B)$' has to be defined for a state involving objects from the underlying theory (the possibilities include classical correlation, a shared quantum state and NS-boxes). It is far from clear whether mutual information can be defined consistently for all nonlocal correlations, nor whether such a definition would be unique.

Pawłowski et al. denote Bob's output by β and quantify the efficiency of Alice's and Bob's strategy by:

$$I \equiv \sum_{k=0}^{N-1} I(a_k : \beta | b = k) \tag{11.28}$$

where $I(a_k : \beta \mid b = k)$ is the Shannon mutual information between a_k and β, computed under the condition that Bob is required to guess the bit $b = k$. They show that if the mutual information $I(\vec{a} : \vec{x}, B)$ for any 'no signaling' theory satisfies three constraints (which are satisfied for quantum information and for classical information, a special case of quantum information):

- consistency with the classical Shannon mutual information when the Alice and Bob subsystems are both classical
- the data-processing inequality: any local manipulation of data can only degrade information, i.e., acting on one subsystem locally by any admissible transformation cannot increase the mutual information
- the chain rule: $I(A : B, C) = I(A : C) + I(A : B \mid C)$, where $I(A : B \mid C)$ is the conditional mutual information

then (i) information causality is satisfied, i.e., $I(\vec{a} : \vec{x}, B) \leq m$, and (ii) $I(\vec{a} : \vec{x}, B) \geq I$. Since $I(\vec{a} : \vec{x}, B) > m$ if $I > m$, it follows that information causality is violated if:

$$I > m \tag{11.29}$$

So if information causality is satisfied, then $I \leq m$, i.e., $I \leq m$ is a necessary condition for information causality. (Note that we could, of course, have $I \leq m$ but $I(\vec{a} : \vec{x}, B) > m$, so (11.29) is not a sufficient condition for information causality.) As the authors emphasize, I is fully specified by Alice's and Bob's input and output bits and is independent of the details of any particular physical theory.

The Shannon mutual information $I(X{:}Y)$ of two random variables is a measure of how much information they have in common: the sum of the information content of the two random variables, as measured by the Shannon entropy (in which joint information is counted twice), minus their joint information:

$$\begin{aligned} I(X : Y) &= H(X) + H(Y) - II(X, Y) \\ &= H(X) - H(X|Y) \end{aligned} \tag{11.30}$$

where $H(X) = -\sum_i p_i \log p_i$ is the Shannon entropy of the random variable X, $H(X, Y) = -\sum_{i,j} p_{i,j} \log p_{i,j}$ is the joint Shannon entropy of the two random variables X, Y representing the joint information, and $H(X / Y)$ is the conditional entropy: $H(X / Y) = H(X, Y) - H(Y)$. Note that $H(X / Y) \leq H(X)$, with equality if and only if X, Y are independent.

So:

$$I \equiv \sum_{k=0}^{N-1} I(a_k : \beta | b = k) = \sum_{k=0}^{N-1} (H(a_k) + H(\beta) - H(a_k, \beta)) \tag{11.31}$$

where the condition $b = k$ has been omitted for ease of reading.

First note that

$$H(a_k|\beta) = H(a_k \oplus \beta|\beta)$$
$$\leq H(a_k \oplus \beta)$$

(11.32)

The first equality follows because only the probabilities of the different alternatives are relevant in the calculation of the entropy. In this case, the probabilities are 0 and 1 and, given that $\beta = 0$, the probability that $a_k = 0$ is the same as the probability that $a_k \oplus \beta = 0$, i.e., that $a_k = \beta$, and the probability that $a_k = 1$ is the same as the probability that $a_k \oplus \beta = 1$, i.e., that $a_k \neq \beta$; and similarly if $\beta = 1$. The second inequality follows because conditioning decreases entropy.

Now:

$$H(a_k \oplus \beta) = h(P_k)$$

(11.33)

so

$$H(a_k|\beta) \leq h(P_k)$$

(11.34)

It follows that:

$$I(a_k : \beta)|b = k) \geq H(a_k) - h(P_k)$$

(11.35)

In the case where the bits in Alice's list are unbiased and independently distributed, $H(a_k) = 1$, so:

$$I(a_k : \beta)|b = k) \geq 1 - h(P_k)$$

(11.36)

i.e.,

$$I \geq N - \sum_{k=0}^{N-1} h(P_k)$$

(11.37)

and since $h(P_k) = \frac{1}{2}(1 + E^n)$, which is independent of k:

$$I \geq N - Nh(P_k)$$

(11.38)

For a PR-box, $E = 1$, $h(P_k) = 0$, and $I = N$. If Bob guesses randomly for all k, then $h(P_k) = 1$, $I = 0$. So in the case where Alice sends m bits of information to Bob, $0 \leq I \leq N$, with a violation of information causality when $I > m$.

If Alice sends Bob one bit of information, information causality is violated if $I > 1$, i.e., if:

$$h(P_k) < 1 - \frac{1}{N} \tag{11.39}$$

or, taking $N = 2^n$, if:

$$h(P_k) < 1 - \frac{1}{2^n} \tag{11.40}$$

which are, respectively, Eqs. 11.18 and 11.19 of Sect. 11.3.

Pawłowski et al. [1, p. 1101] express the condition of information causality as follows:

> Formulated as a principle, information causality states: 'the information gain that Bob can reach about a previously unknown to him data set of Alice, by using all his local resources and m classical bits communicated by Alice, is at most m bits.' The standard no-signalling condition is just information causality for $m = 0$.

Stated in this way, the condition seems trivial: of course, if Alice sends Bob m bits of information, his information gain is at most m bits, and if $m = 0$ his information gain is 0. But implicit in the condition is that Bob's local resources include the marginal probabilities of correlations between Alice and Bob and the values of the correlated variables, and similarly for Alice. The issue concerns the extent to which Alice and Bob can exploit previously established correlations between them in such a way that the m bits of information communicated by Alice to Bob will allow Bob to correctly guess an arbitrarily designated set of bits in Alice's data set, which might contain $N > m$ bits. Of course, without exploiting the correlations, Bob can know some specific, previously agreed upon set of m bits and, exploiting classical correlations, i.e., previously established shared randomness, Bob can know a different specific set of m bits on each occasion that Alice sends him m bits.[5] The relevant insight is that if the correlations are PR-box correlations, then Alice can send Bob a set of m bits chosen on the basis of the Alice-values of the correlated variables, where Alice and Bob select the variables appropriately as the inputs to the PR-boxes, in such a way that Bob can correctly guess *any arbitrary set of m bits in Alice's data set*. In other words, for the case $m = 1$, there is a way of exploiting the PR-box correlations so that the one bit of information can be associated with *any designated bit* in Alice's data set of N bits, for any N (this was pointed out already in [8]).

So in the case where the bits in Alice's data set are unbiased and independently distributed and Alice sends Bob one bit of information, the PR-box correlations can be exploited to achieve $P_k = 1$ for all k, i.e., $h(P_k) = 0$ for all k. The intuition

---

[5] A suitably long shared list of random bits can be used by Alice and Bob to pick a different set of m bits at each round of the guessing game, for some finite set of rounds.

behind information causality is that this is 'too good to be true,' in fact, that the binary entropy should be bounded: $h(P_k) \geq 1 - \frac{1}{N}$. Putting it differently, when the bits in Alice's data set are unbiased and independently distributed, the intuition is that if the correlations can be exploited to distribute one bit of communicated information among the N unknown bits in Alice's data set, the amount of information distributed should be no more than $\frac{1}{N}$ bits, because there can be no information about the bits in Alice's data set in the previously established correlations themselves.

As Pawłowski et al. show, for 'no signaling' correlations, $P_k = \frac{1}{2}(1 + E^n)$, where $N = 2^n$. For classical correlations, $E = \frac{1}{2}$, $h(P_k) \approx .811$ for $n = 1$. For quantum correlations, $E = E_T = \frac{1}{\sqrt{2}}$, $h(P_k) \approx .600$ for $n = 1$, so Alice and Bob can do better exploiting quantum correlations than they can if they are restricted to classical correlations. This is the case for any n, but information causality is always satisfied. The intriguing result by Pawłowski et al. is that information causality is violated *for some value of n* if $E > E_T$. From this perspective, it is misleading to claim that the 'no signaling' condition is 'just information causality for $m = 0$.' If Alice communicates no information to Bob, they have no possibility of exploiting correlations to increase Bob's access to Alice's data set. The condition of information causality concerns the extent to which correlations can be exploited to increase Bob's access to Alice's data set, in the sense of improving Bob's ability to correctly guess any arbitrary bit in Alice's data set.

In fact, the term 'information causality' is suggestive in the wrong sense. The principle really has nothing to do with causality and is better understood as *a constraint on the ability of correlations to enhance the information content of communication in a distributed task*. A more appropriate term would be 'informational neutrality of correlations,' and the principle should be formulated as follows:

> Correlations are informationally neutral: insofar as they can be exploited to allow Bob to distribute information communicated by Alice among the bits in an unknown data set held by Alice in such a way as to increase Bob's ability to correctly guess an arbitrary bit in the data set, they cannot increase Bob's information about the data set by more than the number of bits communicated by Alice to Bob.

So if Alice has a data set of N uniformly and independently distributed bits and sends Bob one bit of information, and Bob can exploit previously established correlations to increase his ability to correctly guess an arbitrary bit in the data set, his information gain about an arbitrary bit in the data set can be no more than $1/N$ bits, i.e., the binary entropy of the probability of a correct guess cannot be less than $1 - 1/N$.

The correlations of a PR-box are not informationally neutral in this sense. While they are logically admissible, they are 'too good to be true' in the way they allow the solution of the following two distributed tasks:

- The 'dating game': Alice and Bob would like to go on a date, but only if they know that they both like each other. In other words, they would like to compute a function that takes the value 1 if they both like each other (i.e., if both inputs to the function are 1), but takes the value 0 if at least one party does not like the other (i.e., if the inputs are both 0, or one input is 0 and the other input is 1). Now, in the real world, there is no way they can do this without revealing information that they both want to keep private: Alice does not want Bob to know that she likes him *if he does not like her*, and similarly for Bob. With a PR-box, they can compute this function, while keeping private the information they want to keep private. Alice and Bob input 0 or 1 into their inputs to the PR-box when they are separate (so neither party sees the other's input). They then come together and share the outputs. If the outputs are different, they know that both inputs were 1, so they happily go on a date. In this case, of course, Alice knows that Bob likes her, and Bob knows that Alice likes him, but that's fine. If the outputs are the same, they know only that either Alice did not like Bob, or that Bob did not like Alice, or that the dislike was mutual. While Alice can infer that Bob does not like her if she likes him, this knowledge is private, so Alice avoids any humiliation; and similarly for Bob.

- 'One-out-of-two' oblivious transfer: Alice has a data set consisting of two bits of information. The constraint on Alice is that she can send Bob one bit of information. The requirement for Bob is that he uses the one bit of communicated information to correctly guess whichever bit he chooses in Alice's data set, in such a way that Alice is oblivious of his choice. Again, there is no way to do this in the real world, but if Alice and Bob have access to a PR-box they can successfully achieve this task. The protocol is the same as the protocol for the $N = 2$ case discussed in Sect. 11.2.

The remarkable result of Pawłowski et al. shows that, while quantum correlations are 'more like' PR-box correlations than classical correlations, insofar as they increase the ability of Alice and Bob to perform distributed tasks relative to classical correlations, they represent the limit of what is possible if correlations are 'informationally neutral,' in the sense that correlations established prior to the choice of a data set can contain no information about such a data set, and hence should not be able to be exploited to allow a party who has no access to the data set to correctly guess any arbitrary bit in the set. This considerably extends related results by van Dam [6, 7], Brassard et al. [5], Linden et al. [9]. Note that there are other results in which nonlocal boxes are exploited to derive the Tsirelson bound. See Skrzypczyk et al. [10], in which a dynamics is defined for PR-boxes and the Tsirelson bound is derived from a condition called 'nonlocality swapping.'

Pawłowski et al. [1, p. 1103–1104] conclude with the following remarks:

In conclusion, we have identified the principle of Information Causality, which precisely distinguishes physically realized correlations from nonphysical ones (in the sense that quantum mechanics cannot reach them). It is phrased in operational terms and in a theory-independent way and therefore we suggest it is at the same foundational level as the no-signaling condition itself, of which it is a generalization.

The new principle is respected by all correlations accessible with quantum physics while it excludes all no-signaling correlations, which violate the quantum Tsirelson bound. Among the correlations that do not violate that bound it is not known whether Information Causality singles out exactly those allowed by quantum physics. If it does, the new principle would acquire even stronger status.

Table 11.2 A deterministic state

a	0		1	
b				
0	$p(00|00) = 1$	$p(10|00) = 0$	$p(00|10) = 1$	$p(10|10) = 0$
	$p(01|00) = 0$	$p(11|00) = 0$	$p(01|10) = 0$	$p(11|10) = 0$
1	$p(00|01) = 1$	$p(10|01) = 0$	$p(00|11) = 1$	$p(10|11) = 0$
	$p(01|01) = 0$	$p(11|01) = 0$	$p(01|11) = 0$	$p(11|11) = 0$

Classical correlations bounded by $E \leq \frac{1}{2}$ can be associated with a polytope, where the vertices represent 'no signaling' deterministic states. For example, in the case considered above for a bipartite system with two binary-valued quantities, the deterministic state in which the values of the two quantities are both zero, for all four possible combinations, is given by Table 11.2.

There are 16 'no signaling' deterministic states (each of which can be represented as a product of local states, an Alice deterministic state and a Bob deterministic state) out of 256 possible deterministic states—the remaining 240 deterministic states allow signaling. The 16-vertex classical polytope is included in a 24-vertex 'no signaling' nonlocal polytope, where the vertices are the 16 'no signaling' deterministic states and 8 additional PR-box states, represented by the probabilities in Table 11.1, or probabilities obtained from Table 11.1 by relabeling the a-inputs, and the A-outputs conditionally on the a-inputs, and the b-inputs, and the B-outputs conditionally on the b-inputs. Quantum correlations bounded by $E = E_T \leq \frac{1}{\sqrt{2}}$ are associated with a spherical convex set with extremal points between the 16-vertex classical simplex and the 24-vertex 'no signaling' nonlocal polytope.

The open question is whether non-quantum correlations represented by points outside the quantum convex set but below the Tsirelson bound can also be excluded by information causality. For a discussion, see Allcock et al. [11].

11.5 Appendix

In [1], the authors prove quite generally that information causality is satisfied for any 'no signaling' theory satisfying three constraints on mutual information (consistency with the classical Shannon mutual information, the data-processing inequality, and the chain rule), hence for quantum information, which satisfies the constraints. It follows that information causality is satisfied at and below the Tsirelson bound.

The following is a simple direct proof (see Sect. 11.3) that if $E = E_T = \frac{1}{\sqrt{2}}$, then:

$$h(\frac{1}{2}(1 + E^n) \geq 1 - \frac{1}{2^n} \tag{11.41}$$

i.e.,

$$-\frac{1}{2}(1+E^n)\log\left(\frac{1}{2}(1+E^n)\right) - \frac{1}{2}(1-E^n)\log\left(\frac{1}{2}(1-E^n)\right) \geq 1 - \frac{1}{2^n} \quad (11.42)$$

After a little algebra, this can be expressed as:

$$\log(1-E^{2n}) + E^n\log\frac{1+E^n}{1-E^n} \leq \frac{1}{2^{n-1}} \quad (11.43)$$

Note that the logarithms are to the base 2.
Now, if $-1 \leq x \leq 1$:

$$\log_e(1+x) = x - \frac{1}{2}x^2 + \frac{1}{3}x^3 - \frac{1}{4}x^4 + \cdots \quad (11.44)$$

$$\log_e\frac{1+x}{1-x} = 2\left(x + \frac{x^3}{3} + \frac{x^5}{5} \cdots + \frac{1}{2m-1}x^{2m-1} + \cdots\right) \quad (11.45)$$

So

$$\log_e(1-E^{2n}) + E^n\log_e\frac{1+E^n}{1-E^n} = E^{2n} + \frac{1}{6}E^{4n} + \frac{1}{15}E^{6n}\cdots$$
$$+ \frac{1}{m(2m-1)}E^{2mn} + \cdots \quad (11.46)$$

Substituting $E = E_T = \frac{1}{\sqrt{2}}$, this becomes:

$$\left(\frac{1}{2}\right)^n + \frac{1}{6}\cdot\left(\frac{1}{2}\right)^{2n} + \frac{1}{15}\cdot\left(\frac{1}{2}\right)^{3n}\cdots + \frac{1}{m(2m-1)}\cdot\left(\frac{1}{2}\right)^{mn} + \cdots \quad (11.47)$$

Since $\log_2 x = \log_2 e\cdot\log_e x$, it follows that $\log(1-E^{2n}) + E^n\log\frac{1+E^n}{1-E^n}$, where the logarithms are to the base 2, can be expressed as the following infinite series:

$$\log_2 e \cdot \left(\frac{1}{2^n} + \frac{1}{6}\frac{1}{2^{2n}} + \frac{1}{15}\frac{1}{2^{3n}} + \cdots + \frac{1}{m(2m-1)}\frac{1}{2^{mn}} + \cdots\right) \quad (11.48)$$

so the inequality (11.43) we are required to prove becomes:

$$\frac{1}{2^n} + \frac{1}{6}\frac{1}{2^{2n}} + \frac{1}{15}\frac{1}{2^{3n}}\cdots + \frac{1}{m(2m-1)}\cdot\frac{1}{2^{mn}} + \cdots \leq \log_e 2 \cdot \frac{1}{2^{n-1}} \quad (11.49)$$

or

$$\frac{1}{2} + \frac{1}{6} \cdot \frac{1}{2^{n+1}} + \frac{1}{15} \cdot \frac{1}{2^{2n+1}} \cdots + \frac{1}{m(2m-1)} \cdot \frac{1}{2^{(m-1)n+1}} + \cdots \leq \log_e 2 \approx .693147$$

(11.50)

This is clearly the case. The largest value of the series is obtained for $n = 1$, when the first term is .5. The remaining terms affect only the second and later decimal places.

Alternatively, from (11.44) we have:

$$\log_e 2 = 1 - \frac{1}{2} + \frac{1}{3} - \frac{1}{4} + \cdots$$

(11.51)

so, subtracting the series on the left hand side of the inequality (11.50) from the series for $\log_e 2$, what has to be proved is that, for any n:

$$\left(\frac{1}{3} - \frac{1}{6} \cdot \frac{1}{2^{n+1}}\right) - \left(\frac{1}{4} + \frac{1}{15} \cdot \frac{1}{2^{2n+1}}\right) + \left(\frac{1}{5} - \frac{1}{28} \cdot \frac{1}{2^{3n+1}}\right)$$
$$- \left(\frac{1}{6} + \frac{1}{45} \cdot \frac{1}{2^{4n+1}}\right) + \cdots \geq 0$$

(11.52)

This is obvious by inspection, since each negative term in parenthesis is smaller than its positive predecessor, for any n.

Acknowledgements This paper was written during the tenure of a University of Maryland semester RASA award.

References

1. Pawłowski, M., Patarek, T., Kaszlikowski, D., Scarani, V., Winter, A., Żukowski, M.: A new physical principle: information causality. Nature **461**, 1101–1104 (2009)
2. Popescu, S., Rhorlich, D.: Quantum nonlocality as an axiom. Foundations Phys. **24**, 379 (1994)
3. Bell, J.S.: On the Einstein-Podolsky-Rosen Paradox. Physics **1**, 195–200 (1964). Reprinted in John Stuart Bell, Speakable and Unspeakable in Quantum Mechanics, Cambridge University Press, Cambridge (1989)
4. Clauser, J.F., Horne, M.A., Shimony, A., Holt, R.A.: Proposed experiment to test hidden variable theories. Phys. Rev. Lett. **23**(880–883), 16 (1969)
5. Brassard, G., Buhrman, H., Linden, N., Méthot, A.A., Tapp, A., Unger, F.: Limit on nonlocality in any world in which communication complexity is not trivial. Phys. Rev. Lett. **96**, 250401–250404 (2006)
6. van Dam, W.: Nonlocality and communication complexity. Ph.D. thesis, Oxford (2000)
7. van Dam, W.: Implausible consequences of superstrong nonlocality. arxiv: quant-ph/0501159v1 (2005)

8. Wolf, S., Wullschleger. J.: Oblivious transfer and quantum nonlocality. In: International symposium on information theory (IEEE Proceedings, ISIT 2005), pp. 1745–1748. IEEE. arxiv:quant-ph/0502030v1 (2005)
9. Linden, N., Popescu, S., Short, A.J., Winter, A.: Quantum nonlocality and beyond: limits from nonlocal computation. Phys. Rev. Lett. **99**, 180502–180505 (2007)
10. Skrzypczyk, P., Brunner, B., Popescu, S.: Emergence of quantum mechanics from non-locality swapping. arXiv: quant-ph/0811.2937v1 (2008)
11. Allcock, J., Brenner, N., Pawlowski, M., Scarani, V.: Recovering part of the boundary between quantum and nonquantum correlations from information causality. Phys. Rev. A **80**, 040103(R)–040107(R) (2009)

Chapter 12
Three Attempts at Two Axioms for Quantum Mechanics

Daniel Rohrlich

Abstract The axioms of nonrelativistic quantum mechanics lack clear physical meaning. In particular, they say nothing about nonlocality. Yet quantum mechanics is not only nonlocal, it is twice nonlocal: there are nonlocal quantum correlations, and there is the Aharonov-Bohm effect, which implies that an electric or magnetic field *here* may act on an electron *there*. Can we invert the logical hierarchy? That is, can we adopt nonlocality as an axiom for quantum mechanics and *derive* quantum mechanics from this axiom and an additional axiom of causality? Three versions of these two axioms lead to three different theories, characterized by "maximal nonlocal correlations", "jamming" and "modular energy". Where is quantum mechanics in these theories?

12.1 Introduction

Among Itamar Pitowsky's many admirable qualities, I admired most his capacity for thoroughly exploring incompatible points of view, approaches and theoretical frameworks. We tend to ignore approaches that are incompatible with our own. It is a natural tendency. It takes work to overcome it. Itamar worked hard to understand all points of view, which led to another of his admirable qualities, his comprehensive knowledge. He was a true *philosopher* – love of knowledge and understanding animated him. As a result, whether in a seminar on campus or at a demonstration on a street corner, he most often understood all the points of view better than anyone else did. The Talmud (ברכות סד, א) expresses this capacity in a adage, "תלמידי חכמים מרבים שלום בעולם" – "scholars bring peace to the world" – which Rabbi Avraham Kook (1865–1935) explained as follows:

D. Rohrlich (✉)
Physics Department, Ben Gurion University of the Negev, Beersheba 84105, Israel
e-mail: rohrlich@bgu.ac.il

Y. Ben-Menahem and M. Hemmo (eds.), *Probability in Physics*, The Frontiers Collection, 187
DOI 10.1007/978-3-642-21329-8_12, © Springer-Verlag Berlin Heidelberg 2012

Some people mistakenly believe that world peace demands uniformity of views and practices. When they see scholars studying philosophy and Torah and arriving at a plurality of views and approaches, they see only controversy and discord. But – on the contrary – true peace comes to the world only by virtue of its plurality. The plurality of peace means appreciating each view and approach and seeing how each has its own place, consistent with its value, context and content.

(From עולת ראיה ח"א ע' של; translations from Hebrew are mine.) One theoretical framework that Itamar explored [1] is the one that I have the honor to present here.   •

Quantum mechanics doesn't supply its own interpretation. The numerous competing interpretations of quantum mechanics testify to the fact that quantum mechanics doesn't supply its own interpretation. About one of these interpretations, due to Everett [2] and Wheeler [3], Bryce DeWitt said the following:

Everett took the quantum theory the way it was and didn't impose anything on it from the outside. No classical realm. Does the theory produce its own interpretation? If so, how? And he, in my view he's the only one who's answered properly in the affirmative that it provides its own interpretation [4].

Yet, more than half a century after Everett's bold and provocative paper, the Everett-Wheeler interpretation remains exactly that – one more competing interpretation, with no consensus in sight. This irony, as well, testifies to the fact that quantum mechanics doesn't supply its own interpretation. What would it take to show that, on the contrary, quantum mechanics does supply its own interpretation? It would take an interpretation so natural and satisfactory as to induce consensus. But how could such an interpretation ever spring from the opaque axioms of quantum mechanics? If we seek a theory that has clear, unambiguous physical meaning, let us derive it from axioms that have clear, unambiguous physical meaning. This paper argues that the way towards a natural and satisfactory interpretation of quantum mechanics passes through new and physically meaningful axioms for the theory, and reports attempts to define such axioms.

The axioms of quantum mechanics are notoriously opaque. Without trying or needing to be comprehensive, let us mention a few of them. "Any possible physical state corresponds to a ray in a Hilbert space." "Physical observables correspond to Hermitian operators." Do these axioms tell us something about the physical world? Or do they merely list mathematical structures useful for describing the world? If they merely list useful mathematical structures, what is it about the underlying physics that makes these structures useful? What we want, after all, is the physics. "If $P(a, \psi)$ is the probability that a measurement of a physical observable A, corresponding to a Hermitian operator $\hat{A}$, on a system in the normalized state $|\psi>$, yields a, then $P(a, \psi) = <\psi| \Pi_a |\psi>$, where Π_a projects onto the subspace of eigenstates of $\hat{A}$ having eigenvalue a." This axiom, too, is opaque; it offers no hint as to whether the probability $P(a, \psi)$ is intrinsic or derives from an underlying determinism. It is more of an algorithm than an axiom.

Long ago, Yakir Aharonov [5] drew an analogy that is at once amusing and penetrating. The special theory of relativity, we know, follows elegantly from two axioms: first, the laws of physics are the same for observers in all inertial reference frames; second, the speed of light is a constant of nature. These axioms have clear physical meanings: the first specifies a fundamental space-time symmetry and the second specifies a physical constant. Suppose we had, instead of these two standard axioms, three "nonstandard" axioms: first, physical objects contract in the direction of their motion (FitzGerald contraction); second, this contraction and a "local time" (lacking clear physical significance) combine in "Lorentz transformations" that form a group; third, simultaneity is subjective. Even supposing we could deduce the special theory of relativity from these three axioms, would there be consensus about its interpretation? What kind of consensus *could* there be, as long as we were stuck with the wrong axioms? Analogously, perhaps consensus about how to interpret quantum mechanics is elusive because we are stuck with the wrong axioms for the theory. Perhaps we have grasped quantum mechanics by the tail, or by the hind legs, instead of by the horns.

If this analogy seems strained, let us note that all three of these "nonstandard" axioms had proponents (not including A. Einstein) already in 1905. Moreover, as late as 1909, H. Poincaré delivered a series of lectures in Göttingen, culminating in a lecture on "La mécanique nouvelle". At the basis of Poincaré's "new mechanics" were three axioms, the third being the FitzGerald contraction: "One needs to make still a third hypothesis, much more surprising, much more difficult to accept, one which is of much hindrance to what we are currently used to. A body in translational motion suffers a deformation in the direction in which it is displaced." [6] In Poincaré's new mechanics, the FitzGerald contraction was an axiom in itself, not a consequence of other axioms, hence Poincaré found it "much more difficult to accept" than we do. Well, aren't all the axioms of quantum mechanics surprising, difficult to accept, and of *much hindrance* to what we are used to?

We can take the analogy further [5]. The logical structure of the special theory of relativity is exemplary. From its two axioms we can deduce all the kinematics of the theory, and with scarcely more input we can deduce all the dynamics as well. How can two axioms be so efficient? If we look at the axioms with a Newtonian mindset, we see clearly that the two axioms contradict each other. Well, of course they don't contradict each other. But they come so close to contradicting each other, that a unique mechanics reconciles them: Einstein's special theory of relativity. Now in the quantum world, as well, we have two physical principles that come close to contradicting each other. One is the principle of causality: *relativistic* causality, also called "no signalling", is the principle that no signal can travel faster than light. The other principle is nonlocality. Quantum mechanics is nonlocal, indeed twice nonlocal: it is nonlocal in two inequivalent ways. There is the nonlocality of the Aharonov-Bohm [7] and related [8] effects, and there is the nonlocality implicit in quantum correlations that violate Bell's [9] inequality. Let us briefly comment on each of these.

One often reads that the Aharonov-Bohm effect proves that the scalar and vector potentials of electromagnetism have a degree of reality in quantum physics that they do not have in classical physics. This distinction is valid, but it can mislead us into thinking that the Aharonov-Bohm effect is local, because an electron diffracting around a shielded solenoid or capacitor interacts locally with the vector potential of the magnetic field in the solenoid, or with the scalar potential of the electric field in the capacitor. However, there is no way to measure these potentials. What is measurable are the magnetic and electric fields. The dependence of the electron diffraction pattern on these fields, i.e. the Aharonov-Bohm effect, must be nonlocal, because the electron does not pass through them. Yet the fields are produced locally; so one might try (i.e. by adjusting the capacitor or the current in the solenoid) to send a superluminal signal that would show up in the electron diffraction pattern. It is indeed possible to send signals that show up in the interference pattern, but they are never superluminal.

Bell's inequality implies that quantum correlations can be nonlocal in the following sense. Let Alice and Bob, shown in Fig. 12.1, be two experimental physicists, in their respective labs. Note that each is equipped with a black box. From spacelike-separated measurements on their respective black boxes, Alice and Bob obtain correlations that violate Bell's inequality. They may suppose that the correlations already existed (as local "plans" or programs) in the black boxes before their measurements, but this explanation is what Bell ruled out. Another explanation – that the black boxes send superluminal signals to each other – might suggest to them that they, too, could send superluminal signals to each other. But they cannot [10]. Quantum correlations can violate Bell's inequality but they are useless for signalling.

The coexistence of nonlocality and causality in the Aharonov-Bohm effect and in nonlocal quantum correlations is remarkable. If there are nonlocal effects, what stops us from signalling with them? Quantum mechanics must be an extraordinary theory if it can do this trick – if it can make nonlocality and causality coexist. Shimony [11, 12] speculated that quantum mechanics might be the only theory that can do so, i.e. that quantum mechanics might follow uniquely from the two axioms of nonlocality and causality. Aharonov [5] even suggested a part of the logical structure. If we look at the details of just how nonlocality coexists with causality, we discover they always involve quantum uncertainty. Indeed, quantum mechanics has to be a probabilistic theory, because if nonlocal influences were certain, how could they *not* violate causality? Aharonov's suggestion thus inverts the

Bob

Alice

Fig. 12.1 Alice and Bob (drawn by Tom Oreb © Walt Disney Co) with their respective black boxes

conventional logical hierarchy of quantum mechanics: instead of making probability an axiom and deriving nonlocality as a quantum effect, he makes nonlocality an axiom and derives probability (uncertainty) as a quantum effect.

Inspired by Aharonov's suggestion, I will now describe three attempts [13–15] to define more precisely the axioms of "causality" and "nonlocality" and to derive quantum mechanics from them. The three attempts fall under the headings "Maximally nonlocal correlations" (or "PR boxes"), "Jamming", and "Modular energy".

12.2 Maximally Nonlocal Correlations

Our goal is to derive quantum mechanics, but we have still to decide whether we should try to derive nonrelativistic or relativistic quantum mechanics. We have two reasons to decide for nonrelativistic quantum mechanics. First, it is a simpler theory with a superior formalism. Relativistic quantum mechanics naturally allows creation of particle-antiparticle pairs; it is more complicated in that the number of particles is not fixed. Even if we artificially fix the number of particles, the formalism of relativistic quantum mechanics is not satisfactory: not all of its Hermitian operators correspond to physical observables [16]. That is, not all of its Hermitian operators are measurable in practice. Relativistic causality imposes constraints on what is measurable in practice. However, these constraints do not apply to the Hermitian operators of nonrelativistic quantum mechanics; they are all measurable [17]. If our goal is to find the right axioms for quantum mechanics, we have a better chance of finding them for a theory that already has a satisfactory formalism. It is also plausible that if we find the right axioms for the nonrelativistic theory, we will be in a better position to find the right axioms for the relativistic theory, axioms that do not lead to unmeasurable Hermitian operators.

The second reason to decide for nonrelativistic quantum mechanics has to do with the axiom of nonrelativistic causality. According to this axiom, no signal can travel faster than the speed of light, c. In the nonrelativistic limit, c is infinite, so the axiom of relativistic causality appears, initially, to become vacuous. But closer inspection reveals that the axiom of relativistic causality is actually *stronger* in this limit: it tells us that quantum correlations cannot be used to send signals at *any* speed. How so? Suppose that the quantum correlations measured by Alice and Bob could be used to send a signal. These correlations do not depend on whether the interval between Alice's and Bob's respective measurements is spacelike or timelike; indeed, nothing in nonrelativistic quantum mechanics can distinguish between spacelike and timelike intervals, because there is no c. Hence if quantum correlations could be used to send any signal, they could be used to send a superluminal signal. Hence they cannot be used to send any signal. If our choice of an axiom of nonlocality has to do with nonlocal correlations, the axiom of causality implies that quantum correlations are useless for sending signals at any speed. We find ourselves with two axioms with clear physical meaning:

1. No quantum correlations can be used to send signals.
2. Some quantum correlations are nonlocal.

The second axiom means that some quantum correlations violate some version of Bell's inequality. The most suitable version of Bell's inequality for the case of bipartite correlations (measured by Alice and Bob) is due to Clauser, Horne, Shimony and Holt (CHSH) [18]. Let us assume that Alice and Bob make a series of joint measurements on their respective boxes, in which Alice measures either A or A' and Bob measures either B or B' in each joint measurement, where A, A', B and B' are physical observables. In each joint measurement they must each choose which observable to measure – they can never measure both – and the results of their measurements are always 1 or -1. From the results of their joint measurements, they can compute four types of correlations: $C(A, B)$, $C(A', B)$, $C(A, B')$ and $C(A', B')$, where $C(A, B)$ is the correlation between the results of their joint measurements when Alice chooses to measure A and Bob chooses to measure B, and so on. The CHSH inequality states that if these correlations already existed (as local plans or programs) in the black boxes before their measurements, then a certain combination of them is bounded in absolute magnitude:

$$|C(A, B) + C(A, B') + C(A', B) - C(A', B')| \leq 2. \tag{12.1}$$

So if the correlations that Alice and Bob measure satisfy this inequality, they are local correlations. Conversely, if the correlations that they measure do not satisfy the inequality, they are nonlocal correlations. Axiom 2 above, the nonlocality axiom, states that at least for some physical variables A, A', B and B' and some preparation of their black boxes, the correlations they measure violate the CHSH inequality.

Besides the bound 2 of the CHSH inequality, two other numbers, $2\sqrt{2}$ and 4, are important bounds for the left-hand side of Eq. 12.1. If the correlations $C(A,B)$, $C(A', B)$, $C(A, B')$ and $C(A', B')$ were completely independent, the absolute value on the left-hand side could be as large as 4. But if they are quantum correlations, the absolute value on the left-hand side can only be as large as "Tsirelson's bound" [19], which is $2\sqrt{2}$. It can be larger than 2 just as $2\sqrt{2}$ is larger than 2, but it cannot be larger than $2\sqrt{2}$. Now the fact that quantum correlations cannot violate Tsirelson's bound is curious. If they are strong enough to violate the CHSH inequality, why aren't they strong enough to violate Tsirelson's bound?

A plausible answer to this question, in the spirit of Aharonov's suggestion, is that if quantum correlations were any stronger, they would violate causality as well as Tsirelson's bound. Indeed, if quantum mechanics follows from our two axioms, then so does Tsirelson's bound – a theorem of quantum mechanics. Conversely, if Tsirelson's bound does not follow from our two axioms, then certainly quantum mechanics does not. So if we can find "maximally nonlocal" correlation functions $C_{max}(A, B)$, $C_{max}(A', B)$, $C_{max}(A, B')$ and $C_{max}(A', B')$ that obey the two axioms but violate Tsirelson's bound, we must conclude that quantum mechanics does not follow from our two axioms.

It is straightforward to define such correlations. We do it in two steps:

1. In any measurement of A, A', B or B', let the results 1 and -1 be equally likely.
2. Let $C_{max}(A, B) = C_{max}(A, B') = C_{max}(A', B) = 1 = -C_{max}(A', B')$.

The first step insures that whatever Alice chooses to measure will not change the probabilities of ± 1 as results of Bob's measurement – whether he measures B or B' – and vice versa, because in any case the probabilities of all the results equal 1/2. Hence Alice cannot send a signal to Bob by her choice of what to measure, and vice versa. These correlations satisfy the axiom of causality. At the same time, they are so strong – joint measurements of A and B, of A' and B, and of A and B' are perfectly correlated, while joint measurements of A' and B' are perfectly anticorrelated – that the left-hand side of Eq. 12.1 violates the CHSH inequality and Tsirelson's bound maximally.

Our attempt to derive nonrelativistic quantum mechanics from two axioms is apparently a failure, but not necessarily a total failure. We may still hope that, with an additional axiom, we will be able to derive it. Indeed, van Dam [20] showed that in a world containing maximally nonlocal correlations, an important class of communication tasks would become dramatically simpler. Brassard et al. [21] extended this result to nonlocal correlations that are not maximal, indeed not much stronger than nonlocal quantum correlations. More recently, Pawłowski et al. [22] have defined an axiom of "information causality" and shown that any nonlocal correlation violating Tsirelson's bound is incompatible with this axiom. Their results are striking, but the physical meaning of "information causality" is perhaps not sufficiently clear. Maximal nonlocal correlations have even inspired experimental work [23].

12.3   Jamming

There is action at a distance in the Aharonov-Bohm effect. A solenoid acts at a distance on a beam of electrons; the interference pattern of the electrons depends on how the experimental physicist prepares the solenoid. By contrast, there is no action at a distance in quantum correlations. If there were, Alice and Bob could use them to send signals to each other; but in quantum correlations there is only "passion at a distance", as Shimony [11] aptly put it. If we define nonlocality as action at a distance, can we derive quantum mechanics (or some part of it) from nonlocality and causality? We shall see that "jamming", a presumably non-quantum form of action at a distance, could coexist with causality. We shall also discuss an axiom for "modular energy", a quantum form of action at a distance.

Having redefined the axiom of nonlocality, we must now redefine also the axiom of causality. We cannot say, "No action at a distance can be used to send signals," when action at distance is itself a signal. We can only say that the signal cannot be outside the forward light cone of the act of sending it. So we are back to relativistic causality. Our two axioms are then

Bob

Jim the Jammer

Alice

Fig. 12.2 Alice and Bob, with Jim the Jammer

1. There is no superluminal signalling.
2. There is action at a distance.

In the Aharonov-Bohm effect, the act of preparing the solenoid or capacitor produces electromagnetic radiation, which cannot have an effect outside the forward light cone of the preparation. If the effect shows up in the interference pattern, it is always within the forward light cone of the preparation and there is no superluminal signalling.

We return to Alice and Bob, and welcome their friend, Jim the Jammer (in Fig. 12.2). Jim has a large black box of his own, which acts nonlocally on their boxes. If Jim presses the button on his box, it turns their nonlocal correlations into local correlations, i.e. into correlations that obey the CHSH inequality. The probability of each result 1 or -1 does not change – it remains $1/2$ – but the correlations change. Let's assume that their results become completely uncorrelated. So here we have Jim acting at a distance to change the correlations between the black boxes of Alice and Bob. Does this action at a distance obey the no-signalling constraint? Clearly, Jim cannot send a signal to *either* Alice *or* Bob, because his action does not change the probabilities of 1 and -1 as results of their measurements. But can Jim send a signal to Alice *and* Bob? Indeed he can, because when Alice and Bob compare their results and compute correlations, they immediately identify their correlations as local or nonlocal. If Alice, Bob and Jim have arranged in advance that local correlations mean "Yes", and nonlocal correlations mean "No" (in some context), then Jim can signal "Yes" or "No" by choosing to jam, or not to jam, their correlations.

Figures 12.3a and b are spacetime diagrams of jamming (with time on the vertical axis). For simplicity, we reduce jamming to three spacetime events. At j, Jim may press the button on his black box. At a and b, Alice and Bob make their respective measurements. Now, Alice and Bob can compare their results anywhere, and only, in the overlap of their future light cones. Hence, they can receive Jim's signal anywhere, and only, in the overlap of their future light cones. However, if this overlap does not lie entirely within the future light cone of j, Jim's signal can be superluminal. So for jamming to be consistent with the no-signalling axiom,

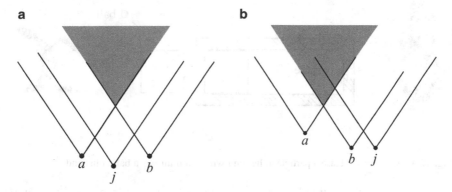

Fig. 12.3 The overlap of the future light cones of a and b either (**a**) lies or (**b**) does not lie entirely within the future light cone of j

jamming must work *only* when the overlap of the future light cones of a and b lies within the future light cone of j.

Hence in Fig. 12.3a Jim *can* jam their correlations, in Fig 12.3b he *cannot*. Then jamming is consistent with our two axioms.

On the face of it, at least, there is nothing like jamming in quantum mechanics. If not, then we have once again shown that causality and nonlocality – now in the guise of action at a distance – do not imply quantum mechanics as a unique theory: at least one other theory, however rudimentary, is consistent with these two axioms. Another failure!

12.4   Modular Energy

Twice, starting with a general axiom of nonlocality, we have failed to derive quantum mechanics. What if we start with an axiom of nonlocality that is tailored to quantum mechanics? Can we then derive quantum mechanics (or part of it)? If we succeed, our success will be less impressive because of the initial tailoring; but at least we may succeed. I will now define an axiom of nonlocality that is, in fact, a dynamical form of quantum nonlocality [24]. The axiom of causality, as well, will be tailored to nonrelativistic quantum mechanics.

We begin with the axiom of causality. The speed of light has no special status in nonrelativistic quantum mechanics. For nonlocal correlations, this fact implies that correlations are useless for any kind of signalling, not just superluminal signalling. Our first try at an axiom of causality was, therefore, that no correlations can be used for signalling. But here we will define nonlocal dynamics instead of nonlocal correlations. Our axiom cannot be that no *dynamics* can be used for signalling; causal relations, including signalling, are inherent in dynamics. If Alice sends a particle to Bob, the particle carries a signal. A more plausible axiom of causality

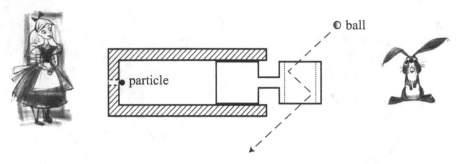

Fig. 12.4 Alice may release a particle at her end while Bob throws a ball at his end

is that signalling is possible only via a material interaction, i.e. only when a particle or other object connects the cause and effect. If Alice's particle mysteriously disappears and then reappears in Bob's lab, *that* particle cannot carry a signal, according to this axiom of causality.

We now define an axiom of nonlocality for a specific physical setting that includes Alice and Bob. Figure 12.4 shows a shaft with a piston in it at Bob's end. The piston can slide without friction in the shaft. Attached to the outer end of the piston is a box with two open sides. Bob throws a ball that ricochets through the box in two elastic collisions, pushing the (stationary) piston in a short distance. Alice's end of the shaft is closed, yet she can release a particle at her end. Our axiom of nonlocality states that the ball and particle – *if* Alice releases a particle – exchange energy *nonlocally* in the time between the two collisions. Energy is conserved, but it can mysteriously disappear from Alice's lab and reappear in Bob's, even before the particle reaches the piston.

On the face of it, this axiom contradicts the causality axiom. If Alice chooses to release a particle into the shaft, Bob will detect a change in the energy of the ball; if she chooses not to, he will detect no change. Thus Alice can signal to Bob via her choice whether or not to release a particle. In the spirit of Aharonov's suggestion, we can infer that Bob's measurements of energy must be uncertain. Even so, there seems to be no way to reconcile the nonlocality axiom with the causality axiom. Let the energies E_A and E_B of the particle and the ball, respectively, be distributed according to probability distributions, so that the energy of each is uncertain. If the particle enters the shaft, we expect the probability distribution of E_B to change. If so, Alice and Bob can still use nonlocal energy exchange to send a signal – if not in one run of the experiment, then in many simultaneous runs on many copies of the experiment: Alice's signal will emerge from the statistics of Bob's measurements.

Yet mathematical analysis [24] reveals an additional possibility. Let E_0 be a parameter with units of energy. For any E_0 and for any energy E, we define an associated quantity $E_{mod} = E$ modulo E_0 (also written $E_{mod} = E \bmod E_0$), which is the energy E minus a multiple of E_0 such that $0 \leq E_{mod} < E_0$. We call this quantity *modular* energy. If the particle and the ball exchange *modular* energy, they exchange at most E_0 in energy. Since the total energy $E_A + E_B$ is conserved in any exchange, so is the total modular energy $E_A + E_B \bmod E_0$, for any E_0. Now if, for any E_0, the distribution of the modular energy $E_B \bmod E_0$ of the ball is flat – i.e. if all values of $E_B \bmod E_0$ between 0 and E_0 are equally likely – then an exchange of energy between the particle and the ball will not change the distribution of $E_B \bmod E_0$, although it will, in general, change the distribution of $E_A \bmod E_0$. This one-way effect occurs if and only if the distribution of $E_B \bmod E_0$ is flat. But if the distribution of $E_B \bmod E_0$ is flat, then the uncertainty in E_B is at least E_0:

$$\Delta E_B \geq E_0.$$

We can say that a nonlocal exchange of energy of up to E_0 between the particle and the ball is consistent with causality, because Bob cannot detect the exchange when $\Delta E_B \geq E_0$.

For how long must the uncertainty ΔE_B be at least E_0? We may reason as follows: ΔE_B must be at least E_0 for the whole time T it takes Alice's particle to reach the piston. This reasoning suggests an inverse relationship between the time T and ΔE_B, since the more energy a particle has, the less time it takes to reach the piston, and the more energy it transfers to the ball. Indeed, let L be the distance from Alice's end of the shaft to the piston and let m_A, m_B and p_A, p_B be the masses and momenta of the particle and the ball, respectively. In the limit of m_A, m_B negligibly small compared to the mass M of the piston, a straightforward classical calculation shows that E_B changes by $4p_A p_B/M$, if and only if Alice releases the particle. Bob can detect a nonlocal transfer of energy only if $\Delta E_B < 4p_A p_B/M$. He cannot detect it if $\Delta E_B > 4p_A p_B/M$. Since $p_A = m_A L/T$, we can eliminate p_A to obtain

$$\Delta E_B > 4m_A L p_B/MT \qquad (12.2)$$

as the condition for Bob not to detect the nonlocal transfer of energy.

Equation 12.2 looks like an uncertainty relation for ΔE_B and T. But T in Eq. 12.2 is the time Alice's particle takes to reach the piston; it is not the minimum time Δt that Bob takes to measure E_B with uncertainty ΔE_B. Moreover, Bob's measurement of E_B is a local measurement. The uncertainty in a local measurement depends only on local variables. The axiom of nonlocality implies that E_B, but not ΔE_B, depends nonlocally on what Alice does. In particular, ΔE_B cannot depend on T if T is not a local variable. If Alice releases a particle with momentum p_A, then E_B changes by $4p_A p_B/M$ (according to the classical calculation), but T depends on L and m_A as well as p_A. Indeed, for a given change in E_B, the time T may be arbitrarily large. So, on the one hand, ΔE_B cannot depend on T. On the other hand, E_0 – which is the maximum nonlocal energy transfer – can certainly depend on T. Such a dependence is quite consistent with the axiom of nonlocality. In brief, what Eq. 12.2 tells us is not how Δt depends on ΔE_B, but rather how E_0 depends on T: Eq. 12.2 tells us that E_0 is inversely related to T.

And now, from this inverse relation, we can infer how nonlocal energy exchange could be consistent with the axiom of causality. Let $E_0 = k/T$, for some constant k, while Δt is the time it takes to measure E_B with uncertainty ΔE_B. If the inequality $\Delta t \geq k/\Delta E_B$ holds, then $\Delta E_B \leq E_0$ implies $\Delta t \geq T$ while $\Delta t < T$ implies

$\Delta E_B > E_0$, and Bob will never detect an exchange of energy before the particle reaches him. Conversely, if $\Delta t \geq k/\Delta E_B$ does not hold, we have $\Delta t\, \Delta E_B < E_0 T$ and Bob can detect the nonlocal exchange of energy by measuring E_B with uncertainty $\Delta E_B < E_0$ in a time $\Delta t < T$. The axiom of causality therefore demands the inequality $\Delta t \geq k/\Delta E_B$; and for this axiom to apply consistently to nonlocal exchange of energy, k must be a universal constant: Alice and Bob must not be able to circumvent the inequality $\Delta t \geq k/\Delta E_B$ by varying parameters of their experiment so as to vary k. Thus k is a universal constant, which we can identify with Planck's constant h.

As noted, in quantum mechanics there is nonlocal exchange of energy. What allows quantum mechanics and causality to coexist, despite this nonlocal exchange, is the uncertainty relation for energy E and time t [24]. We have inverted the logical hierarchy, and from axioms of causality and nonlocality, we have derived a principle of quantum theory: the uncertainty relation for energy and time, $\Delta E\, \Delta t \geq h$.

•

Let us conclude by reviewing our progress towards a derivation of nonrelativistic quantum mechanics from the axioms of causality and nonlocality. All three attempts presented here fall short of a complete derivation. Yet there is reason for optimism. Maximal nonlocal correlations outperform quantum mechanics, but if we take a closer look, we find something quite unreasonable about them. They are so strongly correlated that, for example, Alice can actually *determine* the product of measurements of B and B' by choosing whether to measure A, or A'; for if she measures A, then B and B' are perfectly correlated, while if she measures A', then B and B' are perfectly anticorrelated. Thus Alice could superluminally signal to Bob, if it were not for the (tacit) assumption that Bob cannot measure both B and B' but only one of them, as in quantum mechanics. In quantum mechanics, with its uncertainty relations, this assumption is natural. But in a theory with maximally nonlocal correlations, it is quite *un*natural. Bob can measure B directly and also infer B' indirectly from Alice's measurement. It is true that this method of measuring B and B' does not allow Alice to send a superluminal signal to Bob, but it defeats any uncertainty relation for B and B'. We may still hope, therefore, that if we consider nonlocal correlations as subject to the logic of uncertainty relations, we will arrive uniquely at quantum correlations.

As for jamming, it is notable that the authors of Ref. [14] never proved the incompatibility of jamming with quantum mechanics. It just seemed obvious to them. Yet it is possible to devise a quantum thought-experiment that is equivalent to jamming. Let Alice, Bob and Jim share triplets in the GHZ state [25]:

$$|\Psi_{GHZ}\rangle = \frac{1}{\sqrt{2}}\{|\uparrow\rangle_{Alice}|\uparrow\rangle_{Bob}|\uparrow\rangle_{Jim} - |\downarrow\rangle_{Alice}|\downarrow\rangle_{Bob}|\downarrow\rangle_{Jim}\}, \qquad (12.3)$$

Equation 12.3 does not show the spatial wave functions of Alice, Bob and Jim, but it shows the combined state of an ensemble of three spin-1/2 atoms distributed among them. Now suppose Jim measures either the z-component or the x-component of the spin of his atom. If he measures the z-component of the spin, he leaves the atoms of Alice and Bob in a mixture of product states; if he measures the x-component of the spin, he leaves their atoms in a mixture of entangled states. Jim announces the results of his measurements, but he does not announce *what* he measured; nevertheless Alice and Bob can *deduce* what he measured if they compare the results of their measurements. They can do so, however, only in the future light cone of his announcement (which we can identify with the future light cone of j in Fig. 12.3) because they need the results of Jim's measurements, as well, to deduce what he measured. This is jamming, within quantum mechanics! Hence the question of whether quantum mechanics is the unique theory reconciling causality and action at a distance remains open after all.

These optimistic thoughts arose in the course of writing this paper, and I hope to discuss them further in a separate work.

References

1. Pitowski, I.: Geometry of quantum correlations. Phys. Rev. **A77**, 062109 (2008)
2. Everett III, H.: 'Relative state' formulation of quantum mechanics. Rev. Mod. Phys. **29**, 454 (1957)
3. Wheeler, J.A.: Assessment of Everett's "relative state" formulation of quantum theory. Rev. Mod. Phys. **29**, 463 (1957)
4. Interview of B. DeWitt and C. DeWitt-Morette by K. W. Ford on February 28, 1995, Niels Bohr Library & Archives, American Institute of Physics, College Park. http://www.aip.org/history/ohilist/23199.html (1995), accessed 24 August 2011
5. Aharonov Y.: unpublished lecture notes. See also Aharonov, Y., Rohrlich, D.: Quantum Paradoxes: Quantum Theory for the Perplexed, Chaps. 6 and 18. Wiley-VCH, Weinheim (2005)
6. Poincaré, H.: Sechs Vorträge aus der Reinen Mathematik und Mathematischen Physik (Leipzig: Teubner), 1910, trans. and cited in A. Pais, "Subtle is the Lord...": the Science and Life of Albert Einstein, pp. 167–168. Oxford University Press, New York (1982)
7. Aharonov, Y., Bohm, D.: Significance of electromagnetic potentials in the quantum theory. Phys. Rev. **115**, 485 (1959). See also Aharonov, Y., Rohrlich, D.: op. cit., Chap. 4
8. Aharonov, Y., Pendleton, H., Petersen, A.: Modular variables in quantum theory. Int. J. Theor. Phys. **2**, 213 (1969); Aharonov, Y.: In: Proceedings of the International Symposium on the Foundations of Quantum Mechanics, Tokyo, p. 10 (1983); Aharonov, Y., Casher, A.: Topological quantum effects for neutral particles. Phys. Rev. Lett. **53**, 319 (1984). See also Aharonov, Y., Rohrlich, D.: op. cit., Chaps. 5, 6 and 13
9. Bell, J.S.: On the Einstein Podolsky Rosen Paradox. Physics **1**, 195 (1964). See also Aharonov, Y., Rohrlich, D.: op. cit., Chap. 3
10. Ghirardi, G.C., Rimini, A., Weber, T.: A general argument against superluminal transmission through the quantum mechanical measurement process. Lett. Nuovo Cimento **27**, 293 (1980)
11. A. Shimony, in *Foundations of Quantum Mechanics in Light of the New Technology*, S. Kamefuchi et al. eds. (Tokyo: Japan Physical Society), (1984) p. 225
12. Shimony, A.: In: Penrose, R., Isham, C. (eds.) Quantum Concepts of Space and Time, p. 182. Clarendon, Oxford (1986)
13. Popescu, S., Rohrlich, D.: Quantum nonlocality as an axiom. Foundations Phys. **24**, 379 (1994); Rohrlich, D., Popescu, S.: In: Mann, A., Revzen, M. (eds.) The Dilemma of Einstein,

Podolsky and Rosen 60 Years Later (Annals of the Israel Physical Society, 12) p. 152. Institute of Physics Publishing, Bristol (1996)

14. Grunhaus, J., Popescu, S., Rohrlich, D.: Jamming nonlocal quantum. correlations. Phys. Rev. **A53**, 3781 (1996); Popescu, S., Rohrlich, D.: In: Cohen, R.S., Horne, M., Stachel, J. (eds.) Quantum Potentiality, Entanglement, and Passion-at-a-Distance, pp. 197–206. Kluwer Academic Publishers, Boston (1997); Popescu, S., Rohrlich, D.: In Hunter, G., Jeffers, S., Vigier J.-P. (eds.) Causality and Locality in Modern Physics: Proceedings of a Symposium in Honour of Jean-Pierre Vigier, pp. 383–389. Kluwer Academic, Boston (1998)

15. Aharonov, Y., Rohrlich, D.: op. cit., Sects. 18.5-6

16. Landau, L., Peierls, R.: Erweiterung des Unbestimmtheitsprinzips fuer die relativistische Quantentheorie. Z. Phys. **69**, 56 (1931); trans. In: ter Haar, D. (ed.) Collected Papers of L. D. Landau, pp. 40–51. Pergamon Press, Oxford (1965); Aharonov, Y., Albert, D. Z.: States and observables in relativistic quantum field theories. Phys. Rev. **D21**, 3316 (1980): Aharonov Y., Albert, D.Z.: Can we make sense out of the measurement process in relativistic quantum mechanics? Phys. Rev. **D24**, 359 (1981); Aharonov, Y., Albert, D.Z., Vaidman, L.: Measurement process in relativistic quantum theory. Phys. Rev. **D34**, 1805 (1986); Popescu, S., Vaidman, L.: Causality constraints on nonlocal quantum measurements. Phys. Rev. **A49**, 4331 (1994). See also Aharonov, Y., Rohrlich, D.: op. cit., Chap. 14

17. Aharonov, Y., Susskind, L.: Observability of the sign change of spinors under 2π rotations. Phys. Rev. **158**, 1237 (1967); Aharonov, Y., Susskind, L.: Charge superselection rule. Phys. Rev. **155**, 1428 (1967). See also Aharonov, Y., Rohrlich, D.: op. cit., Chap. 11

18. Clauser, J.F., Horne, M.A., Shimony, A., Holt, R.A.: Proposed experiment to test local hidden-variable theories. Phys. Rev. Lett. **23**, 880 (1969)

19. Tsirelson, B.S. (Cirel'son): Quantum generalizations of Bell's inequality. Lett. Math. Phys. **4**, 93 (1980); Landau, L.J.: On the violation of Bell's inequality in quantum theory. Phys. Lett. **A120**, 52 (1987)

20. van Dam, W.: *Nonlocality and Communication Complexity*, Ph.D. Thesis, Oxford University (2000); van Dam, W.: Implausible consequences of superstrong nonlocality. Preprint quant-ph/0501159 (2005). See also Popescu, S.: Why isn't nature more non-local? Nat. Phys. **2**, 507 (2006)

21. Brassard, G., et al.: Limit on nonlocality in any world in which communication complexity is not trivial. Phys. Rev. Lett. **96**, 250401 (2006). See also [20] Popescu, S.: op. cit.

22. Pawłowski, M., et al.: Information causality as a physical principle. Nature **461**, 1101 (2009)

23. Tasca, D.S., Walborn, S.P., Toscano, F., Souto Ribeiro, P.H.: Observation of tunable Popescu-Rohrlich correlations through post-selection of a Gaussian state. Phys. Rev. **A80**, 030101(R) (2009)

24. Aharonov, Y.: In: Kamefuchi, S., Ezawa, H., Murayama, Y., Namiki, M., Nomura, S., Ohnuki, Y., Yajima, T. (eds.) Foundations of Quantum Mechanics in Light of New Technology. (Proceedings of the International Symposium, Tokyo, Aug 1983), pp. 10–19, Physical Society of Japan, Tokyo (1984); Aharonov Y., Rohrlich, D.: op. cit., Chap. 6

25. Greenberger, D.M., Horne, M., Zeilinger, A.: In: Kafatos, M. (ed.) Bell's Theorem, Quantum Theory, and Conceptions of the Universe (Proceedings of the Fall Workshop, Fairfax, Oct 1988), pp. 69–72. Kluwer Academic, Dordrecht (1989)

Chapter 13
Generalized Probability Measures and the Framework of Effects

William Demopoulos

Abstract This paper is dedicated to the memory of Itamar Pitowsky. It develops the idea that the generalized probability measures of quantum mechanics are the probabilities of "effects." As explained below, effects are included among measurement outcomes but are not exhausted by them. They also differ in key respects from propositions which attribute dynamical properties to the systems that are probed by measurements. These differences are elaborated, and an interpretation of the implicit probability theory of quantum mechanics in terms of effects is outlined. A central feature of this interpretation is that it supports a form of realism that accommodates the no hidden variable theorem of Kochen and Specker, and it does so without appealing to any notion of contextuality.

Generalized measures were introduced by Gleason [2] in the context of his characterization of the measures definable on the closed linear subspaces of Hilbert space. The analysis of the three-dimensional case proved to be fundamental. For this case, a *generalized measure* is a map f from the closed linear subspaces of H^3 to the closed unit interval satisfying the conditions

$$fa + fb \leq 1$$

This paper was intended for presentation to the December 2008 Jerusalem conference in Itamar's honor; unfortunately, circumstances prevented me from attending the conference. The paper is a sequel to "Effects and propositions" [1], to which the reader is referred for a more complete list of references and the historical context of some of the views presented here. A very early version of this paper was given as a talk to Chris Fuchs's Perimeter Institute seminar on the foundations of quantum mechanics in May of 2008. Many thanks to Chris for numerous discussions and comments on subsequent drafts. I am indebted to the Social Sciences and Humanities Research Council of Canada and to the Killam Foundation for support of my research.

W. Demopoulos (✉)
Department of Philosophy, University of Western Ontario, London, ON, Canada
e-mail: wgdemo@uwo.ca

for $a \perp b$, and

$$fa + fb + fc = 1$$

for any three rays a, b, c which are mutually orthogonal. A *generalized two-valued measure* takes values in $\{0, 1\}$.[1] The interpretation of such measures as *probability measures* arises when the rays are taken to represent propositions; and a generalized two-valued probability measure is a *generalized truth-value assignment* when 0 and 1 are interpreted as Truth and Falsity. It is evident that generalized two-valued measures and generalized truth-value assignments are formally interchangeable with one another, whatever the conceptual differences between probability measures and truth-value assignments.

The focus of this paper is a particular feature of the statistical behavior of elementary particles, simple composite systems of them and the quantum probability theory to which their behavior gives rise. This feature was given its canonical formulation by Kochen and Specker [4] in the course of an investigation of the problem of hidden variables. It is captured by their principal theorem (Kochen and Specker [4], Theorem 1) and the discussion of it in Section IV of their paper; and it consists in the fact that there exist simple systems of particles and finite combinations of propositions "belonging to them" for which no generalized two-valued measures are possible, where a proposition *belongs to a particle* if its constituent dynamical property is a possible property of the particle. The assumption of such propositions expresses the idea that the systems which they describe are characterizable by systems of properties which are uncovered when the systems are probed. Hence the notion of a proposition belonging to a particle supports the idea that measurements reveal a particle's dynamical properties. I will argue that the significance of the existence of systems of the sort that underlie the Kochen-Specker construction is to show that the generalized probability measures that arise in quantum mechanics are not naturally interpretable as the probabilities of propositions belonging to particles. (The notion of a proposition belonging a particle is developed further below in conjunction with the explanation of the notion of a natural interpretation.) The idea I will develop is that quantum probabilities are probabilities of "effects," probabilities of the traces of particle-interactions with objects and processes that are epistemically accessible to us in a sense which I will explain. I hope to make it clear that such a view is not committed to anti-realism about the micro-world and that it illuminates at least one otherwise paradoxical feature of quantum mechanics.

I should emphasize that the focus of this paper is not the notion of an effect as it pertains to the study of effect algebras, but the question whether the probabilities that arise in quantum mechanics should be understood to apply to propositions which express the *properties* of particles or whether they should be understood to

---

[1] These definitions are taken from Pitowsky [3].

apply to propositions which express the *effects* of particles. The issue I intend to address is not whether the probabilities of quantum mechanics concern propositions in *any* sense, but whether they concern propositions belonging to particles. I will argue that they do not, and I will explain the kind of proposition with which they are concerned. The terminology of effects offers a simple mnemonic device with which to mark this different kind of proposition.

It is true that classical systems are themselves composed of particles whose quantum probability measures have the peculiarities just noted. And it is also true that quantum mechanics is essential to the correct theoretical description of classical systems. However, I wish to defer the questions 'How should classically described systems be subsumed under quantum mechanics?' and 'How do classically described systems enter into the measurement process?' A burden of the discussion to follow is to clarify the view that there is a basic conceptual difference between classical states and quantum states, between what is represented by a point in phase space and a vector in Hilbert space. The premature consideration of the measurement problem and the quantum theory of classically described systems has a tendency to mask this conceptual difference. For these and other reasons, the conceptual issues raised by these questions are best taken up after the impossibility of two-valued measures has been considered. Although I will not argue directly for this thesis here, the discussion which follows is intended to support the view that the ψ–function represents a state of belief about a system rather than its physical state.[2]

The discussion of measurement and the issues it raises can be deferred, as can the discussion of the relationship between classical and quantum states, since even if quantum mechanics is a theory of the fundamental constituents of matter, the evidence for the theory can perfectly well come from our experience with things for which we do not possess a quantum-theoretical account. And in fact the phenomena relevant to the present discussion were either known prior to the theory's discovery and elaboration or are easily elicited with only a very modest contribution from developments that the theory initiated. Our problem is how to understand the real possibility of physical systems for which there are finitely many direction-dependent propositions,

$(P_\alpha{}^*)$ The square of the spin in the direction $\alpha \neq 0$,

which are so related that there is no generalized two-valued measure definable on them. These direction-dependent propositions are arranged in families of Boolean algebras of which the largest families are generated by three atomic propositions, each associated with one of three mutually orthogonal directions x, y, z of ordinary physical space. (The propositions $P_\alpha{}^*$ are "co-atoms," the Boolean complements of algebraically atomic propositions.) To each such family there corresponds an operational procedure which is interpretable as offering a means of detecting

---

[2] This thesis has been advanced and defended by Itamar in Pitowsky [5, 6]. See also Fuchs [7] for a recent statement and defense from a different "Quantum Bayesian" perspective.

which of the propositions of the Boolean algebra are true and which are false. This is not merely a theoretical possibility, but one that comes close to being actually realizable in the laboratory. Kochen and Specker [4] show this for an atom of orthohelium whose total angular momentum is given by its spin. The atom is a spin-1 system whose *spin* components in three mutually orthogonal directions of space are not commeasurable, but whose *square of the spin* components in any three mutually orthogonal directions are commeasurable.

The canonical operational procedure for the measurement of a component of spin is a Stern-Gerlach magnet which splits a beam of spin-1 systems into three groups, each corresponding to one of the values −1, 0 or +1. The ideal operational procedure for the direct measurement of a square of the spin component is wholly different from that used for a measurement of a component of spin; it must employ an electric rather than a magnetic field, since it is only in the absence of a magnetic field that the Hamiltonian of the system preserves certain of its essential symmetry properties. Such a direct measurement of a square of the spin component distinguishes spin values of 0 from spin values of +1 or −1, but does not distinguish between the latter two possible values, which is what accounts for the way the propositions $P_\alpha{}^*$ are formulated.

The ideal measurement procedure for the square of the spin produces an electric field with the crystalline form of an octahedron. Such a field occurs naturally in a crystal of nickel Tutton salts consisting of an ion surrounded by an octahedron of water molecules. An ideal measurement procedure for the atom of orthohelium thus emulates the nickel ion's environment in the salt crystal by subjecting the orthohelium atom to an external electric field of the same rhombic symmetry as the field inside the crystal. As noted earlier, the use of an electric rather than a magnetic field is important for preserving the spin-Hamiltonian. In the case of nickel Tutton salts it is standardly assumed "that, in the absence of a magnetic field, [the Hamiltonian of the crystal exhibits] rhombic symmetry [; i.e.,] it is possible to choose rectangular co-ordinates Ox, Oy and Oz such that the Hamiltonian is invariant under rotations through π about Ox and Oy" (Stevens [8], p. 238). An ideal test procedure for the square of the spin of the atom of orthohelium would allow one to probe experimentally the behavior of the orthohelium atom as the apparatus is rotated, and the external field turned off and on, by observing the shifts in the electromagnetic spectrum of the atom. One infers the directional properties associated with the dynamical magnitude, *square of the spin in the direction* α, for mutually orthogonal directions α, from the spectral shifts which result as the atom is subjected to the electric field by the measurement device.

By contrast with actual experiments with a nickel ion in a salt crystal, in this idealized experimental situation involving the atom of orthohelium, the field generated by the measurement device can be applied at orientations chosen at the discretion of the experimenter, with each orientation corresponding to a different orthogonal triple of axes of symmetry. The application of every such test procedure determines that exactly one of the propositions $P_\alpha{}^*$, for $\alpha = x, y, z$ is false, and exactly two are true. But on the hypothesis that the families of propositions are related in the way specified, one discovers as one considers all the triples of directions appealed to in the proof of the Kochen-Specker theorem that it is logically impossible that there should be an assignment of truth-values to all these propositions that respects this observation.

Now it may be that a *complete* understanding of the conceptual innovation occasioned by the discovery of systems whose behavior shares this feature of the behavior of an atom of orthohelium will require invoking the quantum theory of the measuring instrument and other classically describable systems. But this does not affect the point that the interpretive problem raised by the statistical behavior of orthohelium is conceptually separable from any such account. The interpretive puzzle depends on the possibility of forcing an interaction with an electric field of specified character and then noting how the interaction gives rise to changes in the atom's electromagnetic spectrum. All of this is characterizable at a pre-quantum-mechanical level of description. In light of these considerations, it seems reasonable to conclude that the problem posed by the Kochen-Specker theorem— the problem of understanding the significance of systems whose propositions do not admit generalized truth-value assignments—is not conceptually dependent on the provision of a quantum theory of measurement.

My plan in the balance of the paper is to argue in support of a conceptual framework that provides a solution to this interpretive problem, not by providing a hidden variable theory that is a counter-example to the theorem, but by providing a framework which yields a natural interpretation of the impossibility of such truth-value assignments, where, by a *natural interpretation*, I mean one that does not violate any of the following three desiderata:

Determinacy: Every proposition which attributes a possible dynamical property to a particle (i.e., every proposition which, in the special sense noted earlier, *belongs to a particle*) is determinately true or false. In particular, if P is a disjunction of propositions, each disjunct of which attributes a possible point value of a dynamical variable, and if the disjuncts exhaust all possible point values, then if P is true, exactly one of its disjuncts must be true. The intuition that supports determinacy is that while it makes perfect sense, when thinking of a fictional world, to treat some propositions belonging to its inhabitants as neither true nor false, this is precluded when we are concerned not with fiction, but with reality. This is because the failure of determinacy is one of the marks that separates our concept of a fictional world from the real one.

Objectivity: Dynamical properties are indicated by a variety of experimental conditions and operational criteria. The methodological basis for objectivity rests on two considerations: (a) every property requires a clear physical criterion for saying when it holds and when it fails to hold; (b) an objective property must have some degree of conceptual independence from the procedures for determining its presence or absence, since the same property must be accessible in different measurement contexts and by alternative measurement procedures. For an interpretation of the theory to be based on the attribution of dynamical properties to physical systems, it is necessary that there should be a conceptual gulf between properties and the procedures which probe their presence or absence. Accessibility in a variety of experimental contexts is not only largely constitutive of what we mean by the objectivity of properties, it is also presupposed by standard forms of counterfactual reasoning about them. For example it is presupposed when we ask whether a property would have obtained had a different operational procedure been applied, or when we ask whether the property would have obtained had the presence of another property been investigated. The possibility of such reasoning is an essential component of the objectivity we associate with physical properties. In the extreme case, where each property is tied to a single operational procedure, this aspect of our concept of objectivity is given up, since admitting only a single operational procedure is tantamount to abandoning the idea that the *same* property may be presented differently. The connection between objectivity and the existence of a variety of operational indicators suggests that there are degrees of objectivity, corresponding to the multiplicity of different operational procedures that are indicative of a property's presence or absence. As we will see, interpretations may differ on the degree of objectivity which they accord the dynamical properties of physical systems.

Observer independence: The reality that attaches to particles is an observer-independent reality; this holds as well for their physically important properties. The observer independence of particles is so closely tied to the objectivity of their properties and the determinacy of propositions involving them that it is generally assumed to be undermined when these desiderata are violated.

I will assume without further argument that an interpretation of the impossibility of generalized truth-value assignments is successful to the extent that it is a natural interpretation in the sense just explained. I intend to show that if the generalized probability measures of quantum mechanics are understood to apply to propositions belonging to particles, it is not possible to frame a natural interpretation of the theory. I will argue that there is an alternative account of the domain over which such probabilities are defined that leads to a natural interpretation. This is the interpretation of probabilities as probabilities of effects. The burden of this paper is to show that such an alternative interpretation does not violate determinacy or objectivity, and that it also satisfies observer independence. The role of the notion of a natural interpretation in the following analysis is therefore a dialectical one: it is used to show that the framework of propositions belonging to particles cannot support an account of the absence of two-valued measures without compromising determinacy or objectivity. As a consequence, such a propositional framework undermines observer independence. By contrast, the framework of effects interprets the absence of two-valued measures as a simple failure of determinism without calling into question the determinacy of physical propositions, and without compromising our conception of the objectivity of physical properties. But the true measure of the success of an analysis in terms of effects turns on its account of observer independence. This issue is taken up at the end of the paper. Let me begin by considering more closely the desiderata of objectivity and determinacy.

Objectivity, in the sense considered here, is motivated by the idea of *contextuality*, which has figured especially prominently in discussions of the Kochen-Specker theorem. The issues surrounding contextuality are particularly clear in the case of directional properties and the direction-dependent propositions of which they are constituents. Consider two orthogonal triples of directions in E^3, (x, y, z) and $(\theta x, \theta y, \theta z)$, where $\theta x = x$ but $\theta y \neq y$ and $\theta z \neq z$. The direction x in E^3 is evidently independent of the family of orthogonal triples—(x, y, z) or $(x, \theta y, \theta z)$—to which it belongs. But in the case of direction-dependent *propositions*, it is not clear a priori that the identity of a proposition is independent of the other direction-dependent propositions with which its truth is evaluated. The constituent directional properties of the propositions might be associated with distinct measurement procedures, and this might be sufficient to justify distinguishing direction-dependent propositions that are associated with the same direction in space.

For example, for directions $\alpha = x, \ldots, \theta z$, consider the direction-dependent propositions,

(P_α) The square of the spin in the direction $\alpha = 0$,

where the P_α are algebraic atoms and are the Boolean complements of the $P_\alpha{}^*$, considered earlier. ($P_\alpha = P_\alpha{}^{**}$, when * is understood as the operation of complementation.) Suppose that the families of directions (x, y, z) and $(x, \theta y, \theta z)$ are associated with distinct ideal measurement procedures, one involving the triple of directions, x, y, z, the other involving the triple of directions, $x, \theta y, \theta z$. The first operational procedure decides the propositions P_x, P_y, P_z, while the second decides the propositions, $P_x, P_{\theta y}, P_{\theta z}$, but there is no measurement procedure that simultaneously decides all the P_α. That is, we have that P_x is comeasurable with P_y and P_z, and with $P_{\theta y}$ and $P_{\theta z}$, but P_y and P_z are not comeasurable with $P_{\theta y}$ and $P_{\theta z}$. Then contextuality concerns the bearing of measurement procedures on the identity of propositions: Is P_x the same proposition when the operational procedure by which the presence or absence of its constituent property is decided is one that measures P_x in conjunction with P_y and P_z as when the operational procedure is one that measures P_x in conjunction with $P_{\theta y}$ and $P_{\theta z}$? It is certainly *possible* that the difference between these two measurement procedures is sufficient to show that P_x splits into two propositions, each with its own constituent property, one decided by an operational procedure associated with (x, y, z), the other by one associated with $(x, \theta y, \theta z)$.

Now it is simply a fact about quantum mechanics that its statistical states are such that the probability measures they generate are *non*-contextual. In the present case, this means that quantum states do not distinguish direction-dependent propositions any more finely than Euclidean geometry distinguishes directions of space. But a contextual hidden variable theory is characterized by the fact that it allows for the possibility that propositions are distinguished more finely by the "hidden" (i.e., two-valued) measures such theories introduce than they are by the probability measures of quantum mechanics. The issues raised by the possibility of such theories center on whether it is justifiable to require of the hidden measures they introduce that they too should be non-contextual. Since the hidden measures are mathematically interchangeable with truth-value assignments, this is equivalent to the question whether the constituent properties of the propositions to which truth-values are assigned vary with the measurement context.

Turning to composite systems, there is also an important sense in which the statistical states of quantum mechanics are *local*. This is not a tendentious remark since it does not contradict the fact that there are quantum states that are non-local in the sense that they violate the inequalities discovered by Bell [9]. The claim that quantum states are local is a simple consequence of the observation that locality is a special case of non-contextuality, the case that concerns the invariance of the probability measures of the theory when one leaves unchanged the local measurement context for one system while varying the test procedure for the system with which it is paired. Locality makes it difficult to invoke the modification of measurement procedures as a justification for distinguishing propositions, since one would have to distinguish propositions belonging to one system on the basis of what properties one chooses to detect by performing a measurement on the spatially separated system with which it is correlated.

For direction-dependent properties, conformity with what is allowed by the geometry of the associated rays of E^3 is a natural measure of objectivity, and by this criterion, quantum mechanics accords the P_α a maximum degree of objectivity since they are individuated exactly as finely as the directions of E^3. Hence quantum mechanics permits a much more inclusive class of measurement procedures for a directional property than any of its contextualist rivals. I claim that it is a desirable property of an interpretation that it should preserve this feature of the theory. Although it is not a decisive objection against an interpretation that it requires a multiplicity of propositions P_x on the basis of contextual considerations involving their constituent properties, it does show that within such a framework, the preservation of determinacy necessitates some sacrifice of objectivity. Relativity to the measurement context secures the determinacy of propositions belonging to particles only by compromising the objectivity of some of their constituent directional properties.

It might seem that one could confine the properties that are contextually individuated to a small subset of the directional properties we have been considering. However it is possible to show that one cannot fix in advance those properties which must be more finely individuated than the directions of space with which they are associated. To be sure, Kochen and Specker's argument isolates a particular orthogonal triple of propositions and shows how the assumption of a truth-value assignment is inconsistent with the assumption that exactly one of the propositions of the triple is true and the others false. But the argument can also be run backwards in the sense that we can choose a different triple and proceed to construct a Kochen and Specker orthogonality graph so that the argument concludes by applying to the selected triple the observation that exactly one of the propositions P_α is true, and the others false.[3] Since the choice of orthogonal triple which conflicts with this observation is completely arbitrary, in order to maintain determinacy *any* square of the spin property may have to be represented by a multiplicity of properties, one for each operational procedure corresponding to a relevant triple of directions. This consequence is a "paradox" of sorts when one considers the realist motivation for securing determinacy together with the fact that for realism the objectivity of physical properties is as fundamental a requirement as the determinacy of the propositions which contain them. That the same property may be presented as the property indicated by a number of different measurement procedures, and that counterfactual reasoning in association with a multiplicity of measurement procedures is legitimate, are no less indispensable to our concept of the objectivity of physical properties than determinacy is to our concept of reality.

To summarize our discussion thus far, in the context of generalized probability measures like those exhibited by quantum mechanics, the desiderata of determinacy and objectivity cannot be maximally satisfied within a framework of propositions belonging to particles; the satisfaction of determinacy involves some sacrifice in objectivity since a maximum degree of objectivity is incompatible with determinacy. As a result, the suggestion that one must give up the idea that particles have an observer independent reality has exercised a powerful appeal over both physicists and philosophers of physics.

By way of articulating an alternative to giving up determinacy, objectivity or observer independence, let me begin by separating "eternal" properties of particles from dynamical properties. Eternal properties are never lost: an electron is *always* a spin-1/2 particle, photons are *always* spin-1, etc. The possession of such properties is not brought into question by the interpretive problems that are raised by the Kochen-Specker theorem. Rather, it is the ascription of dynamical properties and the notion that they are the subject of the theory's probability assignments that poses difficulties for a natural interpretation of the theory. The resolution of these difficulties that I will outline gives up the framework of dynamical properties of particles and the notion of propositions belonging to them and replaces it with the framework of effects. Effects constitute the domain of the algebraic structure over which probabilities regarding the behavior of particles are defined. Effects constitute the evidential basis for all our theoretical assertions about particles and simple combinations of them. They are to be thought of as the traces of particle interactions on systems for which we have "admissible" theoretical descriptions in terms of their dynamical properties. (I will return to the notion of admissibility in a moment.)

---

[3] The possibility of running an argument like Kochen and Specker's backwards is exploited by a theorem of Pitowsky [3] showing that given any two noncomeasurable propositions P_x and P_y represented by rays in H^3, we can always find a *finite* set Γ of rays of H^3 which contains the representatives of P_x and P_y and has an orthogonality structure that forces any generalized two-valued measure on Γ to assign them both 0. More generally, one can show that either the probability of any two noncomeasurable P_x and P_y is 0, or at least one of them has a probability strictly between 0 and 1.

Such systems are epistemically accessible to an extent that systems which are characterized only in terms of their eternal properties and their effects are not.

To see why treating probabilities as probabilities of effects allows for an interpretation of the Kochen-Specker Theorem that leaves intact the desiderata of determinacy and objectivity, recall that the problem of interpreting the theorem arose because effects were implicitly taken to be indicative of a particle's dynamical properties. This meant that the algebraic structure of the theory was interpreted as an algebra of propositions belonging to particles, and the probabilities of the theory were understood to be the probabilities of such propositions. The *objectivity* of properties then demanded that had a different effect been elicited, it would have revealed that the particle had a different dynamical property. Proceeding through Kochen and Specker's sub-algebra of possible propositions, we were led to a contradiction with the observation that for every orthogonal triple of directions x, y, z, exactly one of the propositions P_α is true for $\alpha = x$, y, z. By moving to the framework of effects we give up the idea that probabilities are assigned to propositions containing a particle's dynamical properties, and focus instead on the effects to which particles give rise. Effects are determinate independently of the determinacy of propositions involving a particle's dynamical properties, and their objectivity does not depend on counterfactual reasoning involving such properties. In a framework in which generalized probability measures are defined on effects, the problems posed by determinacy and objectivity are avoided since there is nothing in the concept of an effect to require that effects should obtain in the absence of the interactions in which they are found. The effects framework has no analogue of a state comprised of the totality of dynamical properties as there is in a classical picture of particles and their effects. In particular, there is no assumption of classical trajectories underlying the attribution of an observer independent reality to particles. The effects framework is agnostic about all such classical pictures.

Despite its agnosticism on questions of ontology, the effects framework offers a subtle account of the nature of the conceptual shift from classical to quantum mechanics. The transition to an effects framework consists in replacing the characterization of a particle by a list of its dynamical properties with one according to which a particle's characterization has the logical form of a function: when presented with an experimental idealization of some naturally occurring situation, particles are characterized not by changes in their dynamical properties, but by the effects they produce. The fact that these effects cannot be anticipated with certainty is understood within the effects framework as the unsolvability of the following *Problem of Determinism*: Given a particle and a class of experimental procedures, to predict particle-effects with perfect knowledge, i.e., to predict with probability 0 or 1, uniformly and without foreknowledge of the experimental procedure to which a particle will be subjected, the answer to every question regarding the occurrence of a possible effect. The no hidden variable theorem of Kochen and Specker (together with the related theorems inspired by the work of Bell) shows that the quantum probabilities of such effects are not compatible with the existence of a two-valued measure which solves the Problem of Determinism. Applied to the example of an atom of orthohelium, this means that within the framework of effects the atom is represented not by a collection of dynamical properties but by a function which, when presented with an orthogonal triple of directions associated with the axes of symmetry of an electric field, produces an effect consisting of a shift in its spectrum. In the propositional framework, this shift is taken to be indicative of the truth of exactly two of the propositions,

$$(P_\alpha{}^*)\ \text{The square of the spin in the direction } \alpha \neq 0,$$

attributing dynamical properties to the atom. But this is precisely the interpretive step that is resisted by the effects framework: So far as probability assignments are concerned, there are only effects—in the present case, shifts in the spectrum of the atom—which the atom's interaction with the field induces.

It is important to see how an interpretation in terms of effects bears on realism in view of the fact, already noted, that such an interpretation does not situate the theory within an "ontology." But before turning to the question of observer independence, let me review an analogy which may clarify the status of realism in the present approach.

Imagine that we are concerned to construct a model of past events for which there is very little basis to assume that they resemble the events with which we are familiar. Let us also assume that the traces of these events are accessible to us only in fragments that can be examined one at a time, that the information contained in any one fragment is insufficient to determine a complete account of the events which produced the traces which comprise it, and that the traces are themselves continually changing. Assume further that the fragmentary traces do not combine to give a single consistent story regarding the events at this earlier time. Now suppose it is discovered that although the past is in this way "hidden" from us, our epistemic situation with respect to its traces is systematic and even susceptible of a relatively simple representation. Although systematic, the representation of available traces not only fails to facilitate the reconstruction of the past state of the world, but actually precludes the possibility of a consistent reconstruction on its basis. Under such circumstances we might cease looking for a representation of a past state in terms of the properties that hold of it because we will have come to recognize that there can be no convergence from present or future traces to such a representation. We might then dispense with the search for a theory of such states and focus instead on understanding the distribution of present traces, their relevance to one another, and the task of predicting their likely evolution. This would be a theory of past events of a sort, but not what we had originally imagined such a theory would be like. In particular, it would not aim to *model* the past, but to anticipate its present and future traces. To recover quantum mechanics from the analogy, replace traces with particle-effects, and past states of the world with lists of dynamical properties of particles. Then two things are worth noting: neither such a theory of traces of the past nor the quantum theory of effects contravenes the thesis of determinacy, and where the one theory accepts the reality of the past, the other accepts the reality of the micro-world. In each case, one has merely abandoned a familiar style of theorizing and the modeling associated with it.

Our discussion of the Kochen-Specker theorem, and our resolution of the conceptual difficulties it poses for determinacy and objectivity, is predicated on the idea that the probability assignments of the theory are not interpretable as probabilities of propositions belonging to particles; rather, there is a domain of effects which are epistemically accessible to us in a way in which dynamical properties of particles have been shown not to be, and it is these effects that are the proper subject of the theory's probability assignments. This leaves a large residual issue that we must now address.

I have said that effects are marks or traces particles leave on certain physical systems, and these systems and the traces of their interactions constitute the epistemic basis for our evaluation of our quantum mechanical descriptions of the behavior of single particles and simple composite systems of them. But what is the status of the systems which record the effects of particles? Since they are merely complex systems of large numbers of particles, on the assumption that quantum mechanics is a truly universal and fundamental theory, shouldn't they also fall within the purview of the theory?

The issue these questions raise is a familiar one: it is the issue the early founders of the theory addressed with the doctrine of the indispensability of classical concepts and the necessity of locating the "cut" between physical systems and observers. For them, both the question of the location of the cut and the indispensability of classical concepts arose because of epistemological considerations which had their source in the special status they supposed quantum mechanics assigns observers. They argued that the role of the observer in quantum mechanics is utterly unlike the situation in classical mechanics where it is possible to proceed on the assumption that all physical systems fall within the range of the theory, and where it is possible to treat observers as altogether absent from the application of the classical framework, except insofar as they may happen to occur among the physical systems the theory encompasses.[4] But if the very notion of an effect requires reference to observers and to a preferred, extra-quantum-theoretical, characterization of the physical systems accessible to them, an interpretation of the theory in terms of effects can hardly be advanced as one that restores observer independence to the theory's interpretation. So although the notion of an effect may provide solutions to the problems of determinacy and objectivity, a more elaborate argument is needed to show that the notion has anything new to offer regarding the problem of observer independence.

---

[4] See Camilleri [10] for discussions of the views of Heisenberg, Bohr and Pauli and the similarities and divergences among them. Fuchs's Quantum Bayesian approach to the theory preserves the primacy of the observer that was emphasized by Pauli. Observers in Fuchs's framework are called "agents." In conformity with Fuchs's understanding of the purely epistemic character of the quantum state, effects belong to agents and consist in modifications of their subjective probability judgements.

Let me address this objection by considering first the use of classical concepts. There is a general observation which it is easy to lose sight of. It is that any description of the phenomena we wish to explain is *admissible* just in case reasoning in accordance with generally recognized methodological norms, we are able to reach agreement on the correctness of its application in any particular case. This is simply the non-operationalist core of the methodological framework of Einstein's analysis of simultaneity. Provided our descriptions meet this admissibility condition, there need be nothing methodologically questionable about the continued, or even exclusive, use of classical concepts for the description of the phenomena we are interested in explaining. Our discussion of the atom of orthohelium showed that there is a family of descriptions of the relevant phenomena that are expressed in terms of classical concepts that satisfy this admissibility condition. But to concede that classical mechanics is a descriptive framework that supplies admissible descriptions of the phenomena we seek to explain does not preclude the possibility that we may uncover a framework that revises these descriptions and is in some sense more "fundamental" than the classical one. All that is required to justify the classical framework in this evidentiary role is that there should be consensus about the application of its descriptions. If this observation about admissible descriptions is accepted, the doctrine of the indispensability of classical concepts raises at least two questions: (i) In what sense, if any, is the quantum mechanical framework *more fundamental* than the classical one? (ii) What feature distinguishes the descriptive framework of classical mechanics and makes it not just well-suited but *indispensable* to the provision of admissible descriptions of at least some of the phenomena quantum mechanics is used to explain? The answer to this second question will direct us to an answer to the first, so let us begin with it.

The classical framework involves both dynamical properties and effects of the systems with which it deals. This ability to encompass a system's properties as well as its effects is a consequence of the deterministic character of the classical framework. Here, as before, by the determinism of the classical framework, I mean that feature of it that admits the presence of dispersion-free pure states in the form of two-valued measures on the totality of propositions belonging to a classical mechanical system. The mathematical fact that two-valued probability measures are interchangeable with truth-value assignments entails that it is always possible to represent the state of a classical system in terms of the totality of its dynamical properties: these are the constituents of the propositions which a truth-value assignment (on the Boolean algebra of propositions belonging to the system) maps to Truth. In the case of quantum mechanical systems of particles and simple combinations of them, the absence of such measures led us to interpret the theory's probability assignments as probabilities of their effects, rather than of propositions involving their dynamical properties. But to describe the effects of particles we need a framework whose systems *are* represented by their dynamical properties, since it is the properties of these systems that constitute particle-effects.

From the perspective of the effects framework, the conceptual dependence of quantum mechanics on classical mechanics is a result of the conceptual dependence of descriptions of the systems which record effects on descriptions involving the dynamical properties of these systems. This kind of conceptual dependence, does not preclude the application of quantum mechanics to systems that record effects. Although it is largely a matter of convenience which systems are, and which are not, taken to record effects, it is not wholly a matter of convenience. The development of the quantum theory shows that there are systems that are resistant to a satisfactory classical description—this is the content of Kochen-Specker—and as a result, such systems lack admissible descriptions of the kind we require for the description of the phenomena the quantum theory is invoked to explain.

As for the cut between classical and quantum systems, this also is mandated by the fact that quantum mechanics is a theory devised for the explanation of the behavior of systems that are inherently indeterministic in the sense that (i) they are represented in the theory by their non-dynamical properties and their effects, and (ii) their effects are such that they do not admit a solution to the Problem of Determinism. So far as the conceptual issues raised by the interpretation of the theory are concerned, the basis for the cut lies in the methodological demand for admissible descriptions of the appearances we hope to save.

Although the framework of effects assumes that there are two distinct kinds of system, this is compatible with the thesis that reality is unitary and quantum-mechanical. The reason for this compatibility is that the notion of a system is relative to a theoretical representation. There can be both classical and quantum mechanical systems because to assert that there are classical systems is to claim that theoretical representations expressed in the framework of classical mechanics yield what I earlier characterized as *admissible descriptions*. Nothing in this formulation precludes the possibility that a representation hitherto formulated within classical mechanics might be replaced by a quantum mechanical representation. Whether reality is "captured" by classical mechanics is a separate question, one whose answer may be 'No' compatibly with the descriptions of classical mechanics being admissible.

Since the effects framework locates the classical-quantum cut at the level of differences in theoretical representation, there is no incompatibility between the thesis that there are admissible classical mechanical descriptions and the thesis that reality is quantum mechanical. According to the framework of effects, the world presents us with appearances that we attempt to "save" and that we represent as classical systems. The framework leaves open the empirical question of why it is that the world appears to be amenable to descriptions that are expressible in classical mechanics. But whatever the character of its appearance, supposing quantum mechanics is true, reality itself has all the peculiarities quantum mechanics says it has. This does not mean that quantum mechanics presents us with a "picture of reality"—no theory does—only that quantum mechanics has identified salient aspects of reality, aspects that are missed by classical mechanics. Bohr is supposed to have said: "There is no quantum world. There is only an abstract quantum mechanical description."[5] From the perspective of the framework of

---

[5] Petersen [11]. Thanks to Hilary Putnam for bringing this remark to my attention.

effects, the situation is rather that there is no classical world, only an abstract classical mechanical description.

Appendix: Itamar on Locality as a Special Case of Non-contextuality

This paper was close to its final form when I sent it to Itamar for his reaction to it. I had intended to incorporate a suggestion of his into the paper before submitting the final version for publication. With his death, it occurred to me to simply quote from an e-mail in which he commented at some length on the paper. I think the e-mail conveys some of the remarkable mix of warmth and generosity—greatness of soul—that were characteristic of him and that were so highly valued by his friends.

The context of Itamar's letter is my discussion of contextuality and locality which begins on p. 207. Toward the conclusion of this discussion (beginning on p. 208) I give an argument, based on Kochen and Specker's proof of their principal theorem, for why effects rather than propositions should be understood as the proper subject of the probability assignments of quantum mechanics. Itamar's letter supplements this argument with another which he traces to a thought experiment of Vaidman. Our two arguments are linked by the connection Itamar draws between what I call "propositions belonging to particles" and EPR's elements of reality. Itamar and I argue that both notions should be rejected in favor of effects.

I am very much indebted to Jeffrey Bub for an extended e-mail correspondence which led to the clarification of Itamar's argument and to the suggestion that Vaidman [12] is a plausible choice for the paper of Vaidman's that Itamar had in mind. What we think is the relevant passage from Vaidman is quoted after Itamar's letter. In quoting from Itamar's letter, I have made some very minor stylistic changes which I've left invisible. More significant changes which correct or slightly elaborate Itamar's remarks are enclosed in square brackets.

From Itamar, September 28, 2009:

> ...Back to your paper and to the discussion of locality as a special case of [non-] contextuality, I think that's exactly right and it brings into focus the question about the relation between EPR's "elements of reality" and the concepts of proposition and effect. As you recall, something is an element of reality if its existence can be predicted with certainty. The EPR argument is built upon assigning elements of reality, by this criterion, to properties whose existence is never actually measured. They never leave a trace, and that's where they fail. I think that EPR's criterion is at best necessary but insufficient, and proper elements of reality should also be required to leave a trace that can in principle be retrodicted, at least for a short time after the effect (in fact according to quantum mechanics if there is an effect the wave function changes, while unmeasured "elements" don't change the wave-function). So EPR's mistake is like what you describe for the Kochen-Specker case of assigning truth values to propositions, in particular the value "true" to a contradiction. GHZ's (or Mermin's) version of EPR shows that the two arguments, Kochen-Specker and Bell's, are the same in this respect. The lesson of EPR is not about locality but about how their criterion of reality is insufficient.

There is a paper by Vaidman who shows the shortcomings in the EPR criterion (although Vaidman does not see it that way and does not use the result for this purpose). He creates a simple thought experiment with a system that has no locality issues (no tensor products), and considers a measurement on the system at t_1 and a subsequent measurement (of some other observable) at t_2. Now he considers a hypothetical measurement of property A at a t between t_1 and t_2 and asks: If we measured A at t, would we have discovered that the system had property A? By construction the results of the [measurements at t_1 and t_2] force the answer YES with certainty. However if we use [the measurement at t_1 and] the later measurement at t_2 and ask the same question about a hypothetical measurement [of B at the earlier time t, the answer is also YES with certainty. But A and B are contrary propositions, and therefore cannot be true together. Hence the answer to whether A is true] is NO with probability one. So the property A cannot be assigned a truth value that will be consistent across time.

From (Vaidman [12], pp. 134–135):

A peculiar example of time symmetric counterfactuals is the *three box paradox* [...]. Consider a single particle prepared at time t_1 in a superposition of being in three separate boxes:

$$|\Psi_1\rangle = 1/\sqrt{3}(|A\rangle + |B\rangle + |C\rangle).$$

At a later time t_2 the particle is found in another superposition:

$$|\Psi_2\rangle = 1/\sqrt{2}(|A\rangle + |B\rangle - |C\rangle).$$

For this particle, a set of counterfactual statements, which are *elements of reality* according to the ... definition,

[If we can *infer* with certainty that the result of measuring at time t of an observable O is o, then, at time t, there exists an element of reality $O = o$,]
is:

$$P_A = 1,$$

$$P_B = 1.$$

Or, in words: if we open box A, we find the particle there for sure; if we open box B (instead), we also find the particle there for sure.[6]

---

[6] The definition given in square brackets is from Vaidman [12, p. 133]. The italics are Vaidman's and are intended to emphasize that the definition depends on the atemporal notion of inferring rather than the temporally directed notion of predicting. The three box paradox goes back at least as far as [13]. It is discussed in several other of Vaidman's papers. See for example Vaidman [14] for a more elaborate discussion and for an explicit statement of the principle which underlies this reasoning, namely, the Aharonov-Bergmann-Lebowitz rule for calculating probabilities for the results of an intermediate measurement performed on a pre- and post-selected system.

References

1. Demopoulos, W.: Effects and propositions. Found. Phys. **40**, 368–389 (2010)
2. Gleason, A.M.: Measures on the closed subspaces of Hilbert space. J. Math. Mech. **6**, 885–893 (1957)
3. Pitowsky, I.: Infinite and finite Gleason's theorem and the logic of indeterminacy. J. Math. Phys. **39**, 218–228 (1998)
4. Kochen, S., Specker, E.P.: The problem of hidden variables in quantum mechanics. J. Math. Mech. **17**, 59–87 (1967)
5. Pitowsky, I.: Betting on the outcomes of measurements: a Bayesian theory of quantum probability. Stud. Hist. Philos. Mod. Phys. **34**, 395–414 (2003)
6. Pitowsky, I.: Quantum mechanics as a theory of probability. In: Demopoulos, W., Pitowsky, I. (eds.) Physical Theory and Its Interpretation: Essays in Honor of Jeffrey Bub, pp. 213–240. Springer, Dordrecht (2006)
7. Fuchs, C.A.: QBism, the perimeter of quantum Bayesianism. arXiv:1003.5209v1 [quant-ph] 26 Mar 2010 (2010)
8. Stevens, K.W.H.: The spin-Hamiltonian and line widths in nickel Tutton salts. Proc. R. Soc. Lond. A Math. Phys. Sci. **214**, 237–246 (1952)
9. Bell, J.S.: On the Einstein-Podolsky-Rosen paradox. Physics **1**, 195–200 (1964)
10. Camilleri, K.: Heisenberg and the Interpretation of Quantum Mechanics: The Physicist as Philosopher. Cambridge University Press, Cambridge (2009)
11. Petersen, A.: The philosophy of Niels Bohr. Bull. At. Sci. **19**(7), 8–14 (1963)
12. Vaidman, L.: Counterfactuals in quantum mechanics. In D. Greenberger et al. (eds) Compendium of Quantum Physics: Concepts, Experiments, History and Philosophy. Springer-Verlag, pp. 132–136. Berlin and Heidelberg (2009)
13. Aharonov, Y., Vaidman, L.: Properties of a quantum system during the time interval between two measurements. Phys. Rev. A **41**, 11–20 (1990)
14. Vaidman. L.: Lorentz-invariant "elements of reality" and the joint measurability of commuting observables. Phys. Rev. Lett. **70**, 3369–3372 (1993)

Chapter 14
Infinitely Challenging: Pitowsky's Subjective Interpretation and the Physics of Infinite Systems

Laura Ruetsche and John Earman

Abstract On Itamar Pitowsky's subjective interpretation of quantum mechanics, "the Hilbert space formalism of quantum mechanics [QM] is just a new kind of probability theory" (2006, 213), one whose probabilities correspond to odds rational agents would accept on the outcomes of gambles concerning *quantum* event structures. Our aim here is to ask whether Pitowsky's approach can be extended from its original context, of quantum theories for systems with an finite number of degrees of freedom, to systems with an infinite number of degrees of freedom, such as quantum field theory and quantum statistical mechanics in the thermodynamic limit. An impediment to generalization is that Pitowsky adopts the framework of event structures encoded by *atomic* algebras, whereas the algebras typical of QM for infinitely many degrees of freedom are usually non-atomic. We describe challenges to Pitowsky's approach deriving from this impediment, and sketch and assess strategies Pitowsky might use to meet those challenges. Although we offer no final verdict about the eventual success of those strategies, a testament to the worth of Pitowsky's approach is that attempting to extend it engages us in provocative foundational issues.

14.1 Introduction

For Itamar Pitowsky, "the Hilbert space formalism of quantum mechanics [QM] is just a new kind of probability theory" [1, 213]. Understanding a probability theory to consist in (1) an algebra encoding the structure of events to be assigned

L. Ruetsche (✉)
Department of Philosophy, University of Michigan, Ann Arbor, MI, USA
e-mail: ruetsche@umich.edu

J. Earman
Department of History and Philosophy of Science, University of Pittsburgh, Pittsburgh, PA, USA
e-mail: jearman@pitt.edu

Y. Ben-Menahem and M. Hemmo (eds.), *Probability in Physics*, The Frontiers Collection, 219
DOI 10.1007/978-3-642-21329-8_14, © Springer-Verlag Berlin Heidelberg 2012

probabilities, and (2) a measure over that algebra providing those probabilities, Pitowsky argues that the algebra encapsulating the structure of quantum events is the lattice $\mathcal{P}(\mathfrak{B}(\mathcal{H}))$ of subspaces of a Hilbert space $\mathcal{H}$. To introduce a measure over this event structure, Pitowsky considers what odds rational agents would accept on the outcomes of "quantum gambles" concerning events so structured, and discovers that measure to be one corresponding to standard quantum probabilities. On Pitowsky's view probability is degree of belief and, thus, there is no "BIG" measurement problem. The BIG problem, he contends, arises only for those who think that the quantum state is a real physical state, and consists in identifying a real physical reason for the *sui generis* change of state that is measurement collapse. Asserting the quantum state to be a bookkeeping device for subjective degrees of belief, Pitowsky brands the BIG measurement problem illusory.

Let us observe parenthetically that we are skeptical of the degree of belief interpretation of quantum probabilities and the dismissal of the BIG measurement problem. Pitowsky's general thesis, however, is interesting and attractive. Our aim here is to ask whether Pitowsky's approach can be extended from its original context, of quantum theories for systems with an finite number of degrees of freedom, to systems with an infinite number of degrees of freedom, such as quantum field theory (QFT) and quantum statistical mechanics (QSM) in the thermodynamic limit. Let us lump these latter theories together under the heading of "QM_∞." An obvious impediment to generalization is that Pitowsky adopts the framework of event structures encoded by *atomic* algebras, whereas the algebras typical of QM_∞ are usually non-atomic. Section 14.2 explicates this impediment; Sect. 14.3 assuages the immediate worries it suggests. Section 14.4 identifies, lurking in atomless algebras, deeper challenges to Pitowsky's approach, and sketch strategies Pitowsky might use to meet those challenges. We offer no final verdict about the eventual success of those strategies. But a testament to the worth of Pitowsky's approach is that assessing it as an approach to QM_∞ engages us in some of the most provocative foundational issues that theories of QM_∞ have to offer. Section 14.5 contains some critical remarks on Pitowsky's proposed resolution of the small measurement problem that remains after the BIG problem has been dissolved.

14.2 Atomlessness

The crux of Pitowsky's interpretation is the move from the characteristic quantum event structure provided by $\mathcal{P}(\mathfrak{B}(\mathcal{H}))$ to standard Born rule probabilities *understood as degrees of belief*. Pivotal to this move is a rational agent, presented with "quantum gambles" in the form of sets of possible bets on the outcomes of measurements performed on a system subject to a public preparation procedure. Given an algebra of events with the logical structure $\mathcal{P}(\mathfrak{B}(\mathcal{H}))$, Pitowsky argues, *Gleason's theorem* forces the rational agent's hand. According to Gleason's theorem, provided $\dim(\mathcal{H}) > 2$, there is a one-to-one correspondence between countably additive measures $\mu : \mathcal{P}(\mathfrak{B}(\mathcal{H})) \rightarrow [0, 1]$ and density operators W on $\mathcal{H}$ such that $Tr(WE) = \mu(E)$ for all $E \in \mathcal{P}(\mathfrak{B}(\mathcal{H}))$. Supposing the gambler's probability assignments to $\mathcal{P}(\mathfrak{B}(\mathcal{H}))$ satisfy *prima facie* reasonable constraints, Pitowsky shows that they must take the form of measures characterized by Gleason's theorem. Her betting odds conform to the Born rule, as expressed by the trace prescription. He comments, "The remarkable feature exposed by Gleason's theorem is that the event structure dictates the quantum mechanical probability rule" (222).

For later purposes it will be helpful to restate this seminal result in the language of von Neumann algebras and lattices of projection operators. A *von Neumann algebra* $\mathfrak{M}$ acting on a Hilbert space $\mathcal{H}$ (assumed here to be separable) is a *-algebra of bounded operators with the defining property that $\mathfrak{M}$ is closed in the weak topology or, equivalently, $\mathfrak{M} = (\mathfrak{M}')' := \mathfrak{M}''$ (where $''$ denotes the commutant) (see Sunder [2] for an overview of von Neumann algebras). Corresponding to $\mathfrak{M}$ there is a lattice $\mathcal{P}(\mathfrak{M})$ of projections; in fact this lattice is a complete orthomodular lattice. Conversely, a von Neumann algebra is generated by its projections, i.e. $\mathfrak{M} = \mathcal{P}''$. A *state* on a von Neumann algebra $\mathfrak{M}$ is an expectation value functional, i.e. a normed positive linear mapping $\omega : \mathfrak{M} \rightarrow \mathbf{C}$. A state ω is *pure* iff it cannot be written as a non-trivial convex linear combination $\omega = \lambda\omega_1 + (1 - \lambda)\omega_2$ where $0 < \lambda < 1$ and $\omega_1 \neq \omega_2$. A *mixed state* is a non-pure state. A *vector state* is a state for which there is a $|\psi\rangle \in \mathcal{H}$ such that $\omega(A) = \langle\psi|A|\psi\rangle$ for all $A \in \mathfrak{M}$. Vector states are among the *normal states,* i.e. states ω generated by a density operator W, i.e. $\omega(A) = Tr(WA)$ for all $A \in \mathfrak{M}$.

Ordinary QM is the case where $\mathfrak{M} = \mathfrak{B}(\mathcal{H})$—the full algebra of bounded operators— and $\mathcal{P}(\mathfrak{B}(\mathcal{H}))$ is the full lattice of projections onto all of the closed subspaces of $\mathcal{H}$. For this special case Gleason's theorem is the result that, when $\dim(\mathcal{H}) > 2$, any countably additive probability measure μ on $\mathcal{P}(\mathfrak{B}(\mathcal{H}))$ has a unique extension to a normal state ω_μ on $\mathfrak{B}(\mathcal{H})$. Below we will consider generalizations of Gleason's theorem to more general types of von Neumann algebras encountered outside of ordinary QM.

One of the axioms by which Pitowsky characterizes the quantum event structure requires it to include "maximally informative propositions" which "in the quantum case, . . . correspond to one-dimensional subspaces of the Hilbert space" (218). To appreciate the correspondence Pitowsky draws here, consider a von Neumann algebra $\mathfrak{M}$. If a subalgebra $\mathfrak{N}$ of $\mathfrak{M}$ is abelian (that is, commutative) then the elements of the associated lattice $\mathcal{P}(\mathfrak{N})$, understood as propositions, can be assigned truth values obedient to classical truth tables. If $\mathfrak{N}$ is maximal abelian— that is, such that $\mathfrak{M}$ has no abelian algebra of which $\mathfrak{N}$ is a proper subset—then $\mathcal{P}(\mathfrak{N})$ corresponds to a *largest* set of propositions in $\mathfrak{M}$ that can be attributed classical truth values simultaneously.

An atom of a projection lattice $\mathcal{P}(\mathfrak{M})$ is a projection operator $E \in \mathcal{P}(\mathfrak{M})$ with the property that no non-zero projection operator in $\mathcal{P}(\mathfrak{M})$ has a range that's a proper subspace of E's range. In other words, the atoms of a projection lattice are its minimal elements. Now each atom (if such exists!—see below) of a maximal abelian $\mathfrak{N} \subset \mathfrak{M}$ determines a truth evaluation on the lattice $\mathcal{P}(\mathfrak{N})$.[1] It is in this sense that an atom is a maximally informative proposition (see Bell and Machover [3] for details). When $\mathfrak{M} = \mathfrak{B}(\mathcal{H})$, its atoms are projection operators whose ranges are one-dimensional. Hence, taking $\mathcal{P}(\mathfrak{B}(\mathcal{H}))$ to give the quantum event structure, Pitowsky takes maximally informative propositions to correspond to one-dimensional subspaces of $\mathcal{H}$. The axiom on the quantum event structure imposing this demand he names "atomicity."

We come now to a provocative fact. The von Neumann algebras $\mathfrak{B}(\mathcal{H})$ of ordinary QM are the natural setting of Pitowsky's interpretation. But in QM_∞, one encounters von Neumann algebras not isomorphic to $\mathfrak{B}(\mathcal{H})$, algebras whose projection lattices aren't isomorphic to $\mathcal{P}(\mathfrak{B}(\mathcal{H}))$ Indeed, one encounters von Neumann algebras that lack atoms. We will call such algebras "atomless." (Atomless algebras often appear in the literature under different descriptions. In terms of the Murray and von Neumann classification of von Neumann algebras, $\mathfrak{B}(\mathcal{H})$ is a Type I factor algebra; Types II and III factor algebras are atomless. It follows that every maximal abelian subalgebra of a Type II or III factor algebra is also atomless.)

A fairly homely and intuitive example of an atomless von Neumann algebra, an example to which Sect. 14.4 returns, is the algebra of continuous functions on the closed interval [0,1]. Let $\mathcal{H}$ be the separable Hilbert space $L_2([0,1])$ of square integrable functions on the unit interval [0,1] equipped with the Lebesgue measure. Where f is a bounded measurable function on [0,1], let M_f be the operator on L_2 corresponding to multiplication by f. The collection $\{M_f\} = \mathfrak{D}_Q$ (with addition and multiplication defined pointwise) is a von Neumann algebra acting on $\mathcal{H}$ [4, Example 5.1.6]. $\mathfrak{D}_Q$ is atomless.

To see why, begin with a characterization of the projection operators in $\mathfrak{D}_Q$. For each Borel subset X of [0,1], the characteristic function $\chi_X : \chi_X(p) = 1$ if $p \in X$; $\chi_X(p) = 0$ otherwise, is a projection in $\mathfrak{D}_Q$, and every projection in $\mathfrak{D}_Q$ is the characteristic function for some Borel subset of [0,1] (see Kadison and Ringrose 1997, Ex. 2.5.12, 117). $\mathfrak{D}_Q$ is atomic only if it contains a non-zero χ_S such that no non-zero projection has as its range a proper subspace of χ_S's range. Now, consider what it takes for χ_S to be different from the zero operator. If S is a set of measure 0, the vector norm of $\chi_s|\psi\rangle$ for an arbitrary $\psi \in L^2([0,1])$ is given by

$$|\chi_s|\psi\rangle| = \int_0^1 \psi^*(x)\chi_s\psi(x)dx = \int_s \psi^2(x)dx = 0 \qquad (14.1)$$

If S is a set of measure 0, χ_S maps every element of $L^2([0,1])$ to the zero vector.[2] Thus χ_Y is a non-zero projection operator only if the set Y is measurable. But if Y is

---

[1] The truth evaluation determined by E maps $\mathcal{P}(\mathfrak{N})$ to TRUE if $EF = F$; and to FALSE otherwise.

[2] It turns out that $\mathfrak{D}_Q$'s projection lattice is isomorphic to $\mathcal{B}([0,1]/\mathcal{N})$, the Boolean algebra of equivalence, up to measure 0, classes of Borel subsets of [0,1]. See Halvorson 2001 for details and an illuminating discussion of their significance.

measurable, χ_Y can't be an atom in $\mathfrak{D}_Q$. Every measurable set Y has a measurable proper subset. Let X be a measurable proper subset of Y. $\chi_X \neq 0$, because X is measurable. And χ_X has as its range a proper subspace of χ_Y's range. This spoils χ_Y's claim to be an atom.

As straightforward as the foregoing example may seem, it conceals a remarkable fact. If $\mathfrak{M}$ is atomless, *none of its pure states are normal and perforce none of its pure states correspond to countably additive measures on* $\mathcal{P}(\mathfrak{M})$! It may be tempting to ignore this remarkable fact on the grounds that atomless von Neumann algebras are idle mathematical curiosities. But they are not. Rather, atomless von Neumann algebras are typical of quantum theories of systems with infinitely many degrees of freedom. Here are some examples.

One version of axiomatic algebraic QFT (see [5]) associates a von Neumann algebra $\mathfrak{M}(\mathcal{O})$ of observables with each open bounded region of $\mathcal{O} \subset \mathcal{M}$ of Minkowski space-time $\mathcal{M}$. This association is assumed to have the net property that if $\mathcal{O}_1 \subset \mathcal{O}_2$ then $\mathfrak{M}(\mathcal{O}_1)$ is a subalgebra of $\mathfrak{M}(\mathcal{O}_2)$. The algebra of observables for entirety of Minkowski space-time is the quasilocal algebra $\mathfrak{M}(\mathcal{M})$, obtained as the closure of the union over subregions of their local algebras. For the Minkowski vacuum state of the mass $m \geq 0$ Klein-Gordon field, if $\mathcal{O}$ is a region with non-empty space-like complement, the standard axioms for algebraic QFT imply that $\mathfrak{M}(\mathcal{O})$ is an atomless von Neumann algebra, indeed a Type III factor [6]. Results by Buchholz et al. [7] indicate that the Type III character of local algebras holds not only for free scalar fields but quite generically for quantum fields of physical interest.

Atomless von Neumann algebras are also typical for the thermodynamic limit of QSM, which one reaches by letting the number of constituents of the system and its volume tend to infinity while its density remains finite. In this limit the so-called KMS condition explicates the notion of equilibrium (for a brief exposition, see Sewell [8, pp. 49–51]). In the thermodynamic limit of QSM, KMS states at finite temperatures correspond to Type III (and so atomless) factors for a wide variety of physically interesting systems, including Bose and Fermi gases, the Einstein crystal, and the BCS model (see [9, pp. 139–140]; Bratelli and Robinson [10], Corr. 5.3.36). The exceptions are KMS states at temperatures at which phase transitions occur (if there are any for the systems in question); then the relevant algebras are direct sums/integrals of Type III factors. Systems in equilibrium at *infinite* temperatures are also of interest in QSM. Such systems occupy chaotic states. Chaotic states in the thermodynamic limit of QSM correspond to Type II$_1$ factors, which are also atomless (see [11, Vol. III, Sec. XIV. 1]).

14.3 The Initial Worry Addressed

We have framed an initial impediment to generalizing Pitowsky's approach beyond ordinary QM, whose observable algebras take the richly atomic form $\mathfrak{B}(\mathcal{H})$. The projection lattices and observable algebras encountered in QM$_\infty$ may lack atoms.

This impedes the extension of Pitowsky's approach, because that approach appears to require quantum event structures to contain atoms. But not to worry! The main tools Pitowsky uses are Gleason's theorem and Lüders' Rule and, as we will see, these can be extended to very general von Neumann algebras.

As noted above, Pitowsky uses Gleason's theorem to reconstruct Born rule probabilities in terms of rational odds for quantum gambles. And as intimated above, Pitowsky uses Lüders' Rule to dismiss the BIG measurement problem:

> In our scheme quantum states are just assignments of probabilities to possible events, that is, possible measurement outcomes. This means that the updating of the probabilities during a measurement follows the Von Neumann-Lüders projection postulate and not Schrödinger's dynamics. Indeed, the projection postulate is just the formula for conditional probability that follows from Gleason's theorem. So the BIG measurement problem does not arise. (232)

Let us elaborate for the case of ordinary QM, i.e. in the language of von Neumann algebras the case where the algebra is $\mathfrak{B}(\mathcal{H})$. Suppose that $\mu : \mathcal{P}(\mathfrak{B}(\mathcal{H})) \to [0,1]$ is a countably additive measure. By Gleason's theorem if $\dim(\mathcal{H}) > 2$ there is a unique density operator W on $\mathfrak{B}(\mathcal{H})$ that extends μ to a normal state on $\mathfrak{B}(\mathcal{H})$. Let $E \in \mathcal{P}(\mathfrak{B}(\mathcal{H}))$ be such that $\mu(E) \neq 0$. Then the density operator W_E given by

$$W_E(A) := \frac{Tr(W_E A W_E)}{Tr(W_E)} \text{ for all } A \in \mathfrak{B}(\mathcal{H}) \qquad [\text{LR}]$$

defines a normal state. This state has the property that

(**L**): For any $F \in \mathcal{P}(\mathfrak{B}(\mathcal{H}))$ such that $F \leq E$, $W_E(F) = \mu(F)/\mu(E)$.

where $F \leq E$ holds just in case the subspace onto which F projects is a subspace of the subspace onto which E projects. Furthermore, W_E is the unique density operator with the property (L) (see [12]). These properties of W_E are taken to motivate its interpretation as giving a conditional quantum probability. The rule **LR** for quantum conditionalization is commonly referred to as Lüders' Rule.

Let E be a one-dimensional projection operator (that is, an atom) in $\mathcal{P}(\mathfrak{B}(\mathcal{H}))$. Suppose we subject a system concerning which our degrees of belief are coded by the normal state W to an E measurement, which yields the outcome 1. Invoking Lüders' Rule to update our degrees of belief conditional on this outcome, we find that the Lüders conditionalized state W_E coincides with E *no matter what state W was* (provided, of course, $Tr(W_E) \neq 0$, which is a condition for obtaining the outcome 1 to begin with). The state encoding our updated degrees of belief is the state E into which, on the postulate of measurement collapse, the measured system collapses upon the occasion of an E measurement yielding the outcome 1. There is no BIG measurement problem, Pitowsky claims, because on the subjective interpretation of probability, the collapse of the wave function simply reflects rational updating.

Gleason's theorem and Lüders' rule do the heavy lifting for Pitowsky's interpretation. Do their analogs hold for von Neumann algebras not isomorphic to $\mathfrak{B}(\mathcal{H})$, including the atomless von Neumann algebras of QFT and the thermodynamic limit of QSM?[3] And do their analogues hold when the probability measure on the projection lattice of the algebra is merely finitely additive? In the case of Gleason's theorem the answer is positive. In fact the generalization is quite broad:

> *Generalized Gleason's Theorem* [14, 15, Ch. 5] Let $\mathfrak{M}$ be a von Neumann algebra acting on a separable Hilbert space $\mathcal{H}$ and let $\mu : \mathcal{P}(\mathfrak{M}) \rightarrow [0,1]$ be a finitely additive probability measure. If $\mathfrak{M}$ does not contain any summands of Type I_2,[4] there is unique state ω_μ on $\mathfrak{M}$ that extends μ, i.e. $\omega_\mu(E) = \mu(E)$ for all $E \in \mathcal{P}(\mathfrak{M})$. Moreover, ω_μ is a normal state iff μ is countably additive.[5]

In what follows we will concentrate on countably additive probability measures and their corresponding normal states. But we add in passing that there does not seem to us to be a good motivation for the tradition in the philosophical literature of ignoring finitely additive probability measures and their corresponding non-normal states, especially if probability is given a subjective interpretation (for why shouldn't degrees of belief be merely finitely additive?). One could push from the other end by noting that, according to physics lore, it takes an infinite amount of energy to prepare a non-normal state on the von Neumann algebra associated with a finite, bounded region of spacetime.[6] But such pushing seems to require treating the quantum state as a real physical state, contrary to the subjective interpretation favored by Pitowsky. In any case, what is important is the uniqueness of the extension ω_μ of μ, which makes it possible to think of ω_μ as a bookkeeping device. That ω_μ is not implemented by a density operator when μ is merely countably additive and, thus, does not conform to the familiar form of the Born rule is neither here nor there.

So much for generalizing Gleason's theorem. As for Lüders' Rule, its motivation carries over to all von Neumann algebras for which the generalized Gleason theorem applies. Indeed, the generalized Lüders' Rule can be seen as an adjunct of this theorem. Let ω_μ be the unique extension to a state on $\mathfrak{M}$ of a probability measure μ on $\mathcal{P}(\mathfrak{M})$. Consider the measure μ_E that results from conditionalization on $E \in \mathcal{P}(\mathfrak{M})$ such that $\mu(E) \neq 0$, i.e. $\mu_{E(F)} := \mu(F \wedge E)/\mu(E)$ for any $F \in \mathcal{P}(\mathfrak{M})$ such that $EF = F$ (in which case $(F \wedge E) = EFE$). The unique extension of ω_{μ_E} of μ_E to a state on $\mathfrak{M}$ is the natural candidate for the conditionalized state. But ω_{μ_E} is nothing other than the state obtained by Lüders' Rule.

---

[3] For a treatment of these issues more extensive than the one offered here, see Ruetsche and Earman [13].

[4] An exemplary Type I_2 von Neumann algebra is the algebra of complex-valued 2×2 matrices. A von Neumann algebra is of Type I_2 just in case its identity I is the sum of 2 equivalent abelian projections. For details, Kadison and Ringrose [4].

[5] The proof is highly non-trivial. It proceeds in two stages. First, it is shown that a probability measure on the projection lattice of a von Neumann algebra is uniformly continuous. Next, continuity is used to show that the measure can be extended to a positive linear functional on the algebra.

[6] Ruetsche [16] discusses a quartet of other reasons to be normal.

In sum, the tools essential to Pitowsky's approach continue to function in the setting of QM_∞, notwithstanding the absence from the algebras appropriate to that setting of atoms. Perhaps this should not be a surprise: it is *quantum* mechanics Pitowsky sets out to make sense of, and according to his analysis, the essential ingredient of *quantum* mechanics, the axiom that separates classical from quantum domains, is what he calls the *irreducibility* of the quantum event structure, that is, the presence in that event structure of sets of events that aren't all possible outcomes of the same experiment (218). So for Pitowsky, shedding the axiom of atomicity is consistent with maintaining a distinctly quantum event structure, calling forth a new theory of probability.

14.4 A Further Worry: State Preparation

Having assuaged the initial worry that the very absence of atoms from the observable algebras of QM_∞ derails Pitowsky's approach to interpreting those theories, we turn to other challenges atomless observable algebras pose for Pitowsky. One challenge is to make sense of *state preparation*. Pitowsky faces the challenge because the first step of the quantum bets central to his interpretation is the production of "a *single* physical system ... prepared by a method known to everybody" (223). His key result is that the Born Rule should govern the degrees of belief of agents considering such bets. But the Born Rule provides agents facing quantum bets with rational guidance *only if they are warranted in attributing a particular quantum state* to the system the bets are about. The challenge for Pitowsky is to explain how a system comes to be in such a condition that a particular quantum state ω encode the odds rational agents would accept for gambles on the outcomes of future measurements performed on the system. Put more briefly, the challenge is to account for state preparation.

Although Pitowsky doesn't address this challenge explicitly, the following passage indicates how he might meet it. "If ... we have formed a belief about the state of the system at time $t = 0$ (as a result of a previous measurement, say)," he writes, "we automatically have a probability distribution over the set of all possible outcomes of all possible measurements at each t" (236). We automatically have that distribution because we are entitled to Lüders conditionalize on the outcome of the measurement. As observed in the previous section, if that outcome corresponds to an atom $E \in \mathcal{P}(\mathfrak{B}(\mathcal{H}))$, Lüders conditionalization transforms the degrees of belief encoded by an arbitrary pre-measurement state to degrees of belief corresponding to the specific normal state encoded by the density operator E.

Successfully following Pitowsky's preparation recipe, we start with a system with respect to which an arbitrary, and generally unknown, quantum state ϕ (assumed to be normal) encodes rational degrees of belief, and wind up with a system for which a specific, and known, quantum state ω encodes rational degrees of belief. Lüders' rule is not the only ingredient essential to the preparation recipe. Another is the presence in $\mathfrak{M}$ of what we will call *filters*. In general, and to first approximation, a filter for a state ω on Neumann algebra $\mathfrak{M}$ is a projection operator $E_\omega \in \mathfrak{M}$ such that for any normal state ϕ on $\mathfrak{M}$ that assigns E_ω non-zero probability, ϕ Lüders conditionalized on E_ω coincides with ω on $\mathfrak{M}$. Pitowsky's preparation recipe works in the special case of $\mathfrak{M} = \mathfrak{B}(\mathcal{H})$ because normal pure states on $\mathfrak{B}(\mathcal{H})$ (which are, of course, vector states) have filters in the form of atomic projections.

Generalizing from the particular von Neumann algebra $\mathfrak{B}(\mathcal{H})$, the question of which states on an arbitrary von Neumann algebra $\mathfrak{M}$ have filters is well-posed and has a simple answer. Only pure normal states on $\mathfrak{M}$ have filters.[7] But it was disclosed in Sect. 14.2 that if $\mathfrak{M}$ lacks atoms, it lacks pure normal states as well. Thus, if $\mathfrak{M}$ is atomless, no state on it can be prepared by means of filtration. This means that *even if Lüders' Rule holds in general,* if $\mathfrak{M}$ is atomless, an ingredient crucial to Pitowsky's state preparation recipe is missing.

The challenge for Pitowsky's approach, then, is this: if $\mathfrak{M}$ lacks atoms, it stymies an account of state preparation along familiar lines. Without an account of state preparation, Pitowsky can't forge a link, crucial to his account, between the betting quotients of rational agents and Born rule probabilities implicit in specific quantum states. If Pitowsky can't forge this link, his project of explicating quantum probabilities in the terms of quantum bets collapses.

One way Pitowsky can rise to the challenge is, when confronted with an atomless von Neumann algebra, to justify changing the subject to a Type I von Neumann algebra, an algebra with atoms. In particular, local relativistic QFT may have a way to compensate for the atomlessness of its typical Type III algebras (Dieks 2000). As Sect. 14.2 remarked, the local algebra $\mathfrak{M}(\mathcal{O})$ associated with an open bounded region $\mathcal{O}$ of space-time is generically Type III, therefore atomless, and so filterless. Thus, there can be no local preparation procedure for a normal state on $\mathfrak{M}(\mathcal{O})$ that consists in measuring a filter in $\mathfrak{M}(\mathcal{O})$. Fortunately, however, the standard axioms for local relativistic QFT imply that the funnel property holds for suitable space-time regions in certain models. The net of local algebras $\mathfrak{M}(\mathcal{O})$ have the *funnel property* if and only if for any open bounded $\mathcal{O}$ there is another open bounded region $\hat{\mathcal{O}} \supset \mathcal{O}$ and a Type I factor $\mathfrak{N}$ such that $\mathfrak{M}(\hat{\mathcal{O}}) \supset \mathfrak{N} \supset \mathfrak{M}(\mathcal{O})$. The funnel property entails that normal states on $\mathfrak{M}(\mathcal{O})$ which are the restriction to that algebra of pure normal states on the Type I algebra $\mathfrak{N} \supset \mathfrak{M}(\mathcal{O})$ have filters in that Type I algebra. This guarantees that a local preparation procedure is possible, albeit in an expanded sense of "local" (see [17]).

It is debatable whether it is desirable to thus expand our sense of "local".[8] Clifton [18] furnishes grounds against: given nesting spacetime regions $\hat{\mathcal{O}} \supset \mathcal{O}$ and their atomless von Neumann algebras $\mathfrak{M}(\hat{\mathcal{O}}) \supset \mathfrak{M}(\mathcal{O})$, the choice of an interpolating Type I factor is quite arbitrary. Rather than adjudicate this debate, we will indicate how Pitowsky can weigh in. We'll start with the sample atomless von Neumann

---

[7] See Ruetsche and Earman [13] for a formal definition of 'filter' and an argument for this claim.

[8] Earman and Ruetsche [23] convey a taste of the debate.

algebra introduced in Sect. 14.2, the algebra $\mathfrak{D}_Q$ whose projections correspond to events of being located in subsets Δ of the unit interval.

There is reason to think that the very feature of Pitowsky's account that requires he make sense of state preparation enables him to change the subject from $\mathfrak{D}_Q$ to an atomic algebra. The key point is that "a proposition describing a possible event in a probability space is of a rather special kind. It is constrained by the requirement that there should be a viable procedure to determine whether the event occurs, so that a gamble that involves it can be unambiguously decided" (217). The outcomes subject to quantum bets, that is, the elements of quantum event structure, have to be the sort of thing we can resolve bets about. Given certain assumptions about our perceptual limitations, events of being localized in *arbitrarily small* regions aren't the kinds of things we can place *resolvable* bets on: if Δ is so teeny that we are incapable of determining whether a system's position measurement locates it in Δ, the event χ_Δ can't be subject to a quantum bet. Because not every event in the non-atomic algebra $\mathfrak{D}_Q$ can be the subject of a resolvable bet, $\mathfrak{D}_Q$ doesn't give a quantum event structure of the sort Pitowsky's subjective interpretation requires.

Pitowsky can cite this as reason to concentrate instead on an algebra encoding an event structure of *discriminable* events: events corresponding to, say, being located in one of some set of *finitely extended* regions, where the extent of those regions is keyed to the resolving capacity of our perceptual and technical apparatus. It is plausible that an algebra of such events will be atomic—it may simply be the algebra of eigenprojections of a discretized position observable. It is less plausible that the directive to concentrate on discriminable events picks out a *unique* atomic subalgebra of $\mathfrak{D}_Q$, but we are not inclined to regard this non-uniqueness as a dire problem. It comes with a handy explanation: any coarse-graining that issues in a set of events on which we can wager will deliver the goods of an atomic event structure; insofar as wagering is a practical activity constrained by the unavoidably vague requirement that the objects of wagers be 'recordable,' we shouldn't expect every set of wagers to be indexed to the same coarse-graining.

We will note in passing that Halvorson (2001) makes a move similar to the one we've just offered on Pitowsky's behalf. Halvorson declines to regard every elements of $\mathfrak{D}_Q$ as a 'real' event on broadly operationalist grounds conjoining the claims that (1) quantum probabilities are probabilities for measurement results and (2) measurements are typically approximate. What Pitowsky can add to this maneuver is motivation for the operationalist orientation, in the form of a *subjective* theory of probability requiring quantum events to be such as to settle bets rational agents make.

Now, one can imagine the move being criticized for dismissing some quantum probabilities from physical relevance. Where Δ is a region so subliminal as to be an inappropriate object of a quantum bet, the criticism goes, and ω a quantum state, $\omega(\chi_\Delta)$ is a legitimate quantum probability. But of $\omega(\chi_\Delta)$ Pitowsky's interpretation gives no account. But Pitowsky can stand firm in the face of this criticism. He hasn't promised to explicate every expression anyone has ever regarded as a quantum probability. He has rather offered a proposal about which sorts of things *are* quantum probabilities. He takes this proposal to be part of a package that's our best shot at making sense of the quantum world. According to the proposal, $\omega(\chi_\Delta)$ just isn't a quantum probability. It's a casualty of making the best progress we can on interpreting QM.

We hope to have at least established that Pitowsky has a feasible strategy for precipitating out of the atomless von Neumann algebra $\mathfrak{D}_Q$ a Type I von Neumann algebra abounding in atoms, and so susceptible to accounts of state preparation by filtration. Now let us return to the QFT case. Can the coarse-graining strategy be adapted? What Pitowsky needs is an account of how the appeal to discriminable events, the kind of events that might settle bets, justifies turning attention from atomless local algebras to one (or even some) of the Type I factors interpolating between them. Progress on this question awaits a cogent account of the nature of measurements in QFT. As this is one of the most daunting and unsettled topics in the interpretation of that theory, we take it to be an open, and significant, question whether Pitowsky can account for state preparation in the context of QFT.

It is, however, clear that the QFT stratagem for securing preparation by filtration cannot be adapted to the setting of the thermodynamic limit of QSM. What one would like to be able to prepare is a state of a superconductor or a ferromagnet, that is, a state of the *entire* quasi-local algebra itself. That algebra will typically be atomless, and no funnel property can be invoked to embed it in a Type I algebra. But perhaps in these cases, Pitowsky has another state preparation recipe available. In QSM, the most interesting states are the equilibrium ones. And those we can prepare by waiting.

14.5 The Small Measurement Problem

Believing his account to be proof against the BIG measurement problem, Pitowsky thinks it nevertheless faces "The small measurement problem ... why is it hard to observe macroscopic entanglement, and what are the conditions in which it might be possible?" (233). He gestures to decoherence as giving part—but only part—of the answer.

> Decoherence is a dynamical process and its exact character depends on the physics of the situation. I would like to point to a possibly more fundamental, purely combinatorial reason which is an outcome of the probabilistic structure: *the entanglement of an average ray in a multiparticle Hilbert space is very weak.* (233)

We suspect that Pitowsky has asked the wrong question. The small measurement problem should be: "Why is it hard to observe macroscopic superpositions (e.g. superpositions of live and dead cats)?" The line offered by the decoherence mob is that entanglement of the system we are observing with its environment gives the answer. As it stands this answer is incomplete—entanglement must be combined with the proper semantics of value assignments. For, even at its best, decoherence delivers us Professor Schrödinger's cat described by a reduced density operator W_{CAT} diagonal in an eigenbasis of the cat bio-observable. W_{CAT} is obtained from the density operator $W_{CAT+ENVIR}$ for the composite CAT + ENVIRONMENT system by "tracing out" the environmental degrees of freedom. If, as is typically the case, $W_{CAT+ENVIR}$ is given by the projector onto the ray spanned by a vector that fails to be an eigenstate of the cat bio-observable, the orthodox eigenvalue-eigenstate link of conventional quantum semantics will decline to assign the cat bio-observable a determinate value, no matter how diagonal W_{CAT} is in its eigenbasis. What is needed to rescue Professor Schrödinger's cat from the measurement problem is not only a diagonalized W_{CAT} delivered by (best-case) decoherence but also a semantic rule licensing us to understand a cat attributed W_{CAT} to be either determinately alive or determinately dead. And if this combination works, it does provide not just part but the whole answer to the small measurement problem. But this is not the place to go into details about whether decoherence works the way it is supposed to.

It remains the case that the observability or detection of entanglement is an interesting issue. But again, we are not convinced that Pitowsky has raised it in the right form. He poses the issue in terms of witnesses—operators on the tensor product space of a multiparticle system. For Pitowsky, an entanglement witness N is an observable whose expectation value on non-entangled states is always strictly less than 1, so that any state ω such that $\omega(N) > 1$ is therefore entangled. The more $\omega(N)$ exceeds 1, the easier ω's entanglement is to detect; the closer $\omega(N)$ is to 1, the more likely it is that experimental noise will interfere with detecting its entanglement. Pitowsky introduces a gauge of a state ω's entanglement in the form of ω's best witness: the observable N_ω in the collection N of witness for ω for which $\omega(N_\omega)$ most exceeds 1. ω's best witness is the observable with the best shot of establishing ω's entanglement, notwithstanding the vagaries of experimental noise.

Now consider the n-fold tensor product von Neumann algebra $\oplus_n \mathcal{B}(\mathcal{H})$ where each $\mathcal{H}$ is the linear vector space $\mathbf{C}^2$ appropriate for describing a spin $\frac{1}{2}$ system. The maximum value a state ω on $\oplus_n \mathcal{B}(\mathcal{H})$ assigns its best witness is $\sqrt{2^n}$. Roughly put, Pitowsky's weak entanglement conjecture—the conjecture with whose fate he takes the fate of the small measurement problem to rest—is that, as n grows, states whose best witnesses are this good shrink to a set of measure 0 in the space of pure states on $\oplus_n \mathfrak{B}(\mathcal{H})$. Pitowsky interprets this as a constraint on our capacity to observe macroscopic entanglement: if macroscopic systems are such that entangled states whose best witnesses are good enough to be detectable in spite of experimental noise are exceedingly rare, evidence of macroscopic entanglement will be exceedingly rare as well.

We are skeptical of Pitowsky's witness approach to the question of whether entanglement is observable because what we typically observe is not the multiparticle system in its entirely but some subsystem. And what we would like to know is whether subsystems carry signatures of their entanglement with the larger system.

The decoherence people would say that the classical appearance of a macro-subsystem is a signature of their entanglement with the environment! But this is contentious. Returning to QFT, we encounter less contentious signatures of entanglement, and encounter them everywhere. A QFT-adapted case that entanglement is generally undetectable would, we contend, make it plausible that states on these local algebras rarely, if ever, bear signatures of entanglement. But there is reason to despair of such a solution. In QFT, the correlations instituted across spacelike separated regions by *typical* states have the capacity to violate Bell inequalities, even maximally (for results, see [19, 21]; for an account aimed at philosophers, see [20]). Just as restricting the entangled spin singlet state to the algebra of observables pertaining to a single subsystem eventuates in a mixed state on the subsystem algebra, so too restricting typically entangled global QFT states to local regions results in mixed states. The local signature of the global entanglement is the Type III character of the local algebras: *all* of their normal states are mixed states; none of them are pure. That the global state restricted to appropriate sub-regions is a KMS state at finite temperature (e.g. in the Unruh effect) can be taken as a signature of entanglement.

It follows that, in the setting of QFT, to believe that the local state is a normal state on the local algebra (which Pitowsky's agents are presumably beholden to do) is to believe that that state carries the signature of entanglement.

14.6 Conclusion

Pitowsky's approach is nuanced and resourceful. While we believe the impediments we've presented to extending it to QM_∞ are serious, we do not claim that they are insurmountable. We do claim that the project of bringing Pitowsky's subjective interpretation into contact with theories of QM_∞ has a payoff that makes the project worth betting on: the contact exposes features both of Pitowsky's interpretation and of these physical theories that deserve further attention.

References

1. Pitowsky, I.: Quantum mechanics as a theory of probability. In: Demopoulos, W., Pitowsky, I. (eds.) Physical Theory and Its Interpretation: Essays in Honor of Jeffrey Bub, pp. 213–239. Springer, Dordrecht (2006)
2. Sunder, V.: An Invitation to von Neumann Algebras. Springer, Berlin (1986)
3. Bell, J.L., Machover, M.: A Course in Mathematical Logic. North-Holland, Amsterdam (1977)
4. Kadison, R.V., Ringrose, J.R.: Fundamentals of the Theory of Operator Algebras. Vol. 2: Advanced Theory, American Mathematical Society, Providence (1997)
5. Haag, R.: Local Quantum Physics, 2nd edn. Springer, New York (1996)
6. Araki, H.: Type of von Neumann algebra associated with free field. Prog. Theor. Phys. **32**, 956–965 (1964)
7. Buchholz, D., D'Antoni, C., Fredenhagen, K.: The universal structure of local algebras. Commun. Math. Phys. **111**, 123–135 (1987)
8. Sewell, G.: Quantum Theory of Collective Phenomena. Oxford University Press, Oxford (1986)
9. Emch, G.: Algebraic Methods in Statistical Mechanics and Quantum Field Theory. Wiley, New York (1972)
10. Bratelli, O., Robinson, D.W.: Operator Algebras and Quantum Statistical Mechanics 2, 2nd edn. Springer, New York (1997)

11. Takesaki, M.: Theory of Operator Algebras, vols. 2 and 3. Springer, Berlin (2003)
12. Bub, J.: Von Neumann's projection postulate as a probability conditionalization rule. J. Philos. Logic **6**, 381–390 (1977)
13. Ruetsche, L., Earman, J.: Interpreting probabilities in quantum field theory and quantum statistical mechanics. In: Beisbart, C., Hartman, S. (eds.) Probabilities in Physics. Oxford University Press, Oxford (2010)
14. Hamhalter, J.: Quantum Measure Theory. Kluwer, Dordrecht (2003)
15. Maeda, S.: Probability measures on projections in von Neumann algebras. Rev. Math. Phys. **1**, 235–290 (1989)
16. Ruetsche, L.: "Why be Normal?" Studies In History and Philosophy of Modern Physics **42**, 107–115 (2011)
17. Buchholz, D., Doplicher, S., Longo, R.: On Noether's theorem in quantum field theory. Ann. Phys. **170**, 1–17 (1986)
18. Clifton, R.: The modal interpretation of algebraic quantum field theory. Phys. Lett. A **271**, 167–177 (2000)
19. Halvorson, H., Clifton, R.: Generic bell correlation between arbitrary local algebras in quantum field theory. J. Math. Phys. **41**, 1711–1717 (2000)
20. Butterfield, J.: Vacuum correlations and outcome independence in algebraic quantum field theory. Ann. N. Y. Acad. Sci. **755**, 768–785 (1995)
21. Summers, S., Werner, R.: Maximal violation of Bell's inequalities is generic in quantum field theory. Commun. Math. Phys. **110**, 247–250 (1987)
22. Halvorson, H.: On the Nature of Continuous Physical Quantities in Classical and Quantum Mechanics. J. Philos. Logic **30**, 27–50 (2001)
23. Earman, J., Ruetsche, L.: Relativistic Invariance and Modal Interpretations. Philosophy of Science **72**, 557–583 (2005)

Chapter 15
Bayesian Conditioning, the Reflection Principle, and Quantum Decoherence

Christopher A. Fuchs and Rüdiger Schack

Abstract The probabilities a Bayesian agent assigns to a set of events typically change with time, for instance when the agent updates them in the light of new data. In this paper we address the question of how an agent's probabilities at different times are constrained by Dutch-book coherence. We review and attempt to clarify the argument that, although an agent is not forced by coherence to use the usual Bayesian conditioning rule to update his probabilities, coherence does require the agent's probabilities to satisfy van Fraassen's [1984] *reflection principle* (which entails a related constraint pointed out by Goldstein [1983]). We then exhibit the specialized assumption needed to recover Bayesian conditioning from an analogous reflection-style consideration. Bringing the argument to the context of quantum measurement theory, we show that "quantum decoherence" can be understood in purely personalist terms—quantum decoherence (as supposed in a von Neumann chain) is not a physical process at all, but an application of the reflection principle. From this point of view, the decoherence theory of Zeh, Zurek, and others as a story of quantum measurement has the plot turned exactly backward.

This article is dedicated to the memory of Itamar Pitowsky. Already 10 years ago, he saw the potential in a Bayesian understanding of quantum probabilities and had a respect for our efforts like no one else [1–3]. The respect was mutual. For you, Itamar.

C.A. Fuchs
Perimeter Institute for Theoretical Physics, Waterloo, ON, Canada
e-mail: cfuchs@perimeterinstitute.ca

R. Schack (✉)
Department of Mathematics, Royal Holloway, University of London, Egham, Surrey, UK
e-mail: r.schack@rhul.ac.uk

Y. Ben-Menahem and M. Hemmo (eds.), *Probability in Physics*, The Frontiers Collection,   233
DOI 10.1007/978-3-642-21329-8_15, © Springer-Verlag Berlin Heidelberg 2012

15.1   Introduction

At the center of most accounts of Bayesian probability theory [4] is the procedure of
Bayesian conditioning. By this we mean the following. Assume a Bayesian agent,
at some time $t = 0$, has assigned probabilities $P_0(E)$, $P_0(D)$ and $P_0(E, D)$ to events
E and D and their conjunction. As long as $P_0(D) \neq 0$, the conditional probability of
E given D is then

$$P_0(E|D) = \frac{P_0(E,D)}{P_0(D)} \; . \tag{15.1}$$

Now assume that, at a later time $t = \tau$, the agent learns that D is true and updates
his probability for E. We denote the agent's updated probability by $P_\tau(E)$. Standard
Bayesian conditioning consists of setting

$$P_\tau(E) = P_0(E|D). \tag{15.2}$$

The rule Eq. 15.2 can be viewed as a possible answer to the general question of
how an agent's probabilities at two different times should be related. We will
address this question from a personalist Bayesian perspective [4–9], according to
which probabilities express an agent's uncertainty, or degrees of belief, about future
events and acquire an operational meaning through decision theory [4]. Although
they are not determined by agent-independent facts, personalist probability
assignments are not arbitrary. Dutch-book coherence [5, 8, 9] as a normative
principle requires an agent to try her best to make her numerical belief assignments
conform to the usual rules of the probability calculus. When coupled with the
agent's overall belief system, this is a powerful constraint [10]. Personalist Bayesian probability is at the heart of Quantum Bayesianism, a radical new approach to
the foundations of quantum mechanics developed by Caves, Fuchs, Schack,
Appleby, Barnum, and others. (See [11, 12] for an extensive reference list.) The
motivation for the present investigation is to explore the relevance of Bayesian
conditioning to the Quantum Bayesian program.

It was first pointed out by Hacking [13] that there is no coherence argument that
compels the agent to take into account the earlier probabilities $P_0(E|D)$ when
setting the later probabilities $P_\tau(E)$. Similar points have been made by other authors
(see, e.g., [14]). Hacking was writing about the standard *synchronic* Dutch book
arguments, but the above statement remains true even for the *diachronic* Dutch
book arguments, originally due to Lewis and first reported by Teller [15]. Without
further assumptions, diachronic coherence does not compel the agent, at $t = \tau$, to
use the Bayesian conditioning rule (15.1).

The way diachronic coherence arguments connect probability assignments at
different times is more subtle. It is expressed elegantly through van Fraassen's
reflection principle [16], which itself entails the related constraints of Shafer [17]

and Goldstein [18]. The key idea behind the reflection principle is to consider the agent's beliefs about his own future probabilities, i.e., to consider expressions such as $P_0(P_\tau(E) = q)$. Shafer [17] put the point very nicely,

> This interpretation is based on the assumption that a person has subjective probabilities for how his information and probabilities may change over time. This means we are concerned not with how the person *should* or *will* change his beliefs, but rather with what he believes about how these beliefs will change. [Emphases added.]

The same idea underlies the approach this paper takes towards Bayesian conditioning and quantum decoherence.

In the next section, we present a detailed example where the agent appears justified to depart from the Bayesian conditioning rule. In Sect. 15.3 we review the standard, synchronic, Dutch book arguments and show why they do not imply Bayesian conditioning. Section 15.4 introduces diachronic coherence and presents a derivation of the reflection principle. In Sect. 15.5 we show that the Bayesian conditioning rule can be understood as a variant of the reflection principle valid for a particular class of situations. Section 15.6 addresses an argument that has been advanced against the reflection principle and shows that it is based on a misconception of the role that coherence considerations have in probability assignments. Finally, in Sect. 15.7 we give a natural application of the reflection principle to decoherence in quantum mechanics from a Quantum Bayesian perspective.

15.2 Example: Polarization Data

Consider a physicist running an experiment to discover the linear polarization of photons coming from a rather complicated optical device which he had built himself. Perhaps he is convinced that every photon is produced the same way, only that he has forgotten which orientation θ he gave to a certain polarization filter deep within the set-up. It might thus be easier to discover θ and recalibrate than to tear the whole thing apart and readjust. A statistical analysis is in order.

Our experimenter will measure polarization for a sequence of n individual photons and carry out a Bayesian analysis of the outcomes consisting of a string, s_n, of zeros and ones. The zeros stand for horizontal polarization and ones for vertical polarization. Before starting, the experimenter records his probabilistic prior, which in the personalist approach to probability adopted in this paper, represents his Bayesian degrees of belief about the measurement outcomes. To be specific we assume that the prior is *exchangeable*, i.e., of the form [8, 19]

$$P_0(s_n) = \int_0^1 q_0(x) x^k (1 - x)^{n-k} dx , \qquad (15.3)$$

where k is the number of zeros in s_n, and $q_0(x)$ is a probability density. If the experimenter is completely ignorant of what orientation he had given the filter, he might assume $q_0(x)$ to be the constant density, but the precise form of $q_0(x)$ is of no great importance to the argument below.

The prior $P_0(s_n)$ is a convex sum of binomial distributions $\{x, 1-x\}$, with $x = \cos^2\theta$. It is symmetric in the sense that it is invariant under permutations of the bits in s_n. By adopting this prior, the experimenter judges that the order in which the photons arrive is irrelevant to his analysis. To a Bayesian, this is in fact the operational meaning that all photons are "produced the same way." A simple consequence of this is that a posterior probability calculated from this prior by Bayesian conditionalization after some number of trials will not depend on the order of the zeros and ones found in those trials.

Suppose now that the experimenter observes 4,000 trials, and finds very nicely that the number of zeros and ones is not very far from 2,000 each. Technically this means that if the experimenter updates his prior to a posterior by Bayesian conditionalization, the posterior for the next n bits will be

$$P_{\text{posterior}}(s_n) = \int_0^1 q'(x) x^k (1 - x)^{n-k} dx , \qquad (15.4)$$

where $q'(x)$ is a function on the interval $[0,1]$ peaked very near $1/2$. If the experimenter were asked to bet on the next bit, this probability distribution would advise him to bet at close to even odds.

But now consider the following fantastic scenario. Suppose the experimenter becomes aware that the string s_n he accumulated is identical to the first 4,000 bits of the binary expansion of π! Any sane person would be flabbergasted. Even though the experimenter built the device with his own hands, he would surely wonder what was up. Perhaps one of his lab partners has played an immense joke on him?

The following question becomes immediate: If the experimenter is rational, how should he bet *now* on the next bit? Sticking doggedly with Bayesian conditioning, he would be advised to use near even odds, just as before. But the number π is too significant to ignore: His heart says to bet with

$$P_\tau(E) = 0.99 \qquad (15.5)$$

on the event E that the next bit equals the 4001th bit of the binary expansion of π. Our experimenter faces a stark choice: He can either ignore his heartfelt belief and use the value $P_\tau(E) \approx 1/2$ obtained from the conditioning rule, or he can ignore the conditioning rule and use the value $P_\tau(E) = 0.99$ representing what he really feels. Both choices are deeply problematic: the second one seems to be incoherent because it is contradicted by the usual understanding of the formalism, whereas the first one seems to be ignoring common sense to the point of being foolish.

Is the second choice really incoherent however? Does it violate a reasonable normative requirement? We will see in the next sections that this is not the case.

15.3 Synchronic Dutch Book

A simple way to give an operational meaning to personalist Bayesian probabilities is through the agent's betting preferences. When an agent assigns probability p to an event E, she regards $\$p$ to be the fair price of a standard lottery ticket that pays $\$1$ if E is true. In other words, an agent who assigns probability p to the event E regards both buying and selling a standard lottery ticket for $\$p$ as fair transactions; for her, the ticket is worth $\$p$.

In the following, we will call a set of probability assignments *incoherent* if it can lead to a sure loss in the following sense: there exists a combination of transactions consisting of buying and/or selling a finite number of lottery tickets which (i) lead to a sure loss and (ii) the agent regards as fair according to these probability assignments. A set of probability assignments is *coherent* if it is not incoherent. We accept as a normative principle that an agent should aim to avoid incoherent probability assignments.

The standard, synchronic, Dutch book argument [5, 8, 9] shows that an agent's probability assignments P_0 at a given time are coherent if and only if they obey the usual probability rules, i.e., $0 \leq P_0(E) \leq 1$ for any event E; $P_0(S) = 1$ if the agent believes the event S to be true; and $P_0(E \vee D) = P_0(E) + P_0(D)$ for any two events E and D that the agent believes to be mutually exclusive.

In this approach, conditional probability is introduced as the fair price of a lottery ticket that is refunded if the condition turns out to be false. Formally, let D and E be events, and let $\$q$ be the price of a lottery ticket that pays $\$1$ if both D and E are true, and $\$q$ (thus refunding the original price) if D is false. For the agent to make the conditional probability assignment $P_0(E/D) = q$ means that she regards $\$q$ to be the fair price of this ticket.

It is then a consequence of Dutch-book coherence that the product rule $P_0(E, D) = P_0(E/D)P_0(D)$ must hold [9]. In other words, conditional probability assignments violating this rule are incoherent. If $P_0(D) \neq 0$, we obtain Bayes's rule,

$$P_0(E|D) = \frac{P_0(E,D)}{P_0(D)} \ . \tag{15.6}$$

It is worth pointing out that Bayes's rule emerges here as a theorem, combining terms that are defined independently, in contrast to the common axiomatic approach to probability theory where Eq. 15.6 is used as the definition of conditional probability.

The above shows that a coherent agent must use Bayes's rule to set the conditional probability $P_0(E/D)$. The value of $P_0(E/D)$ expresses what ticket prices the agent regards as fair at time $t = 0$, i.e., before she finds out the truth value of either D or E. It says nothing about what ticket prices she will regard as fair at some later time $t > 0$.

In particular, assume that, at some time $t = \tau > 0$, the agent learns that D is true and updates her probabilities accordingly. Denote by $P_\tau(E)$ the agent's updated probability of E, meaning that she now regards $\$P_\tau(E)$ as the new fair price of a ticket that pays \$1 if and only if E is true. Nothing in the Dutch book argument sketched above implies that $P_\tau(E)$ should be equal to $P_0(E/D)$ [13]. All probabilities used in the argument are the agent's probabilities at time $t = 0$; they are defined via ticket prices for bets on E, D and their conjunction which the agent regards as fair at $t = 0$. The Dutch book argument leading to Eq. 15.6 is a *synchronic* argument. It does not connect in any way the agent's probability assignments at $t = 0$ and $t = \tau$. In particular, it does not imply that the agent has to use Bayesian conditioning to update her probabilities. In the next section we will see what connection between the agent's probability assignments at different times actually is implied by diachronic Dutch book arguments.

15.4 Diachronic Dutch Book

To set the scene, we consider an investor who today buys 500 shares of some company at a price of \$20 each, which he regards as a fair deal. The next day, his appreciation of the market has changed, and he sells his 500 shares at \$18 each, which now, given the new situation, he again regards as a fair deal. Despite the fact that the investor makes a net loss of \$1,000, he does not behave irrationally. By selling his shares at a lower price on the next day, he simply cuts his losses.

But what if the investor is certain today that tomorrow he will regard \$18 as the fair price for a share? It would then be foolish for him to buy, today, 500 shares for \$20 each, because he is certain that tomorrow he would be willing to sell the shares for \$18 each, leading to a net loss of \$1,000. As a matter of fact, buying shares at any price above \$18 today would be foolish in this situation, as would be selling shares today at any price below \$18.

In the above example, we have assumed that money has the same utility for the investor today and tomorrow, i.e., we have assumed a zero interest rate. This is an assumption we will make throughout the present paper. More precisely, we will assume that the time at which she receives a sum of money is irrelevant to a Bayesian agent.

In probability language, what we have just described is the following. Assume $P_0(E)$ is an agent's probability at $t = 0$ of some event E. The agent buys a lottery ticket that pays \$1 if E is true, for $\$P_0(E)$ which she regards as the fair price. At a later time $t = \tau$, she updates her beliefs. Her probability for E is now $P_\tau(E)$, which happens to be less than $P_0(E)$. At this point, the agent decides to cut her losses by selling the ticket for $\$P_\tau(E)$. Despite the net loss, there is nothing irrational about the agent's transactions.

But now suppose that at $t = 0$ the agent is certain that, at $t = \tau$, her probability of E will be q, where $q \neq P_0(E)$. In the case $q < P_0(E)$, this means that, at $t = 0$, she is willing to buy a ticket for $\$P_0(E)$ although she already knows that later she will be willing to sell it for the lower price $\$q$. In the case $q > P_0(E)$, it means that, at $t = 0$, she is willing to sell a ticket for $\$P_0(E)$ although she already knows that later she will be willing to buy the same ticket for the higher price $\$q$. In both cases, already at $t = 0$ the agent is certain of a sure loss.

This simple scenario contains the main idea of van Fraassen's diachronic Dutch book argument. Similar to the synchronic case discussed in the previous section, we call an agent's probability assignments incoherent if there exists a combination of transactions consisting of buying and/or selling a finite number of lottery tickets at two different times such that (i) *already at the earlier time*, the agent is sure of a net loss; and (ii) each transaction is regarded as fair by the agent according to her probability assignments *at the time* the transaction takes place. We will continue to accept as a normative principle that an agent should aim to avoid incoherent probability assignments.

To turn our simple scenario into the full-fledged diachronic Dutch-book argument, we only need to relax the assumption that at $t = 0$ the agent is *certain* that $P_\tau(E) = q$. Instead, we assume that

$$P_0(P_\tau(E) = q) > 0 , \tag{15.7}$$

i.e., at $t = 0$ the agent believes with some positive probability that at $t = \tau$ her probability of E will be equal to q. We will now show that this implies the agent's probability assignments are incoherent unless

$$P_0(E \,|\, P_\tau(E) = q) = q , \tag{15.8}$$

i.e., unless at $t = 0$ the agent's conditional probability of E, given that $P_\tau(E) = q$, equals q. This is van Fraassen's *reflection principle* [16].

To derive the reflection principle, denote by Q the proposition $P_\tau(E) = q$, i.e., the assertion that at time $t = \tau$, the agent will regard $\$q$ as the fair price for a ticket that pays $\$1$ if and only if E is true. The inequality (15.7) thus becomes $P_0(Q) > 0$. To establish that coherence implies the reflection principle (15.8), one must show that the assumption $P_0(E/Q) \neq q$ leads to a sure loss for an appropriately chosen set of bets no matter what outcomes occur for the events considered.

As a warm-up to gain intuition, suppose that $P_0(E/Q) > q$ and that Q is true. This means that at $t = 0$, the agent is willing to buy a ticket for $\$P_0(E/Q)$ that pays $\$1$ if both Q and E are true, and refunds the ticket price if Q is false. But, because of Q's truth, this ticket will further be equivalent to a ticket that pays $\$1$ if E is true. Finally, the truth of Q also implies that at $t = \tau$ the agent will be willing to sell this ticket for $\$q$, which is less than what she paid for it. In other words, if Q is true, the agent is sure to lose $\$d$, where $d = P_0(E/Q) - q$.

But this simple argument—illustrative though it may be—is not a full-fledged proof of incoherence. To get a full proof we need to show that the agent is sure of a loss not only when Q turns out to be true, but also when Q turns out to be false. For this it is sufficient to consider an alternate scenario where there is a side bet on Q, such that the agent loses some amount if Q is false, and wins less than $\$d$ if Q is true.

Such a side bet may be realized by a lottery ticket that pays $d/2 if Q is true, which the agent is willing to buy for $P_0(Q)d/2.

We have thus the following combination of transactions, each of which the agent regards as fair at the time it takes place:

(i)   to buy, at $t = 0$, for $P_0(E/Q)$, a ticket that pays $1 if both Q and E are true, and refunds the ticket price if Q is false;

(ii) to buy, at $t = 0$, for $P_0(Q)d/2$, a lottery ticket that pays $d/2 if Q is true;

(iii) if Q is true (i.e., if $P_\tau(E) = q$), to sell, at $t = \tau$, for $q, a lottery ticket that pays $1 if E is true.

Already at $t = 0$, the agent knows that these transactions result in a net loss, equal to $(P_0(Q) + 1)d/2 if Q is true, and $dP_0(Q)/2 if Q is false. We have thus shown that the assumption $P_0(E/Q) > q$ implies that the agent's probability assignments are incoherent.

The final piece of a proof is to consider the case $P_0(E/Q) < q$. By reversing the signs of all transactions above, it is easy to see that this case leads to a sure loss in exactly the same way. Putting these two cases together, this completes the full derivation of the reflection principle.

The coherence condition of Shafer [17] and Goldstein [18] follows by a simple application of synchronic coherence along with the reflection principle. Suppose the agent instead of contemplating a single $Q = [P_\tau(E) = q]$ for what she will believe of E at $t = \tau$, contemplates a range of mutually exclusive and exhaustive propositions $\{Q\}$ to which she assigns probabilities $P_0(Q)$. Then, straightforward synchronic coherence requires

$$P_0(E) = \sum_Q P_0(Q)P_0(E|Q) = \sum_q P_0(P_\tau(E) = q) P_0(E \mid P_\tau(E) = q) , \quad (15.9)$$

for which reflection in turn implies

$$P_0(E) = \sum_q P_0(P_\tau(E) = q) q . \quad (15.10)$$

This implication of the reflection principle will turn out to be particularly important for our exposition of quantum decoherence.

15.5   Bayesian Conditioning in Reflectional Terms

The reflection principle is a constraint on an agent's present beliefs about her future probability assignments. It does not directly provide an explicit rule for assigning probabilities, either for the present ones or the future ones. An agent whose probabilities violate the reflection principle is incoherent and should strive to remove this incoherence. The reflection principle does not provide a recipe for how to do this.

One way in which the agent can achieve coherence is by adopting an "updating strategy" [20] based on the Bayesian conditioning rule. We will now explore to what extent the Bayesian conditioning rule follows for such a strategy in a way analogous to the reflection principle—that is, in a way "concerned not with how the person should or will change his beliefs, but rather with what he believes about how these beliefs will change" [17].

Let E and D be events, and let $P_0(E)$, $P_0(D)$ and $P_0(D/E)$ denote the agent's respective probabilities at $t = 0$. Assume that the truth value of D will be revealed to the agent at $t = \tau$. Suppose she now adopts the strategy that, if at $t = \tau$ she learns that D is true, her updated probability of E, denoted by $P_\tau(E)$, will be given by some value q, $0 \leq q \leq 1$.

The above can be phrased in terms of the agent's probabilities at $t = 0$. For her to adopt this strategy simply means that she is certain that, if D turns out to be true, she will make the probability assignment $P_\tau(E) - q$, i.e.,

$$P_0(P_\tau(E) = q|D) = 1 \ . \tag{15.11}$$

This statement about the agent's current belief about her future probability captures the essence of Bayesian conditioning. Together with diachronic coherence it implies that

$$q = P_0(E|D) \ , \tag{15.12}$$

i.e., the Bayesian conditioning rule. Presented in this way, it can be regarded as a variant of the reflection principle, valid whenever the condition (15.11) holds.

To derive Eq. 15.12, we consider again the combinations of bets introduced in Sect. 15.4, but with the event D replacing Q throughout. We assume first that $P_0(E/D) > q$ and define $d = P_0(E/D) - q$. The transactions are

(i)   to buy, at $t = 0$, for \$$P_0(E/D)$, a ticket that pays \$1 if both D and E are true, and refunds the ticket price if D is false;
(ii) to buy, at $t = 0$, for \$$P_0(D)d/2$, a lottery ticket that pays \$$d/2$ if D is true;
(iii) if D is true, to sell, at $t = \tau$, for \$$q$, a lottery ticket that pays \$1 if E is true.

At $t = 0$, the agent is certain that these transactions result in a net loss, equal to \$$(P_0(D) + 1)d/2$ if D is true, and \$$P_0(D)d/2$ if D is false. At $t = 0$, the agent regards (i) and (ii) as fair transactions, and because of Eq. 15.11, she is certain that at $t = \tau$ she will regard (iii) as a fair transaction. We have thus shown that the agent's probabilities are incoherent if $P_0(E/D) > q$. The case $P_0(E/D) < q$ is similar. Thus coherence implies that $P_0(E/D) = q$, as required.

The key assumption in this derivation, expressed by Eq. 15.11, is that the agent can identify an event D that she expects to determine her future beliefs. There are more general updating strategies that are not of this form. Jeffrey's probability kinematics [21] is such an example. Probability kinematics is a coherent updating strategy [20] which does not make use of the Bayesian conditioning rule, but it too can be put in reflectional terms as we did with Bayesian conditioning.

Actually, one can go still further along these reflectional lines if one strengthens the assumption in Eq. 15.11 to also make a direct identification between the possible values for $P_\tau(E)$ and the D, i.e., that there is bijection between them. In such a case, one can say that Bayesian conditioning *follows* *directly* from the reflection principle. For then,

$$P_0(E|D) = P_0(E|D, Q) = P_0(E|Q) \tag{15.13}$$

by standard synchronic logic, and $P_0(E/Q) = q$ by reflection.

The discussion above is entirely in terms of the agent's beliefs at $t = 0$. What if, at $t = \tau$, after learning that D is true, the agent re-analyses the situation, possibly taking into account circumstances she was not aware of at $t = 0$, and concludes that her new probability, $P_\tau(E)$, differs from $P_0(E/D)$. Does this imply that her probability assignment is incoherent? The answer is no. Coherence is a condition about an agent's *current* *beliefs*, including her beliefs about her future probability assignments. In the above scenario, the agent's beliefs at $t = \tau$ are coherent as long as $0 \leq P_\tau(E) \leq 1$. Nothing in the Dutch book argument implies that the agent's actual probabilities at $t = \tau$ are constrained by her probabilities at $t = 0$, which supports the conclusion of Sect. 15.2 that there is no conflict with coherence for an experimenter who assigns $P_\tau(E) = 0.99$ although the Bayesian conditioning rule appears to mandate $P_\tau(E) \approx 1/2$.

15.6 Sirens, Car Keys, and Married Couples

We have seen in the previous section that one way of satisfying the reflection principle and thereby avoiding incoherence is to set your future probabilities in terms of your current probabilities via Bayesian conditioning. The form of the reflection principle, however, suggests a different way of proceeding. Since Eq. 15.8 expresses a constraint on a current probability, conditioned on a future probability assignment, one could take the future probability as given and set the current probability in terms of it, thus reversing the usual direction of Bayesian updating. This can be a useful and legitimate procedure. An important application will be given in Sect. 15.7 below.

If the reflection principle is taken as a rule to set future probabilities in terms of current probabilities, it can lead to decisions that appear irrational [22–25]. A classic example [26] is provided by the story of Ulysses and the Sirens. Ulysses knows that tomorrow, as soon as he is within earshot of the Sirens, he will make a catastrophic decision. Does the reflection principle force him to endorse this catastrophic decision today?

Since analysing this story would involve a discussion of utility, here is another famous example [23]. I know that I will get drunk this evening and that I will assign probability 10^{-6} to the event E of my causing an accident while driving home late at night. Does reflection imply that I must assign probability 10^{-6} to the event E now?

Examples like this have led, e.g., Christensen [23] to the conclusion that the reflection principle is unsound. This conclusion stems from a confusion about the role of coherence arguments, however. The reflection principle can be regarded as a tool to detect incoherence. The Dutch book arguments show that incoherent probability assignments have the potential to lead to catastrophic consequences. This justifies accepting a normative rule that an agent should adhere to the reflection principle in order to avoid incoherence. The reflection principle does not, however, give a prescription for setting probabilities, either today's in terms of tomorrow's or vice versa. There is a range of options for the agent once she has detected an incoherence, as we will now illustrate.

Suppose that, in the example above, my initial conditional probability for an accident if I drive home drunk is 0.01. Suppose further that I am certain that I will get drunk, and that my probability for an accident will then be 10^{-6}. These probabilities violate the reflection principle. My probability assignments are therefore incoherent. One way of avoiding this incoherence would be to decide not to get drunk, which would mean assigning probability 0 to this event and therefore restore coherence. There is another very practical solution, which is to give my car keys to a trusted friend before I start drinking. My probability assignments will still be incoherent, but I will be unable to act on them.

Ulysses's solution, 3,000 years ago, was very similar. He ordered his men to chain him to the mast of his ship. His men were to plug their ears. He accepted incoherence, but prevented himself from acting on his incoherent probability assignments. His men achieved coherence by reducing the probability of hearing the Sirens to zero. Coherence is an ideal one should always strive for. Incoherent probability assignments have the potential to lead to catastrophic consequences. If one can't achieve coherence, one should give up the car keys, plug one's ears or chain oneself to a mast.

In his article contra reflection, Christensen [23] pointed out that the reflection principle is very similar to a related principle which he called *solidarity*. Consider husband and wife who share a bank account. Denote the husband's probabilities by P_h and the wife's by P_w, and consider some event, E. Solidarity is the principle that, given that the wife's probability of E is q, the husband's probability must also be q, i.e.,

$$P_h(E|P_w(E) = q) = q \, . \tag{15.14}$$

Violating solidarity leads to a sure loss for the joint bank account exactly like in the diachronic Dutch book argument for the reflection principle.

Christensen argued that the solidarity principle is clearly absurd. This may be a case of confusion between normative and descriptive rules. The solidarity principle is a normative principle and does not claim that actual agents' probability assignments are always compatible with it. What it says instead is that, to avoid potential catastrophic consequences for their common bank account, husband and wife must strive for coherence. The solidarity principle, or more generally, the reflection principle, provides a tool to detect incoherence. It is then up to the agents how to resolve the incoherence. The husband might give, the wife might give, or they might compromise after debating all the relevant issues. The key point is that deliberation is to their mutual benefit, and coherence is their goal.

15.7 Quantum Decoherence

In the last section, we described how the reflection principle can be used to detect incoherence and thus to avoid catastrophic consequences. In this section, we will see that there is a generic situation in quantum theory where the reflection principle is used directly to set today's probabilities in terms of tomorrow's.

We will look at a standard quantum measurement situation [27] from the perspective of Quantum Bayesianism, according to which all quantum states, pure or mixed, represent an agent's degrees of belief about future measurement outcomes. Assume an agent has, at time $t = 0$, assigned a quantum state (i.e., density operator) ρ_0 to a quantum system. She intends to perform two measurements on the system, the first one at time $t = \tau > 0$, the second one at a still later time $t = \tau'$. She describes the first measurement by a collection of trace-decreasing completely positive maps $\{\mathcal{F}_i\}$, each corresponding to a potential outcome, i, for the first measurement. These completely positive maps determine the agent's probabilities $P_0(i) = \text{tr}[\mathcal{F}_i(\rho_0)]$, at time $t = 0$, for the outcomes i, but they also determine the states she will assign to the system after the measurement: If outcome i, then $\rho_\tau = P_0(i)^{-1} \mathcal{F}_i(\rho_0)$.

To describe the second measurement, it is enough to use a POVM, i.e., positive operator valued measure, $\{E_j\}$, since we will not be considering any further measurements after it. In this description each positive operator E_j corresponds to a potential outcome, j, for the second measurement. If ρ_τ is the agent's system state at time $t = \tau$, then her probabilities, at $t = \tau$, for the outcomes j are given by $P_\tau(j) = \text{tr}(E_j \rho_\tau)$.

Now suppose our agent is confronted at time $t = 0$ with a bet concerning the outcome j at $t = \tau'$. How should she gamble without having yet performed the measurement at $t = \tau$? We can read the answer straight off the reflection principle as written in the form of Goldstein and Shafer, Eq. 15.10—remember here that $P_\tau(j)$ is implicitly dependent upon i:

$$P_0(j) = \sum_i P_0(i)P_\tau(j) = \sum_i \text{tr}[E_j \mathcal{F}_i(\rho_0)] . \tag{15.15}$$

Cleaning this up a bit, we can write:

$$P_0(j) = \text{tr}[E_j \sum_i \mathcal{F}_i(\rho_0)] = \text{tr}(E_j \rho'_0) , \tag{15.16}$$

where

$$\rho'_0 = \sum_i \mathcal{F}_i (\rho_0). \tag{15.17}$$

What we have shown here is that the reflection principle entails that the agent can obtain her probabilities at $t = 0$ for the outcomes of the second measurement from the density operator ρ'_0. The state ρ'_0, which has the form of a "decohered" state, *is* the agent's quantum state at $t = 0$ *as far as the second measurement is concerned.*

These conclusions are valid for any pair of measurements, but a little more can be said if the POVM $\{E_j\}$ is informationally complete, i.e., if the state ρ_τ is fully determined by the probabilities $P_\tau (j)$. In this case ρ'_0 as defined in Eq. 15.17 is the only density operator that gives rise to the probabilities $P_0 (j)$ required by the reflection principle.

Equation 15.17 takes a perhaps more familiar form if the first measurement is a von Neumann measurement and the updating is given by the Lüders rule. In this case the action of the maps $\mathcal{F}_i$ on the state ρ_0 can be written as $\mathcal{F}_i (\rho_0) = \Pi_i \rho_0 \Pi_i$, where the Π_i are projection operators, and Eq. 15.17 becomes

$$\rho'_0 = \sum_i \Pi_i \rho_0 \Pi_i. \tag{15.18}$$

A common attitude about quantum measurement is that it is something that demands a detailed physical explanation. Much of the folklore since the publication of John von Neumann's 1932 book *Mathematical Foundations of Quantum Mechanics* is that a quantum measurement is something that occurs in two steps: First, there is a kind of "pre-measurement" where the quantum system becomes entangled with a measuring device. Secondly, there is a "selection" of one of the entangled state's components; this is what singles out a particular measurement result.

The trouble with this description, however, is that the entangled wave function, with its freedom to be expressed in any bipartite basis, does not have enough structure to specify how it should be decomposed so that such a "selection" can be effected. The theory of quantum decoherence, developed by Zeh, Zurek, and others [28], attempts to overcome this deficiency in the von Neumann story by supplementing it with a further story of interaction between the measuring device and an environment: The idea is that the specific form of the interaction with the environment specifies how the joint state of system plus device ought to be decomposed. In this picture, the decoherence process preceding the "selection" step leads to a state of the form (15.18), or more generally, (15.17). What remains mysterious in this picture, however, is "the selection step" itself. Decoherence theorists usually leave that question aside, implicitly endorsing one variety or another of an Everettian interpretation of quantum mechanics.

In contrast, the Quantum Bayesian view of quantum theory leaves most of the usual von Neumann story aside: Instead of taking quantum states and unitary evolution as the ontic elements to which the theory refers, it takes the idea of an individual agent's decisions and experience as the theory's real subject matter. In this view, the process called "quantum measurement" is nothing other than an agent acting upon the world and experiencing the consequences of her actions. For a Quantum Bayesian, the only *physical* process in a quantum measurement is what was previously seen as "the selection step"—i.e., the agent's action on the external world and its unpredictable consequence for her, the data that leads to a new state of belief about the system.

Thus, it would seem there is no foundational place for decoherence in the Quantum Bayesian program. And this is true. Nonetheless, in the two-time measurement scenario we described above, there is a coherent state assignment at time $t = 0$ for the second measurement that mimics a belief in decoherence. This is simply a consequence of the implications of the reflection principle. The "decohered" state ρ'_0 is not the agent's state after she has made the first measurement (that would have been *one* of the ρ_τ depending upon the i found). It is not the state resulting from the measurement interaction before the selection step takes place as the decoherence program would have it (nothing is so intricately modeled here). It is simply a quantum state the agent uses at time $t = 0$ before the first measurement to make decisions regarding the outcomes of the second measurement.

That is the story of decoherence from the Quantum Bayesian perspective. Decoherence does not come conceptually before a "selection," but rather is predicated on a time $t = 0$ belief regarding the possibilities for the next quantum state at time $t = \tau$. Decoherence comes conceptually *after* the recognition of the future possibilities. In this sense the decoherence program of Zeh and Zurek [28], regarded as an attempt to contribute to our understanding of quantum measurement, has the story exactly backward.

Acknowledgements We thank Lucien Hardy for persisting that the example in Sect. 15.2 *should be* important to us. We thank Matthew Leifer for bringing the work of Goldstein [18] to our attention, which derivatively (and slowly) led us to an appreciation of van Fraassen's reflection principle [16]; if we were quicker thinkers, this paper could have been written 6 years ago. This work was supported in part by the U. S. Office of Naval Research (Grant No. N00014-09-1-0247).

References

1. Pitowsky, I.: Betting on the outcomes of measurements: a Bayesian theory of quantum probability. Stud. Hist. Philos. Mod. Phys. **34**, 395 (2003)
2. Pitowsky, I.: Quantum mechanics as a theory of probability. In: Demopoulos, W., Pitowsky, I. (eds.) Physical Theory and Its Interpretation: Essays in Honor of Jeffrey Bub, pp. 213–240. Springer, Berlin (2006)

3. Bub, J., Pitowsky, I.: Two dogmas about quantum mechanics. In: Saunders, S., Barrett, J., Kent, A., Wallace, D. (eds.) Many Worlds? Everett, Quantum Theory, and Reality, pp. 433–459. Oxford University Press, Oxford (2010)
4. Bernardo, J.M., Smith, A.F.M.: Bayesian Theory. Wiley, Chichester (1994)
5. Ramsey, F.P.: Truth and probability. In: Braithwaite, R.B. (ed.) The Foundations of Mathematics and other Logical Essays, pp. 156–198. Harcourt, Brace and Company, New York (1931)
6. de Finetti, B.: Probabilismo. Logos **14**, 163 (1931); transl., Probabilism. Erkenntnis **31**, 169 (1989)
7. Savage, L.J.: The Foundations of Statistics. Wiley, New York (1954)
8. de Finetti, B.: Theory of Probability, vols. 1 and 2. Wiley, New York (1990)
9. Jeffrey, R.: Subjective Probability. The Real Thing. Cambridge University Press, Cambridge (2004)
10. Logue, J.: Projective Probability. Clarendon, Oxford (1995)
11. Fuchs, C.A., Schack, R.: Quantum-Bayesian coherence. arXiv:0906.2187v1, (2009)
12. Fuchs, C.A.: QBism, the perimeter of quantum Bayesianism. arXiv:1003. 5209v1, (2010)
13. Hacking, I.: Slightly more realistic personal probability. Philos. Sci. **34**, 311 (1967)
14. Howson, C., Urbach, P.: Scientific Reasoning: The Bayesian Approach, 2nd edn. Open Court Publishing, La Salle (1993)
15. Teller, P.: Conditionalization and observation. Synthese **26**, 218 (1973)
16. van Fraassen, B.C.: Belief and the will. J. Philos. **81**, 235 (1984)
17. Shafer, G.: A subjective interpretation of conditional probability. J. Philos. Log. **12**, 453 (1983)
18. Goldstein, M.: The prevision of a prevision. J. Am. Stat. Assoc. **78**, 817 (1983)
19. Caves, C.M., Fuchs, C.A., Schack, R.: Unknown quantum states: the quantum de Finetti representation. J. Math. Phys. **43**, 4537 (2002)
20. Skyrms, B.: Dynamic coherence and probability kinematics. Philos. Sci. **54**, 1 (1987)
21. Jeffrey, R.: The Logic of Decision. McGraw Hill, New York (1965)
22. Maher, P.: Diachronic rationality. Philos. Sci. **59**, 120 (1992)
23. Christensen, D.: Clever bookies and coherent beliefs. Philos. Rev. **100**, 229 (1991)
24. Skyrms, B.: A mistake in dynamic coherence arguments? Philos. Sci. **60**, 320 (1993)
25. Green, M.S., Hitchcock, C.R.: Reflections on reflection: Van Fraassen on belief. Synthese **98**, 297 (1994)
26. van Fraassen, B.C.: Belief and the problem of Ulysses and the Sirens. Philos. Stud. **77**, 7 (1995)
27. Nielsen, M.A., Chuang, I.L.: Quantum Computation and Quantum Information, 10th edn. Cambridge University Press, Cambridge (2010)
28. Schlosshauer, M.: Decoherence and the Quantum-to-Classical Transition. Springer, Berlin (2007)

Chapter 16
The World According to de Finetti: On de Finetti's Theory of Probability and Its Application to Quantum Mechanics

Joseph Berkovitz

Abstract Bruno de Finetti is one of the founding fathers of the subjectivist school of probability, where probabilities are interpreted as rational degrees of belief. His work on the relation between the theorems of probability and rationality is among the corner stones of modern subjective probability theory. De Finetti maintained that rationality requires that degrees of belief be coherent, and he argued that the whole of probability theory could be derived from these coherence conditions. De Finetti's interpretation of probability has been highly influential in science. This paper focuses on the application of this interpretation to quantum mechanics. We argue that de Finetti held that the coherence conditions of degrees of belief in events depend on their verifiability. Accordingly, the standard coherence conditions of degrees of belief that are familiar from the literature on subjective probability only apply to degrees of belief in events which could (in principle) be jointly verified; and the coherence conditions of degrees of belief in events that cannot be jointly verified are weaker. While the most obvious explanation of de Finetti's verificationism is the influence of positivism, we argue that it could be motivated by the radical subjectivist and instrumental nature of probability in his interpretation; for as it turns out, in this interpretation it is difficult to make sense of the idea of coherent degrees of belief in, and accordingly probabilities of unverifiable events. We then consider the application of this interpretation to quantum mechanics, concentrating on the Einstein-Podolsky-Rosen experiment and Bell's theorem.

J. Berkovitz (✉)
Institute of History and Philosophy of Science and Technology, University of Toronto, Toronto, ON, Canada
e-mail: joseph.berkovitz@utoronto.ca

Y. Ben-Menahem and M. Hemmo (eds.), *Probability in Physics*, The Frontiers Collection,   249
DOI 10.1007/978-3-642-21329-8_16, © Springer-Verlag Berlin Heidelberg 2012

16.1 The Background and Motivation

The foundations of this paper were laid in 1988/1989, when I worked on a seminar paper for Itamar Pitowsky's course in the philosophy of probability.[1] The question that motivated the paper was whether subjective probability, and more specifically de Finetti's subjectivist interpretation, could successfully be applied in quantum mechanics (QM). This question, which was raised by Itamar, may seem a bit anachronistic now that the subjective interpretation of quantum probabilities is gaining popularity. But back then this interpretation was undeveloped.[2]

In de Finetti's interpretation, probabilities have no objective reality. They are the expressions of the uncertainties of individuals. Itamar's question was not whether such a radical subjective interpretation could constitute an adequate interpretation of probabilities in quantum mechanics. Rather, it was the question whether de Finetti's interpretation could be reconciled with the apparent non-classical character of these probabilities. We explain this concern in Sect. 16.1.3, and discuss it in more detail in Sect. 16.2. To prepare the ground for this discussion, we now turn to present Bell's theorem and two different interpretations of it. In Sects. 16.3 and 16.4, we introduce the main ideas of de Finetti's theory of probability, and in Sects. 16.5–16.7 we discuss the application of this theory to the quantum realm.

16.1.1 Bell's Theorem and Its Common Interpretation

Recall the Einstein-Podolsky-Rosen (EPR) experiment. Pairs of particles are emitted from the source in opposite directions. When the particles are spacelike separated, they each encounter a measurement apparatus that can measure their position or momentum. The distant measurement outcomes are curiously correlated. Einstein et al. [6] thought that this kind of correlation reflects the incompleteness of QM rather than non-local influences. They argued that the QM state-description is incomplete, and they believed that a more complete description would render the distant measurement outcomes probabilistically independent. The idea is that the correlations between such distant outcomes could be explained away by a local common cause: the complete pair-state at the emission. Given this state, the joint probability of the outcomes would factorize into their single probabilities (see Factorizability below), and so the correlations between them would not entail the existence of non-locality.

---

[1] See Berkovitz [1, 2].

[2] For applications of the subjective interpretation to QM, see for example, Caves, Fuchs and Schack [3–5] and Pitowsky [59]. While these applications appeal to de Finetti's subjective theory of probability, both the interpretation of de Finetti and the focus of its application are substantially different from the ones offered below.

In his celebrated theorem, Bell [7–11] considers models of the kind EPR may have had in mind, but he focuses on Bohm's [12] version of the experiment (henceforth, the EPR/B experiment), where the measured quantities are spins in various directions. These models postulate the existence of "hidden variables" that are supposed to constitute a (more) complete pair's state, and this state is supposed to determine the measurement outcomes or their probabilities in a perfectly local way. Bell's theorem demonstrates that such models are committed to certain inequalities concerning the probabilities of measurement outcomes, the so-called "Bell's inequalities," which are violated by the predictions of QM and (granted very plausible assumptions) actual experimental results. In Clauser and Horne's [13] version, the inequalities are concerned with the probabilities of measurement outcomes of spins in two different directions in each wing of the EPR/B experiment, henceforth the "Bell/CH inequalities" (for details, see Sect. 16.2).

The common view is that Bell's theorem demonstrates that local hidden-variables models cannot reproduce the predictions of QM [7–11, 13–16]. On this view, the derivation of Bell/CH inequalities involves the following premises.

(i) The distribution of the complete pair-state is determined by the QM pair-state, and is independent of the settings of the measurement apparatuses. That is, for any QM pair-state ψ, complete pair-states λ, and setting of the L- and R-apparatus to measure spins in the directions x and y, respectively, we have:

$$(\lambda - \text{independence}) \quad \rho_{\psi xy}(\lambda) = \rho_{\psi}(\lambda);$$

where $\rho_{\psi}(\lambda)$ and $\rho_{\psi xy}(\lambda)$ are the probability distributions of λ given ψ and given $\psi \& x \& y$, respectively. Note that in our notation for conditional probabilities, we place the conditioning events in the subscript rather than after the conditionalization stroke. Unlike Kolmogorov's [17] axiomatization, in this approach conditional probability is not defined as a ratio of unconditional probabilities. Rather, conditional probability may be thought of as a conditional, which does not necessarily entail the corresponding conditional probability *a la* Kolmogorov (for more details, see Sect. 16.3.7). In this concept of conditional probability, the conditioning events ψ and $\psi \& x \& y$ are not part of the probability spaces referred by $\rho_{\psi}(\ )$ and $\rho_{\psi xy}(\ )$, respectively. To highlight this fact, we place them in the subscripts. As we shall see later, this alternative concept of conditional probability is in line with de Finetti's theory of probability. Arguably, it is also a more appropriate representation of the basic idea of conditional probability in other interpretations of probability [18, 19]. Yet, while this representation is important for pedagogical reasons, it is not essential for our analysis of Bell's theorem and the feasibility of interpreting probabilities in the quantum realm along de Finetti's theory.

(ii) For each complete pair-state λ and apparatus settings x and y, the model prescribes probabilities of single and joint measurement outcomes: $P_{\lambda x}(O_x)$, $P_{\lambda y}(O_y)$ and $P_{\lambda xy}(O_x \& O_y)$, where O_x is the outcome "up" in x-spin measurement on the L-particle; and similarly, *mutatis mutandis*, for the outcome O_y in the R-wing.

(iii) The joint probability of distant outcomes given the complete pair-state and apparatus settings factorizes into the single probabilities of the outcomes. The idea here is that the correlation between the distant outcomes are explained by a common cause, i.e. the complete pair-state, so that conditionalization on the common cause renders the outcomes probabilistically independent. More precisely, for any λ, x, y, O_x and O_y:

$$\text{(Factorizability)} \quad P_{\lambda xy}(O_x \& O_y) = P_{\lambda x}(O_x) \cdot P_{\lambda y}(O_y).$$

(iv) The QM probabilities of outcomes are reproduced as statistical averages over the model probabilities of outcomes – namely, as sum-averages over the model probabilities according to the distribution of the complete pair-state. That is, granted λ-*independence*, for any ψ, x and y, we have:

$$P_{\psi x}(O_x) = \int_{\lambda} P_{\lambda x}(O_x)\, d\rho(\lambda),\ P_{\psi y}(O_y) = \int_{\lambda} P_{\lambda y}(O_y)\, d\rho(\lambda),\ P_{\psi xy}(O_x \& O_y)$$
$$= \int_{\lambda} P_{\lambda xy}(O_x \& O_y)\, d\rho(\lambda).$$

Bell's theorem demonstrates that in any model that satisfies (i)-(iv), the probabilities of measurement outcomes in the EPR/B experiment are constrained by the Bell/CH inequalities (see Sect. 16.2). Thus, granted the plausibility of λ-*independence* and the overwhelming evidence for the empirical adequacy of QM (in its intended domain of application), the consensus has it that *Factorizability* fails in this experiment. The failure of this condition is commonly thought of as indicating some type of non-locality (for a recent review of quantum non-locality, see [20] and references therein).

16.1.2   Fine's Interpretation of Bell's Theorem

Following Bell [10], the above analysis of Bell's theorem relies on a principle of causal inference which is similar to Reichenbach's [21] principle of the common cause. That is, it is assumed that non-accidental correlations have causal explanation, and the kind of explanation is as spelled out in (iii) and (iv) above. While this kind of inference is common, there are dissenting views. Fine [22–24] denies that non-accidental correlations must have causal explanation, and he argues that the correlations between the distant measurement outcomes in the EPR/B experiment do not call for causal explanation; and Cartwright [25, Chaps. 3 and 6] and Chang and Cartwright [26] challenge the assumption that common causes must render their joint effects probabilistically independent.

More important to our consideration, Fine [27, p. 294] argues that

(F) What hidden variables and the Bell/CH inequalities are all about are the requirements that make "well defined precisely those probability distributions for non-commuting observables whose rejection is the very essence of quantum mechanics."

The idea is that Bell's theorem focuses on models that presuppose the existence of joint probability over non-commuting spin observables in the EPR/B experiment – a distribution that does not exist according to standard QM. In more detail, Fine [27] argues that:

I. (Corresponding to "Proposition 1") "The existence of a deterministic hidden-variables model is strictly equivalent to the existence of a joint distribution probability function $P(AA'BB')$ for the four observables of the experiment, *one that returns the probabilities of the experiment as marginals.*" [27, p. 291]
II. ("Proposition 2") "Necessary and also sufficient for the existence of a deterministic hidden-variables model is that Bell/CH inequalities hold for the probabilities of the experiment." [27, p. 293]
III. ("Proposition 3") "There exists a factorizable stochastic hidden-variables model for a correlation experiment if and only if there exists a deterministic hidden-variables model for the experiment." [27, p. 293]

Fine believes that (I)–(III) entails (F), and this suggests that the common interpretation of Bell's theorem – namely, that (granted the very plausible assumption of λ-*independence*) the violation of Bell/CH inequalities entails quantum non-locality – is misguided.

16.1.3 Subjective Probability, Joint Distributions and Verifiability

De Finetti held that for degrees of belief to be coherent they have to be probabilities, i.e. they have to satisfy the probability axioms. It is commonly presupposed, albeit implicitly, that a person's coherent degrees of belief concerning all the propositions she considers are to be represented by a joint probability distribution, which returns these degrees of belief as marginals; for notable examples, see Lewis's [28] "A Subjectivist's Guide to Objective Chance" and Carnap's [29] "On Inductive Logic." If the subjectivist interpretation were committed to such an assumption, and the view that the Bell/CH inequalities follow from the assumption of a joint distribution over non-commuting observables in the EPR/B experiment were correct, followers of this interpretation would be bound to have probabilities that are constrained by Bell/CH inequalities, and accordingly incompatible with the predictions of QM.

Indeed, followers of the subjectivist interpretation may agree with Fine's analysis of Bell's theorem, yet reject the view that a person's degrees of belief are to be represented by a single probability distribution. The question is whether they have non-*ad hoc* reasons to reject this view. Based on a reconstruction of de Finetti's probability theory in Sects. 16.3 and 16.4, we shall argue in Sect. 16.5 that followers of de Finetti have such reasons in the context of the EPR/B experiment and Bell's theorem. That is, we shall argue in Sect. 16.4 that de Finetti's notion of coherent degrees of belief embodies a certain verifiability condition. Consequently: (a) Degrees of belief in events that are not verifiable have no definite coherence conditions, and accordingly have no probability. (b) There are no joint probability distributions over events that are not jointly verifiable. (c) The coherence conditions of degrees of belief in events that are not jointly verifiable are weaker than they would have been had the events been jointly verifiable. Thus, the coherence conditions of degrees of belief in events that are not jointly verifiable are weaker than the familiar coherence conditions discussed in the literature on subjective probability. Accordingly, the inequalities that constrain the probabilities of such events are weaker than those that constrain the probabilities of events that are jointly verifiable.

In Sects. 16.5 and 16.7, we shall consider the implications of these consequences for the structure of probabilities in models of the EPR/B experiment in which probabilities are interpreted along de Finetti's theory of probability. These sections reflect the implications of de Finetti's theory, as reconstructed in Sects. 16.3 and 16.4. De Finetti himself struggled to understand the nature of the QM probabilities and their relation to "classical" probabilities. In Sect. 16.6, we shall briefly look at de Finetti's own analysis of the QM probabilities. But first we turn to present the Bell/ CH inequalities and to consider Fine's claim that these inequalities follow from, and are equivalent to the assumption of a joint distribution over non-commuting observables in the EPR/B experiment.

16.2 Joint Distributions, Probabilistic Inequalities and Bell's Theorem

The term "Bell/CH inequalities" is ambiguous. It refers to different kinds of inequalities. The first kind is a theorem of probability theory:

(Bell/CH − prob)
$$- 1 \leq P_\lambda(X\&Y) + P_\lambda(X'\&Y) + P_\lambda(X\&Y') - P_\lambda(X'\&Y') - P_\lambda(X) - P_\lambda(Y) \leq 0.$$

Indeed, this inequality obtains for any joint probability distribution over any four events X, X', Y, Y' (or propositions about them). In the context of the hidden-variables models of the EPR/B experiment, it is natural to think about λ as the complete pair-state, and X (Y) and X' (Y') as referring to spin properties of the particles, or properties that determine their dispositions to spin in measurements. For example, X (Y) may be the event of the L- (R-) particle spinning "up" in the direction x (y), or some other property that determines the disposition of the L- (R-) particle to spin "up" along the direction x (y) in a spin measurement along this direction.

The second and third kinds of Bell/CH inequalities are not theorems of probability theory:

(Bell/CH - phys - λ) $\quad -1 \leq P_{\lambda xy}(O_x \& O_y) + P_{\lambda x'y}(O_{x'} \& O_y) + P_{\lambda xy'}(O_x \& O_{y'})$
$$-P_{\lambda x'y'}(O_{x'} \& O_{y'}) - P_{\lambda x}(O_x) - P_{\lambda y}(O_y) \leq 0$$

(Bell/CH - phys - ψ) $\quad -1 \leq P_{\psi xy}(O_x \& O_y) + P_{\psi x'y}(O_{x'} \& O_y) + P_{\psi xy'}(O_x \& O_{y'})$
$$-P_{\psi x'y'}(O_{x'} \& O_{y'}) - P_{\psi x}(O_x) - P_{\psi y}(O_y) \leq 0$$

where, as before, ψ is the QM pair-state, x (y) is the setting of the L- (R-) apparatus to measure spin in the direction x (y), and O_x (O_y) is the outcome "up" in x- (y-) spin measurement on the L- (R-) particle; and similarly, *mutatis mutandis*, for x' (y') and $O_{x'}$ ($O_{y'}$). (Bell/CH – physics – λ) is an inequality of probabilities of the hidden-variables model, whereas (Bell/CH – physics – ψ) is an inequality of QM probabilities. The latter inequality is derived from the former by integrating over all the complete pair-states λ while assuming λ-independence.

In (Bell/CH – prob) all the probabilities belong to the same probability space, whereas in (Bell/CH – phys – λ) and (Bell/CH – phys – ψ) each of the probabilities belongs to a different probability space. This should be clear from the fact that each of the probabilities in these latter inequalities has a different subscript. Thus, unlike the former inequality, these inequalities cannot be derived purely on the basis of considerations of coherence or consistency.

Indeed, (Bell/CH – phys – λ) and (Bell/CH – phys – ψ) are sometimes represented in terms of conditional probabilities *a la* Kolmogorov with the conditioning events placed after the conditionalization stroke rather than in the subscript, where in each inequality all the probabilities are embedded in one "big" probability space:

(Bell/CH –phys – λ – big)
$$-1 \leq P(O_x \& O_y / \lambda \& x \& y) + P(O_{x'} \& O_y / \lambda \& x' \& y) + P(O_x \& O_{y'} / \lambda \& x \& y')$$
$$-P(O_{x'} \& O_{y'} / \lambda \& x' \& y') - P(O_x / \lambda \& x) - P(O_y / \lambda \& y) \leq 0$$

(Bell/CH – phys – ψ – big)
$$-1 \leq P(O_x \& O_y / \psi \& x \& y) + P(O_{x'} \& O_y / \psi \& x' \& y) + P(O_x \& O_{y'} / \psi \& x \& y')$$
$$-P(O_{x'} \& O_{y'} / \psi \& x' \& y') - P(O_x / \psi \& x) - P(O_y / \psi \& y) \leq 0.$$

Yet, these inequalities are not theorems of probability theory. Unlike (Bell/CH – prob), they cannot be derived from the assumption of a joint distribution over the measurement outcomes, the (QM or complete) pair-state and apparatus settings.

We shall discuss the relationships between (Bell/CH – prob) and (Bell/CH – phys – λ) below and in Sect. 16.3.7.

In hidden-variables models of the EPR/B experiment that postulate the existence of definite values for all the four spin quantities that are involved in the Bell/CH inequalities, it is natural (though not necessary) to suppose a joint probability over these probabilities.[3] Thus, in such models, it is plausible to expect (Bell/CH – prob). But (Bell/CH – prob) is neither necessary nor sufficient for (Bell/CH – phys – ψ) or (Bell/CH – phys – ψ – big). Indeed, unless we make some assumptions about the relationships between the probabilities of the spin quantities in (Bell/CH – prob) and the probabilities of their measurement outcomes, the assumption of joint probability over these quantities will do little to constrain the probabilities of their measurement outcomes. Two natural assumptions are λ-independence and the assumption that the probabilities of spin-measurement outcomes "mirror" the probabilities that the particles' spins have before the measurements: for any spin properties X and Y, apparatus settings x and y to measure these properties, and the corresponding measurement outcomes O_x and O_y,

$$\text{(Mirror)} \quad P_{\lambda x}(O_x) = P_\lambda(X), \quad P_{\lambda y}(O_y) = P_\lambda(Y), \quad P_{\lambda xy}(O_x \& O_y) = P_\lambda(X \& Y).$$

Although these assumptions may seem natural, models of the EPR/B experiment that postulate joint probability over the values of the particles' spin in various directions violate at least one of these assumptions; and their violation bears directly on the question whether the quantum realm involves some kind of non-locality. λ-independence fails in models of the experiment that postulate retro-causal influences from the measurement events to the source at the emission, so that the distribution of the complete pair-state depends on the measured quantities (for recent discussions of such models, see [31–35], and references therein). In such models, the QM statistics may be accounted for by such retro-causal influences rather than non-locality.

Mirror may be violated in various "hidden-variables" theories. For example, it is violated in Bohmian mechanics, if X and X' (Y and Y') are respectively the positions of the L- (R-) particle relative to planes aligned along the directions x and x' (y and y') at the emission. In Bell's [36] "minimal" Bohmian mechanics spins are not intrinsic properties of the particles, and the positions of the particles at the emission influence their spin dispositions, i.e. their behavior in spin measurements: X (Y) determines the spin disposition of the L- (R-) particle in the direction x (y) in a measurement of spin x (y), if the L- (R-) measurement occurs first; and similarly for X' (Y') and $x'(y')$. Yet, due to non-local influences, the distribution of these dispositions is different from the distribution of the outcomes of the corresponding spin measurements. If, for example, at the emission both particles are disposed to spin "up" in a z-spin measurement, and the L-measurement occurs first, this

---

[3] Svetlichny et al. [30] argue that if probabilities are interpreted as infinitely long-run frequencies in random sequences, such a joint probability distribution need not exist.

measurement will change the z-spin disposition of the R-particle: after the L-measurement, it will be disposed to spin "down" on z-spin measurement ([36, 37], Chap. 7, [20], Sect. 5.3.1).

While the joint distribution over the spin quantities of the particle-pair in the EPR/B experiment (the "hidden variables") is neither necessary nor sufficient condition for (Bell/CH – physics – ψ) or (Bell/CH – physics – ψ – big), the question arises whether some other joint distributions are. The most comprehensive, relevant joint probability distribution in the context of these inequalities is a distribution over the QM and complete pair-state, the various relevant apparatus settings and the corresponding measurement outcomes,[4] and such distribution is neither necessary nor sufficient for these inequalities. (Bell/CH – phys – ψ) follows from *Factorizability* and λ-*independence* [13],[5] and as it is not difficult to see these conditions do not presuppose a joint distribution over the pair-state, apparatus settings and measurement outcomes. Similarly, (Bell/CH – phys – ψ – big) follows from factorizability and λ-independence expressed in terms of conditional probabilities *a la* Kolmogorov – for any QM and complete pair-states, λ and ψ, apparatus settings x and y to measure the particles' spins along the directions x and y, and the corresponding measurement outcomes O_x and O_y,

(Factorizability*) $P(O_x \& O_y / \lambda \& x \& y) = P(O_x / \lambda \& x) \cdot P(O_y / \lambda \& y)$

(λ - independence*) $\rho(\lambda / \psi \& x \& y) = \rho(\psi)$

– and these conditions do not presuppose such a joint distribution. Indeed, each particular case of *Factorizability** presupposes a joint distribution over the complete pair-state, two measurement outcomes (O_x and O_y) and two apparatus settings (x and y), and each particular case of λ-*independence** presupposes a distribution over the QM and complete pair-state and two apparatus settings. But these conditions do not presuppose a joint distribution over the QM and the complete pair-state and all the four apparatus settings and four corresponding measurement outcomes that are involved in (Bell/CH – phys – ψ – big). Thus, a joint probability over the QM and complete pair-state, apparatus settings and measurement outcomes is not a necessary condition for (Bell/CH – phys – ψ – big). It is also

---

[4] In fact, one may also add to this list the complete states (the "hidden variables") of the apparatus settings. While such a distribution will be even more comprehensive, it will not change the conclusion of the analysis below.

[5] The derivation of (Bell/CH – phys – ψ) from *Factorizability* and λ-*independence* is straightforward. $-1 \leq a \cdot b + a' \cdot b + a \cdot b' - a' \cdot b' - a - b \leq 0$ obtains for any real numbers $0 \leq a, a', b, b' \leq 1$. Substituting $a = P_{\lambda x}(O_x)$, $a' = P_{\lambda x'}(O_{x'})$, $b = P_{\lambda y}(O_y)$, $b' = P_{\lambda y'}(O_{y'})$ and applying *Factorizability* we have (Bell/CH – phys – λ), and integrating over λ while assuming λ-*independence* we obtain (Bell/CH – physics – ψ).

not sufficient for (Bell/CH – phys – ψ – big), as it is easy to construct such a distribution that violates the inequality.[6]

Fine [27] discusses a fourth kind of Bell/CH inequality, where the probabilities are supposed to be "the observed distributions for each of the four observables involved in the EPR/B experiment plus the joint observed distributions for each of the four compatible pairs" of these observables. (Fine [27], p. 291)

(Bell/CH – Fine)

$$-1 \leq P(O_x \& O_y) + P(O_{x'} \& O_y) + P(O_x \& O_{y'})$$
$$-P(O_{x'} \& O_{y'}) - P(O_x) - P(O_y) \leq 0;$$

where, presumably, P is a probability function that depends on the QM pair-state ψ. (Bell/CH - Fine) follows from the assumption of a joint probability over the measurement outcomes. The question is what could motivate such an assumption. Surely, the probabilities in this inequality need to depend on the apparatus settings, so that they either belong to different spaces (each characterized by different apparatus settings), as in (Bell/CH – phys - ψ), or are in the same probability space but are conditional on the QM pair-state and apparatus settings, as in (Bell/CH – phys - ψ - big). In the first case, the motivation for (Bell/CH – Fine) should probably include assumptions like *Mirror* and *λ-independence*, and as we have seen the violation of such assumptions is relevant to the question whether the quantum realm involves non-locality. In the second case, one may assume a joint distribution for the QM pair-state, apparatus settings and measurement outcomes, but as we argued above such a distribution would not entail (Bell/CH – phys - ψ - big). So in either case, (Bell/CH – Fine) has to be motivated by assumptions about the physical nature of the systems involved in the EPR/B experiment – in particular, assumptions about the state of the particles at the source, the causal relations between this state and the state of the measurement apparatuses during the measurements, and the causal relations between measurements in the two distant wings of the experiment. And granted such assumptions, the violation of (Bell/CH – Fine) will have bearings on the causal relations in the EPR/B experiment in general, and the question of quantum non-locality in particular.

It is also noteworthy that in the derivation of the Bell/CH inequalities, or more precisely (Bell/CH – Fine), Fine [27] in fact presupposes *λ-independence* and some factorizability conditions. That he presupposes λ-independence is clear from the

---

[6] For example, (Bell/CH – phys – ψ – big) fails for any joint distribution that returns the following probabilities as marginals for apparatus settings that satisfy $|x - y| = |x' - y| = |x - y'| = 60°$ and $|x' - y'| = 180°$: $P(\psi \& x \& y) = P(\psi \& x' \& y) = P(\psi \& x \& y') = P(\psi \& x' \& y') = 1/4$, $P(\psi \& x) = P(\psi \& y) = 1/2$, $P(O_x \& O_y \& \psi \& x \& y) = P(O_{x'} \& O_y \& \psi \& x' \& y) = P(O_x \& O_y \& \psi \& x \& y') = 1/32$, $P(O_{x'} \& O_y \& \psi \& x' \& y') = 1/8$, $P(O_x \& \psi \& x) = P(O_y \& \psi \& y) = 1/4$.

fact that he takes the distribution of λ to be the same for all spin measurements; and as it is not difficult to see from equations (2) and (11) in his paper, his characterization of hidden-variables models embody factorizability conditions. Recalling footnote 5, it is not difficult to show that λ-independence and these factorizability conditions are sufficient for the derivation of Bell/CH inequalities. So the question arises as to the role that the assumption of joint distribution plays in Fine's derivation of these inequalities. It may be tempting to argue that such an assumption is necessary for the physical plausibility of the hidden-variables models. But, first, this argument is not open to Fine, who holds that the rejection of such assumption is the very essence of QM. Second, even if we suppose that the assumption of joint distribution were important for the ontological status of the hidden-variables models (an assumption that Bell, Clauser and Horne and many others reject), the violation of this assumption *per se* is not sufficient to vindicate Fine's claim that "what the hidden-variables models and the Bell/CH inequalities are all about are the requirements that make well defined precisely those probability distributions for non-commuting observables." [27, p. 291] Since factorizability fails in the EPR/B experiment, Fine has to appeal to the view that the violation of this condition has no implications for the question of non-locality [22–24]. For if we suppose that the failure of factorizability involves some kind of non-locality, as a broad consensus has it, then the fact that factorizability fails in standard QM as well as in any alternative interpretation or hidden-variables model in which λ-*independence* obtains, will entail that the common interpretation of Bell's theorem is on the right track.

In any case, as we shall see in Sect. 16.5, if probabilities are interpreted along de Finetti's probability theory, (Bell/CH – Fine) cannot be derived from the assumption of joint probability distribution over the measurement outcomes since such distribution does not exist. Similarly, λ-*independence* and *Mirror* do not entail (Bell/CH – phys – ψ) since (Bell/CH – prob) does not hold; for the joint probability distribution over the spin quantities in this latter inequality does not exist. But before turning to discuss the application of de Finetti's theory to the quantum probabilities, we introduce the highlights of this theory in Sects. 16.3 and 16.4.

16.3 De Finetti's Theory of Probability

Our aim here is to offer a new reading of de Finetti's theory of probability and, assuming that quantum probabilities are interpreted along this theory, to study their logical structure – i.e. the inequalities that constrain them. Thus, for lack of space, the presentation of de Finetti's theory will be uncritical.

16.3.1   The Probability Axioms Are Not Merely Formal Conventions

De Finetti held that "probability theory is not merely a formal, merely arbitrary construction, and its axioms cannot be chosen freely as conventions justified only by mathematical elegance or convenience. They should express all that is necessarily inherent in the notion of probability and nothing more." [38, pp. xiii–xiv] He thought of probability as a guide of life under uncertainty. Having been influenced by positivism, he held that probability, like other notions of great practical importance, should have an operational definition, namely "a definition based on criterion which allows us to measure it." [39, p. 76] Also, being a guide of life under uncertainty, de Finetti maintained that probability should be closely related to rational decisions under uncertainty. ([38], pp. xiii–xiv, Chaps. 1 and 2; [39], 76–89) The decision framework that he had in mind is Bayesian, where a person's degrees of belief reflect her uncertainty concerning the things she cares about, her utilities reflect her subjective preferences, and the outcomes of rational decisions are actions that maximize her expected utility.[7] De Finetti thought of probability as reflecting rational degrees of belief, and of coherence as a necessary condition for degrees of belief being rational, and he argued that all the theorems of probability theory could be derived from the coherence conditions of degrees of belief. ([39], pp. x, 72–75, 87–89; [38], Chaps. 1 and 2)

16.3.2   The Domain of Probability Is the Domain of Uncertainty

De Finetti made a distinction between the domain of certainty, i.e. that which one takes as certain or impossible, and the domain of uncertainty, i.e. the range over which one's uncertainty extends. The distinction between these domains is very important and fundamental to de Finetti's philosophy of probability, as his long and detailed discussion of this topic demonstrates [39, Chap. 2]. The domain of uncertainty depends on one's (actual and/or hypothetical) background knowledge [39, pp. 27, 47] and one's reasoning, and thus it may include events that are logically impossible or certain, e.g. complicated contradictions or tautologies that one fails to recognize. The domain of probability is the domain of uncertainty. This domain is supposed to include all the atomic uncertain events (or the propositions that such events occur) and their logical combinations, which may be certain (for example, if A is an uncertain event, the domain of uncertainty will also include the certain event A or not-A). Whether an event is atomic is a pragmatic matter, which does not depend on metaphysical questions. It is noteworthy that for de

---

[7] It is noteworthy that unlike Frank Ramsey [40], another founding father of the modern school of subjective probability, de Finetti held that probability is not strictly related to rational preferences.

Finetti, there is a sharp distinction between being certain about an atomic event, and having a degree of belief one in it. The former belongs to the domain of certainty, whereas the latter belongs to the domain of uncertainty.

16.3.3 Probabilities Are Subjective and Instrumental

Many friends of the subjective interpretation of probability think that coherence is a necessary but not sufficient condition for the rationality of degrees of belief. They hold that for degrees of belief to be rational, they also have to be constrained by knowledge of objective facts about the world. In particular, it is frequently maintained that when objective probabilities are available, they should constrain the corresponding subjective probabilities. Thus, many hold that rationality requires that a person's subjective probability of E given that the objective probability of E is p, and she assumes, believes or knows nothing else about the prospects of E, should be p. An influential expression of this idea is Lewis's [28] "principal principle."

De Finetti rejected the idea that subjective probabilities are supposed to be guesses, predictions or hypotheses about the corresponding objective probabilities, or based on such probabilities or any other objective facts. Indeed, he argued that probabilities are inherently subjective, and that none of the objective interpretations of probability makes sense. He held that objective probability does not exist, and that recognition of its inexistence would constitute a progress in scientific thinking. "The abandonment of superstitious beliefs about the existence of Phlogiston, the Cosmic Ether, Absolute Space and Time, ... , or Fairies and Witches, was an essential step along the road to scientific thinking. Probability, too, if regarded as something endowed with some kind of objective existence, is no less misleading misconception, an illusory attempt to exteriorize or materialize our true [i.e. actual] probabilistic beliefs." [39, p. x][8]

De Finetti [39, pp. x–xi] argued that probability and probabilistic reasoning should always be understood as subjective. Probability only reflects uncertainty, and accordingly no fact could prove or disprove a degree of belief. He did not deny, however, the psychological influence that facts may have on degrees of belief. "I find no difficulty in admitting that any form of comparison between probability evaluations (of myself, or of other people) and actual events may be an element influencing my further judgment, of the same status as any other kind of information ... But, as with any other experience, these modifications would not be governed by a mechanical rule; it is, in each case, my personal judgment that is responsible for giving a weight to the facts (for instance, according to my feelings

---

[8] The addition of the word "actual" in the square brackets is mine, as the translation from Italian seems incorrect. The word "vero" could be translated as "actual" or "true", and it is clear that in this context it should be translated as "actual."

about the success of the other person being due to his skill and competence or merely due to a meaningless chance)." [38, p. 21]

The source of uncertainty is immaterial. "It makes no difference whether the uncertainty relates to unforeseeable future, or to an unnoticed past, or to a past doubtfully reported or forgotten; it may even relate to something more or less knowable (by means of a computation, a logical deduction, etc.) but for which we are not willing or able to make the effort; and so on. ... The only relevant thing is uncertainty – the extent of our knowledge and ignorance. The actual fact of whether or not the events considered are in some sense *determined*, or known by other people, and so on, is of no consequence." [39, pp. x–xi] The important thing for de Finetti is that in all these different states of uncertainty, subjective probability could be useful as a guide. The role of probability is *purely instrumental*, and its value should be determined solely on the basis of its potential to serve as a guide in everyday and science. De Finetti went to great pains in his attempt to show that his subjective theory of probability could serve as such a guide.

As de Finetti's *Philosophical Lectures on Probability* suggest, he was instrumentalist about probabilistic theories [41, pp. 53–54], interpreting their probabilities as subjective, representing nothing but degrees of expectations. [41, p. 52] And he held that distributions brought to us by probabilistic theories, such as Statistical Mechanics and Quantum Mechanics, "provide more solid grounds for subjective opinions." [41, p. 52]

Like other instrumental views, de Finetti thought that subjective probability could play its role as a guide, independently of our metaphysical assumptions about the world. "[P]robabilistic reasoning is completely unrelated to general philosophical controversies, such as Determinism versus Indeterminism, Realism versus Solipsism – including the question of whether the world 'exists', or is simply the scenery of 'my' solipsistic dream." [39, p. xi]

16.3.4 Intuition, Prudence and Learning from Experience

It is common to portray probability in de Finetti's radical subjective interpretation as unconstrained, too permissive and possibly whimsical (see, for example, [42, Sect. 3.5.4]. On the other hand, de Finetti held that assigning or "evaluating" probabilities is an inductive reasoning, and as such it is based on learning from experience; and "to speak about inductive 'reasoning' means, however, to attribute a certain validity to that mode of learning, to consider it not as a result of a capricious psychological reaction, but as a mental process susceptible of an analysis, interpretation and justification." [38, p. 147] Indeed, he warned against superficiality in assigning probabilities, which is frequently associated with subjective probability. In particular, he warns against two common patterns of superficiality. "On the one hand You may think that the choice, being subjective, and therefore arbitrary, does not require too much of an effort in pinpointing one particular value rather than a different one; on the other hand, it might be thought that no mental effort is required, since it can be avoided by the mechanical application of some standardized procedure." [39, p. 179] He recommended various features that must underlie each probability evaluation, like for example to "think about every aspect of the problem...try to imagine how things might go...encompass all conceivable possibilities; and also take into account that some might have escaped attention...identify those elements which, compared to others, might clarify or obscure certain issues...enlarge one's view by comparing a given situation with others...attempt to discover the possible reasons lying behind those evaluations of other people..." [39, pp. 183–4]. This is not surprising given that de Finetti held that "the (subjectivistic) theory of probability is a normative theory, not a descriptive one," and the value of probability theory is "precisely as an aid to the avoidance of plausible and frequently serious shortcomings and errors." [38, p. 151]

De Finetti's philosophy of probability presupposes that people have the intuitive faculty to form reasonable opinions about uncertain events and, with the aid of probability theory, the capacity to form reasonable probabilistic opinions. De Finetti held that people need to develop and refine this faculty, and apply reason to learn to guard it against the tendency to form superficial probabilistic opinions. Yet, he cautioned against the misunderstanding of the role of reason. In particular, he warned that "the tendency to overestimate reason – often in an exclusive spirit – is particularly harmful. Reason, to my mind, is invaluable as a supplement to the other psycho-intuitive faculties, but never a substitute for them. Figuratively, reason is a pole that may keep the plant of intuitive thought from growing crooked, but it is not itself either a plant or a valid substitute for a plant." [38, pp. 147–8]

Learning from experience is important for assigning both "prior" and posterior' probabilities. De Finetti held that every probability is conditional "not only on the mentality or the psychology of the individual involved, at the time in question, but also, and essentially, on the state of information in which he finds himself at the moment," though in many cases there is no need to mention explicitly the background information, and accordingly it is suppressed [39, p. 134]. So both prior and posterior probabilities are conditional probabilities. The prior probabilities are conditional on some prior background information, and they are updated according to the increase or change in background knowledge/beliefs/assumptions. De Finetti makes a distinction between updating and changing opinions. When one conditionalizes on new information, one keeps the same opinion yet updates it to a new situation [41, p. 35]. And when one revises one's probability function, one changes one's opinion. Change of opinion could result from reconsideration of neglected, inaccurate or ambiguous information, or change of mind about the relevance of information, or superficial or careless evaluations, etc. Thus, de Finetti held that realistically the evolution of one's subjective probabilities involves both updating and changing opinions [41, pp. 39–40].

Due to the disparity in subjective evaluations, prior probabilities are expected to vary significantly. Yet, de Finetti held that the effects of "the disparity between the initial judgments of people or of vagueness in the initial judgments of one person are often largely eliminated," if the additional information gathered between the prior and the posterior evaluations is sufficiently revealing and the prior probabilities are "sufficiently gentle or diffuse," i.e. not too opinionated [38, p. 145].

16.3.5   Probabilities Are Coherent Degrees of Belief

Probabilities are not just any degrees of belief. They are coherent degrees of belief in (propositions about) events that belong to the (agent's) domain of uncertainty. The notion of coherent degrees of belief is commonly understood in terms of Dutch books, i.e. bets that results in loss come what may. The idea is that incoherent degrees of beliefs are subjected to Dutch books. ([39, 40, 60], Chaps. 3 and 4) Coherence is thus characterized in a betting framework, where a person is subjected by a clever bookie to series of bets. The person assigns certain odds to these bets according to her degrees of belief, and the bookie prescribes the possible gains and losses according to these odds.

In his later work, de Finetti preferred a different decision-theoretic framework (for the motivation, see Sect. 16.3.6). ([38], Chaps. 1 and 2; [39], Chaps. 3 and 4) In this alternative framework, there is no bookie. Individuals express their degrees of belief, and they are subjected to fixed gains and variable monetary losses, the so-called "loss functions," which are functions of their degrees of belief about events and the occurrence of these events. That is, letting E being any verifiable event, d a degree of belief in E, and $\mathbf{E}$ an indicator function denoting whether E occurs ($\mathbf{E} = 1$ if E occurs, and $\mathbf{E} = 0$ otherwise), the loss L that the individual is subjected to is:

$$(\text{L1}) \quad L = \frac{(\mathbf{E} - d)^2}{k};$$

where k is an arbitrary constant which is fixed in advance and which may differ from one case to another. In the case of multiple degrees of belief, the total loss is the sum of the losses incurred by each degree of belief. For example, the loss function for the degrees of belief d_1, d_2, d_3 in the events E_1, E_2, E_3, respectively, is:

$$(\text{L2}) \quad L = \frac{(\mathbf{E}_1 - d_1)^2}{k_1} + \frac{(\mathbf{E}_2 - d_2)^2}{k_2} + \frac{(\mathbf{E}_3 - d_3)^2}{k_3}.$$

In this alternative decision-theoretic scheme, coherent degrees of belief are explicated in terms of admissible decisions. The "decisions" are the individual's degrees of belief in various events, and they are admissible if they are not dominated by any other decisions, i.e. by any other degrees of belief in the same events; where a set of degrees of belief in events is dominated by another set of degrees of belief in the same events, if it leads to higher losses come what may. A set of degrees of beliefs is coherent just in case it is not dominated by any other set of degrees of belief in the same events.

16.3.6   Measurements of Degrees of Belief

De Finetti assigned a great importance to the measurement of degrees of belief. He thought that since probability is supposed to be a guide of life, it should have a meaning that renders it effective as such. Being influenced by positivism, he held that "in order to give an effective meaning to a notion – and not only an appearance of such in a metaphysical-verbalistic sense – an operational definition is required." By operational definition, he meant "a definition based on a criterion which allows us to measure it." [39, p. 76] His inspiration came from early twentieth century physics. "The notion of probability, like other notions of practical significance, ought to be operationally defined (in the way that has been particularly stressed in physics following Mach, Einstein, and Bridgman), that is, with reference to observations, in experiments that are at least conceptually feasible. In our case, the experiments concern the behavior of an individual (real or hypothetical) facing uncertainty." [38, p. xiv]

The main reason why de Finetti preferred the loss-functions decision-theoretic scheme is that the Dutch-book framework involves a bookie, an "opponent," the presence of whom may intrude with the measurement of degrees of belief. In particular, de Finetti mentioned the possibility that the bookie or the individual take advantage of differences of information, competence or shrewdness [39, p. 93]. The presuppositions of this scheme are that individuals strive to maximize their expected utility, and that utility is linear with money, where k is supposed to warrant this linearity. Granted these assumptions, it is not difficult to show that it is in the best interest of individuals to express their actual degrees of belief; for any other degrees of belief will lower their (subjective) expected utility.

Since de Finetti defines probability in terms of betting or measurement contexts, it may be tempting to interpret him as behaviorist about degrees of belief, holding that degrees of belief, and accordingly probabilities, do not exist outside these contexts [43, 185–9]. This interpretation is particularly suggestive given the inspiration that de Finetti took from Bridgman's [44] operationalism, where theoretical terms are defined in terms of the operational procedures of their measurements. Yet, de Finetti did not intend the betting and the loss-function decision-theoretic frameworks as Bridgman-like operational definitions of degrees of belief. Indeed, he held that degrees of belief exist independently of the contexts of their measurement. "The criterion, the operative part of the definition which enables us to measure it, consists in this case of testing, through the *decisions* of individual (which are observable), his opinions (previsions, probabilities), which are not directly observable." [39, p. 76] Moreover, as Eriksson and Hájek [43, p. 190] point out, de Finetti's worries about the relation between utility and money and about agents who care too much or too little about their bets, do not make sense if degrees of beliefs are interpreted along Bridgman's operationalism. The operational procedure is supposed to provide a reliable measurement of degrees of belief, not a definition of them. Yet, as we shall see in Sect. 16.4, the operational procedure plays an important role in explicating the coherence conditions of degrees of belief and to that extent it plays an important role in defining subjective probabilities.

16.3.7   Conditional Probability

Following Kolmogorov's [17] influential axiomatization of probability, it is common to define conditional probability in terms of unconditional probabilities: $P(B/A) \equiv P(B\&A)/P(A)$. De Finetti rejected this axiomatic approach. He thought that probability theory should be derived from the analysis of the meaning of probability. He held that every probability is conditional "not only on the mentality or the psychology of the individual involved, at the time in question, but also, and essentially, on the state of information in which he finds himself at the moment," though in many cases there is no need to mention explicitly the background information, and accordingly it is suppressed [39, p. 134]. Thus, he maintained that conditional probability is the fundamental object of probability theory, and unconditional probability does not make sense (except when it is a conditional probability in disguise).[9]

In introducing the concept of conditional probability, de Finetti says that "we shall write $P(E|H)$ for the *probability 'of the event E conditional on the event H'* (or even the *probability 'of the conditional event $E|H$'*), which is the probability that You attribute to E if You think that in addition to your present information, i.e. the H_0 which we understand implicitly, *it will become known to You that H is true (and nothing else).*" [39, p. 134] This characterization is ambiguous. On the one hand, conditional probability is characterized as a conditional with a probabilistic consequent, whereas on the other it is likened to unconditional probability of a "conditional event."

The association of conditional probability with a "called-off" bet in the betting decision-theoretic framework, and the loss function for conditional probability in the loss-function decision-theoretic framework both suggest the first interpretation. The loss function for the probability of E given H and the background knowledge H_0 is:

$$(L3) \quad L = \frac{H_0 H (E - d)^2}{k}$$

where d is a degree of belief in E, $\mathbf{E}$ and $\mathbf{H}$ are indicator functions, denoting the truth value of E and H, and $\mathbf{H_0}$ is an indicator function denoting the truth value of H_0. Based on (L3), the proposition that the conditional probability of E given H and H_0 equals p may be characterized by the following conditional:

(CP1) If you have the background knowledge H_0 and you come to know H (and nothing else), then your degree of belief in E will be p.

---

[9] In fact, the idea that conditional probability is the fundamental object of probability theory could also be defended in other interpretations of probability. [18, 19, 45]

The idea is that a person with such a conditional probability is subjected to a loss of $(\mathbf{E} - p)^2/k$ on the condition that she has the background knowledge/beliefs H_0 and she comes to know H and nothing else; and the loss is zero, if she does not have the knowledge/beliefs H_0 or does not come to know H. This is very similar to the idea of a called-off bet, where the probability of E given $H\&H_0$ being p is explicated by a bet in which a person pays pS dollars on the condition that she knows $H\&H_0$ for the opportunity to earn S dollars if E occurs and zero otherwise, and the bet is called off if she does not know $H\&H_0$.

The notion of conditional probability applies not only to cases where one knows H_0 and H, but also to cases where one assumes or believes H_0 and H. We may thus extend the meaning of conditional probability as follows:

(CP2) If you know, believe or assume H_0 and you come to know, believe or assume H (and nothing else), then your degree of belief in E will be p.

Further, the conditioning event and the background knowledge may be counterfactual rather than actual. In such cases, conditional probability may be characterized by the following counterfactual conditional:

(CP3) If you had the background knowledge or beliefs H_0 and you had come to know, believe or assume H (and nothing else), then your degree of belief in E would have been p.

Beware! (CP2) is neither the material nor the strict conditional. It is true if one knows, believes or assumes H and nothing else beside one's background knowledge H_0, and one's degree of belief in E is p; it is false when one has the background H_0 and comes to know, believe or assume H but one's degree of belief in E is not p; and it is indeterminate when one does not have the background H_0 or does not come to know, believe or assume H. (CP3) is not the Stalnaker–Lewis counterfactual conditional, though it may be interpreted as being true in case one's degree of belief in E is p in the most similar relevant worlds or scenarios in which one holds H_0 and H. For a more detailed discussion of these conditionals, see Berkovitz [45].

In order to distinguish the above notion of conditional probability from that of Kolmogorov, we shall place the conditional event in the subscript: $P_{H_0H}(E)$ will denote the conditional probability of E given H and the background knowledge H_0. De Finetti ([38], Chap. 2, [39], Chap. 4) demonstrates that coherence entails that:

$$(\text{C1}) \quad P_{H_0H}(E) \cdot P_{H_0}(H) = P_{H_0}(E\&H);$$

where $P_{H_0}(H)$ and $P_{H_0}(E\&H)$ are respectively the probability of E given H_0 and the probability of $E\&H$ given H_0. When $P_{H_0}(H)$ is definite and non-zero, we obtain Kolmogorov's definition of conditional probability as a coherence condition on degrees of belief.

Recall (Sect. 16.2) the two different ways of representing the Bell/CH inequalities: in terms of conditional probabilities with the conditions (namely, the pair-state and the apparatus settings) in the subscript, as in (Bell/CH – phys – ψ); and in terms of conditional probabilities *a la* Kolmogorov, where the conditions are placed after the conditionalization stroke, as in (Bell/CH – phys – ψ – big). (C1) suggests a way to relate these different representations.

De Finetti's proposal that the probability of E given H may be seen as the probability of the "conditional event" $E|H$ suggests another interpretation of conditional probability. Conditional events (or "tri-events") are in effect three-valued propositions about events, the truth-value of which depends on the condition ([39], p. 139, [46], Appendix, pp. 307–11). In particular, $E|H$ is the proposition that E occurs, but its truth-value depends on whether H occurs. If H occurs, then $E|H$ is true if E occurs and false if E does not occur; and if H does not occur, then $E|H$ has indeterminate truth-value. The idea is then to assign probabilities only to conditional events that are true or false, so that indeterminate conditional events have no probabilities.

We shall return to consider the implications of the two different interpretations of de Finetti's concept of conditional probability in our discussion of his verificationism in Sect. 16.4, and in the application of his theory of probability to QM in Sect. 16.5.

16.3.8 Symmetry and Exchangeability

Judgments of equally probable events, and accordingly of symmetries, are central to all interpretations of probability. In objective interpretations of probability, the symmetries concern the way things are. For de Finetti, the relevant symmetries concern one's opinions and judgments. De Finetti held that any evaluation of equally probable events is based on subjective judgments, and that the notion of exchangeability is central to such judgments. A collection of events is said to be exchangeable if the probability p_h that h of them occur depends only on h and is independent of their order of appearance [38, p. 229]. Followers of de Finetti's interpretation and friends of the Bayesian interpretation of quantum probabilities attribute a great importance to exchangeability. Indeed, the notion of exchangeability, and the related notion of partial exchangeability are bound to play a central role in the interpretation of the quantum probabilities along de Finetti's probability theory. For example, Caves et al. [4] apply de Finetti's work on exchangeability to the interpretation of the notion "unknown quantum states" and the related notion of "unknown quantum probabilities" from a subjectivist Bayesian perspective. Yet, as the notion of exchangeability is not central to our main focus – the study of the coherence conditions of degrees of belief in the context of QM – we postpone its discussion to another opportunity.

16.4   Verifiability, Coherence and Contextuality

De Finetti [39, p. 34] held that the events in probability assignments have to be verifiable. "In general terms, it will always be a question of examining, if, and in which sense, a statement really constitutes an 'event,' permitting in a more or less realistic acceptable form, and in unique way, the 'verification' of whether it is 'true' or 'false'. . . A and B are events (observables), but it is not possible to observe both of them, and, therefore, it is not possible to call the product AB an event (observable)."

An important implication of this view is that the constraints on probabilities of events that are not jointly verifiable are weaker. For example, if A and B are jointly verifiable, their probabilities are subjected to the inequality $P(A)+P(B)-P(A\&B)\leq1$. But if A and B are not jointly verifiable, they have no joint probability, and accordingly their probabilities are not subjected to this inequality.

De Finetti [46, p. 260] acknowledged that verifiability is "a notion that is often vague and illusive" and thought that it is necessary "to recognize that there are various degrees and shades of meaning attached to [it]." He took a pragmatic attitude toward the kind and degree of verifiability that is actually required for events to have a definite probability [46, Appendix]. To simplify things, we shall focus on verifiability in principle, and by "verifiable events" we shall mean events that are verifiable in theory according to one's beliefs.

Unlike probabilities, de Finetti was not antirealist about events. Yet, he held that notions of great practical importance should have "operational definitions," namely definitions based on criteria that render them measurable. If events are not verifiable, they cannot have such an operational definition. Further, the prospects of adequate measurements of degrees of belief in such events are dim, thus undermining the idea that probability should also have an operational definition. The most obvious explanation for de Finetti's verificationism is the influence of positivism. De Finetti [39, p. 76] was worried that events that are not verifiable may appear to be sensical but in fact be meaningless, and accordingly degrees of belief in such events will be useless.

In the context of de Finetti's philosophy of probability, there is a different reason to motivate his verificationism. It is difficult to make sense of the idea of coherent degrees of belief in, and accordingly probabilities of unverifiable events. This is clear in the betting decision-theoretic framework. Bets on events that are in principle unverifiable could never be concluded. Accordingly, no Dutch book could be based on such bets, and the idea that Dutch book could be used to explicate the notion of "coherent degrees of belief" collapses. Things are not so obvious in the loss-function decision-theoretic framework, as this framework appears to provide a way to explicate this notion even in the case of unverifiable events; for the notion of "admissible decision," which is used to explicate coherence in this framework, seems applicable even in the case of unverifiable events. But a little reflection on the nature of probabilities in de Finetti's theory suggests that this appearance is deceptive. In this theory, there are no objectively correct probability assignments. Probabilities are subjective opinions that only reflect uncertainty about things. The value of probabilities reside solely in their instrumental role as a guide for decisions under uncertainty, and this role could only be measured in terms of verifiable "gains" and "losses," or more generally verifiable consequences. In the case of unverifiable events, the instrumental value of probabilities vanishes because the consequences of probability assignments are in principle unverifiable. This lack of instrumental value reflects on the prospects of explicating the notion of coherent degrees of belief. Incoherent degrees of belief in unverifiable events have no verifiable harmful consequences, and so radical subjectivists about degrees of belief, like de Finetti, who deny the existence of objective probabilities, have no incentive to have coherent degrees of belief in such events. Accordingly, like in the betting decision-theoretic framework, the idea that the loss-function decision-theoretic framework could be used to explicate the notion of coherent degrees of belief collapses in the case of unverifiable events.

De Finetti proposes to make the verifiable nature of events explicit by assigning probabilities to "conditional events" $E|H$ (see Sect. 16.3.6) rather than to the events themselves; where H is an observation that enables to verify the event E [46, pp. 266–7, 307–313]. The idea is to assign probabilities only to conditional events $E|H$ with determinate truth-values, so that unverifiable events E have no probabilities. This idea is easily generalized to complex "conditional events," i.e. logical combinations of conditional events. Consider, for instance, $E_{12}|H_{12}$, the conjunction of the conditional events $E_1|H_1$ and $E_2|H_2$; where H_i is an observation that enables to verify whether E_i is true, and E_{12} is the event that denotes the conjunction of the events E_1 and E_2. $E_{12}|H_{12}$ is true if H_{12} and E_{12} are both true, false if H_{12} is true and E_{12} false, and has indeterminate truth-value if H_{12} is false. By restricting probabilities to conditional events, "complex" conditional events (like E_{12}) may fail to have definite probabilities, even when the "atomic" events that constitute them (E_1 and E_2) do. In this approach, a person's probabilities are represented by a "big" probability space with "gaps" in the place of some complex events (henceforth, DF-big-space). The logic and probability of conditional events seem to require some kind of three-valued logic, and indeed de Finetti discussed various three-valued logics that could serve as a basis for such probability theory [46, Appendix, pp. 302–313]. For de Finetti's early thoughts about conditional events and their logic, see De Finetti [47] and Mura [48].

De Finetti also entertained the idea of representing probabilities of verifiable events in terms of classical, two-valued logic. In fact, as we shall see in Sect. 16.6, he preferred such an approach. This alternative approach is in line with our proposal in Sect. 16.3.7 that conditional probability *a la* de Finetti may be characterized as a conditional with a probabilistic consequent. Indeed, this interpretation of de Finetti suggests a natural way of representing probabilities of verifiable events in terms of two-valued events. The main idea is to suppose that the "unconditional" probability of an event E being p has in effect a logical structure of a conditional with a probabilistic consequent: if an observation H that enables to verify E occurs (occurred), the probability of E is (would be) p. Recall (Sect. 16.3.7) that in our suggested notation, this conditional is represented as $P_H(E) = p$, i.e. as a conditional probability with the conditioning event in the subscript; and probabilities with different subscripts, i.e. conditionals with different antecedents, correspond to different probability spaces. That is, we could represent de Finetti's verificationism by supposing that a person's subjective probabilities are represented by multiple probability spaces, in each of which probabilities of events are conditional (implicitly) on an observation that enables to jointly verify all the events in the space. On this view, a person's coherent degrees of belief are represented by many "smaller" probability spaces (henceforth, DF-many-spaces), each contains events that could be jointly verified.

Although the two approaches are different, in de Finetti's philosophy of probability they are closely related. In both approaches, probabilities of events are conditional on observations that enable to verify them. This is not obvious in the DF-big-space, where probabilities appear to be unconditional. But recall (Sect. 16.3.7) that for de Finetti probabilities of "conditional events" are closely connected, if not equivalent, to the corresponding conditional probabilities. The similarity between conditional probability, represented as a conditional with probabilistic consequent, and the corresponding probability of conditional event is hindered by de Finetti's formal notation, which is similar to the common notation for conditional probability *a la* Kolmogorov. Yet, in both cases only verifiable events E have probabilities, and the observations H that enable their verification have no probability, as they are not events in the probability space. To highlight this feature, in our representation of conditional probability as a conditional with a probabilistic consequent, we have placed the conditioning events H in the subscript rather than after the conditionalization stroke, $P_H(E)$; and, as de Finetti [38, p. 104] remarks, the conditional event $E|H$ "must be considered as a whole," and accordingly H is not part of the probability space. Indeed, the inclusion of H in the probability space while maintaining de Finetti's verificationism would lead to an infinite regress, where H would have to be a conditional event, the condition of which would have to be represented by a conditional event, and so forth.

Finally, as represented above de Finetti's verificationism is very stringent. Conditionalizing probabilities of events on observations that enable to verify them would severely restrict the range of events that have probabilities. First, this brand of verificationism restricts probabilities to observational contexts. Second, in various cases the required observations are actually impossible to carry out. Third, it threatens to render de Finetti's philosophy of probability extremely operationalist, as the probability of an event may vary according to the kind of observation that enables to verify it. Yet, it is possible to sustain the main thrust of de Finetti's verificationism while avoiding the above undesired consequences by conditionalizing probabilities of events on the proposition that the events are verifiable in principle, rather than on the proposition that observations that enable to verify them have been performed. In fact, this weaker version of verificationism is what de Finetti seemed to have in mind. We shall discuss the implications of the weaker and the stronger versions of verificationism in the next section.

16.5   Coherent Degrees of Belief for the EPR/Bohm Experiment

The most important implication of de Finetti's verificationism is that the coherence conditions on degrees of belief in events that are *not* jointly verifiable are weaker than they would have been had the events been jointly verifiable. Let's consider again (Bell/CH – prob) (see Sect. 16.2). In de Finetti's theory, (Bell/CH – prob) is a necessary condition for coherent degrees of belief in, and accordingly for probabilities of X, Y, $X\&Y$, $X'\&Y$, $X\&Y'$ and $X'\&Y'$ only *when* these events (propositions) are jointly verifiable. But in various hidden-variable models of the EPR/B experiment, X and X' (Y and Y') are values of non-commuting spin observables, which are not jointly verifiable. Similarly, the measurement outcomes in (Bell/CH – Fine) are not jointly verifiable, and so they are not necessary conditions for coherent degrees of belief, and accordingly for probabilities of the measurement outcomes involved in this inequality. Thus, if probabilities are interpreted along de Finetti's theory, (Bell/CH – prob) and (Bell/CH – Fine) do not apply to the EPR/B experiment.

Recalling (Sect. 16.4) that de Finetti formalizes his verificationism in terms of conditional probabilities, the failure of these inequalities can be manifested in two different ways, corresponding to the two different interpretations of de Finetti's concept of conditional probability. Consider, for example, (Bell/CH – prob). In the DF-big-space approach, probabilities are assigned only to conditional events. In our case, the relevant conditional events are X/H_X, Y/H_Y, $X'/H_{X'}$, $Y'/H_{Y'}$, $X\&Y/H_{XY}$, $X'\&Y/H_{X'Y}$, $X\&Y'/H_{XY'}$, $X'\&Y'/H_{X'Y'}$, $X\&X'/H_{XX'}$ and $Y\&Y'/H_{YY'}$; where, as before, H_i is either a measurement that enables to verify the event i, or the proposition that the event i is verifiable (we shall consider below the differences between these two interpretations of H_i). Since it is impossible in principle to jointly observe X and X' (Y and Y'), individuals who are familiar with this feature of the quantum realm will not assign a determinate truth-value to $X\&X'/H_{XX'}$ ($Y\&Y'/H_{YY'}$), and so $X\&X'/H_{XX'}$ ($Y\&Y'/H_{YY'}$) and any conjunction that includes it has no probability. Consequently, a (Bell/CH - prob)-like inequality is not a necessary condition for the probabilities of the conditional events X/H_X, Y/H_Y, $X\&Y/H_{XY}$, $X'\&Y/H_{X'Y}$, $X\&Y'/H_{XY'}$ and $X'\&Y'/H_{X'Y'}$. In the DF-many-spaces approach, the assumption that X and X' (Y and Y') are not jointly verifiable entails that the events X, X', Y and Y' are not in the same probability space. There are four smaller probability spaces, each contains two of these events: $\{X,Y\}$, $\{X',Y\}$, $\{X,Y'\}$ and $\{X',Y'\}$. So (Bell/CH - prob) is not a necessary condition for coherent degrees of belief in, and accordingly for the probabilities of the events X, Y, $X\&Y$, $X'\&Y$, $X\&Y'$ and $X'\&Y'$. The upshot is that followers of de Finetti, who assume that the spins of a particle in different directions are not jointly verifiable, are not committed to (Bell/CH – prob). Thus, they may assume *Mirror* (e.g. that the probability distribution of spin-measurement outcomes reflects the probability distribution of the particles' spins before the measurements) and λ-*independence* (e.g. that the distribution of the particles' spins is independent of the measurements), yet assign probabilities that are not constrained by (Bell/CH – phys – ψ). Similarly, followers of de Finetti will not see (Bell/CH – Fine) as a necessary constraint on the probabilities of the four spin-measurement outcomes involved in each of the Bell/CH inequalities.

Two challenges may be posed for de Finetti's verificationism. The first is for the DF-many-spaces approach. In this approach, the same event may have different probabilities in different spaces: e.g. event X may have the probability p_1 in the probability space S_1 that is constituted by the "atomic" events X and Y, and p_2, $p_2 < p_1$, in the space S_2 that is constituted by the "atomic" events X and Y'. For recall that the probabilities in S_1 are conditionalized on a measurement H_{XY} that enables to verify whether X and Y occur, and the probabilities in S_2 are conditionalized on a measurement $H_{XY'}$ that enables to verify whether X and Y' occur. If H_{XY} and $H_{XY'}$ are incompatible measurements, there is no Dutch-book argument to dictate that the probability of X should be the same in both probability spaces.

Things are different, however, in our suggested interpretation of de Finetti's verificationism, where events are conditionalized on their verifiability rather than on measurements that enable their verification (see Sect. 16.4). In this version, the probability of X has to be the same in S_1 and in S_2 on pain of a Dutch book, where a bookie offers to sell a bet on X for $\$p_1$ and buy it back for $\$p_2$, thus "pumping" money out of any individual who holds that the probability of X in S_1 is different from the probability of X in S_2. The reasoning is as follows. An individual who holds the above probabilities should consider as fair a bookie's offer to (i) sell a conditional bet on X given that X and Y are jointly verifiable at the price of $\$p_1$, and (ii) buy a conditional bet on X given that X and Y' are jointly verifiable at the price of $\$p_2$. Since in each of these cases the bet is conditional on the relevant events being *verifiable*, rather than on actually being verified by measurements, the two bets could be jointly realized. Thus, if the individual accepts both bets as fair, she is destined to lose money come what may.

The second challenge is for both approaches, and it is related to the Kochen and Specker's (1967)'s no-go theorem. Due to its verificationism, de Finetti's theory of probability prescribes weaker constraints on probabilities in the EPR/B experiment. This provides followers of de Finetti's theory with some flexibility that is lacking in other interpretations of probability. Thus, for example, hidden-variables models of this experiment in which probabilities are interpreted along de Finetti's theory may postulate the existence of definite values for non-commuting spin observables, i.e. values of spins in various directions, even if they assume *Mirror* and *λ-independence*. Yet, Kochen and Specker's theorem and other similar theorems impose heavy constraints on assignments of definite values to such non-commuting observables (for a review of these theorems, see [49]), which substantially limit the scope of such flexibility. The reasoning is as follows.

In their theorem, Kochen and Specker consider a spin-1 particle and triples of the square values of spins in three orthogonal directions, S_x^2, S_y^2, S_z^2. The observables S_x^2, S_y^2, S_z^2 commute and accordingly their values are jointly verifiable (though the observables S_x, S_y, S_z do not commute and so their values are not jointly verifiable).

Kochen and Specker demonstrate that granted the following assumptions, there is no coherent way of distributing the values of spins in 117 directions.

Values: All physical quantities of a quantum system, i.e. all the observables that pertain to it, have definite values at all times.

Non-contextuality: Properties that a system possesses, i.e. the values of the observables that pertain to it, are non-relational to other properties or the measurement context.

More recently proofs involving less observables have been given (for references, see [49]). The upshot is that any "hidden-variables" model that satisfies these assumptions cannot provide a coherent assignment to a particle's spins in more than a limited number of directions. Indeed, the challenge that Kochen and Specker's theorem raises is not particular to the interpretation of probabilities along de Finetti's theory; it is posed for any interpretation of the probabilities of "hidden-variables" models. Yet, these theorems substantially restrict the advantages that de Finetti's interpretation provides.

De Finetti was also verificationist about events (see Sect. 16.4), and his verificationism may provide a way around Kochen and Specker's theorem. The proof of the theorem requires a truth-value assignment to propositions about events that are not jointly verifiable, and given de Finetti's verificationism about events the assignment of truth values to propositions about events that are not jointly verifiable may be more flexible, so as to avoid a Kochen and Specker-like contradiction; for such an assignment may violate *Non-contextuality*. Recall (Sect. 16.4) that de Finetti argued for verificationism on the grounds that the instrumental value of notions depends on their verifiability, and that this reasoning relies heavily on a positivist philosophy. Recall also that in the case of probabilities of events, de Finetti's verificationism can be motivated on different grounds – namely, by the radical subjectivist and instrumental nature of probability in his theory; for due to this nature, it is difficult to make sense of the notion of *coherent* degrees of belief, and accordingly of probabilities of unverifiable events. Such a motivation does not seem to exist in the case of events *per se*, as Finetti was not antirealist about events.

Followers of de Finetti's interpretation of probability who do not wish to adhere to de Finetti's positivism may circumvent Kochen and Specker's theorem by rejecting *Values*. They may for example follow the orthodox interpretation and accordingly reject *Values*; for recall that in this interpretation, the particles in the EPR/B experiment have no definite spins before the measurements. While the rejection of *Values* does not entail the failure of *Mirror*, it is more difficult to motivate the later premise when the former fails. Alternatively, followers of de Finetti may hold *Values* but reject *Non-contextuality*. For instance, they may hold that the values of spin quantities are *relational* to the values of other spin quantities,[10] so that the value of the particle's spin along the direction x relative

---

[10] For an example of interpretation of QM that postulates such relationalism, see Berkovitz and Hemmo's [50] relational modal interpretation.

to the values of its spins in the (mutually) orthogonal directions y and z is different from its value relative to the values of its spins in different (mutually) orthogonal directions y' and z'. Given such contextuality, there exist coherent assignments for the values of all the spin quantities that are involved in the Kochen and Specker theorem. The question whether such contextuality is compatible with *Mirror* is rather delicate and go beyond the scope of our current discussion. But, in any case, the above reasoning seems to suggest that the challenges that the Kochen and Specker theorem poses limit the advantage that de Finetti's interpretation of probability may have over other interpretations of probabilities.

16.6 De Finetti on the Nature of Quantum Probabilities

De Finetti found QM both fascinating and challenging. He dedicated a substantial part of the long appendix of his *Theory of Probability* to the analysis of QM probabilities [46, pp. 302–333]. Unlike his analysis of the foundations of probability, the discussion of the nature of QM probabilities lacks incisiveness and clarity. De Finetti refers frequently to von Nenumann's [51] *Mathematical Foundations of Quantum Mechanics*, Bodieu's [52] *Theorie dialectique des probabilities* and Reichenbach's [53] *Philosophic Foundations of Quantum Mechanics*. He models his analysis as a simplified version of Bodieu's and Reichenbach's. Like Bodieu, de Finetti believes that quantum probabilities are a special case of a general calculus of probability. Yet, he thinks that Reichenbach presents "the questions most lucidly from the logical and philosophical point of view," and he thus uses Reichenbach's comments as guidelines for developing his own analysis of the QM probabilities. The aim of de Finetti's analysis is "finding the logical constructions which will prove suitable for resolving the difficulties we find ourselves" in trying to interpret QM. He believes that "the correct path is straightforward and simple" and "it is obscured precisely by preconceived ideas about what it is that constitutes a necessary prerequisite for any logic," and the key for resolving the difficulties is to recognize that the logic of events should be three-valued [46, p. 303, 305–9].

Reichenbach presented the three truth-values in reference to observations: E is true if the observation H has given the result E; E is false if the observation H has given the result not-E; and E is indeterminate or meaningless if the observation H has not been made. De Finetti [46, p. 307] thinks that Reichenbach's presentation corresponds to his conditional three-valued events, the only difference being that in his framework the third value is called "void." Following Reichenbach, he seems to favor the view that the third truth value lies between true and false; for "[t]his is, in fact, the requirement that must be satisfied if something is to be called a mathematical structure or, in particular, a logical structure." Yet, later, in his philosophical lectures on probability, he [41, p. 169] explicitly rejects this view when he says that denoting the third truth-value by "$1/2$" instead of "$\emptyset$" "is not appropriate because it somewhat suggests that it is an intermediate value between true and false." This later view of the indeterminate truth-value is more in line with our interpretation of de Finetti, where indeterminate conditional events have no determinate truth-value and accordingly have no probability.

In any case, de Finetti [46, p. 308] thinks that all the logical construction of Reichenbach's three-valued logic "could be expressed in terms of two-valued logic", so as to avoid "creating a number of symbols and names of operations and consequent rules (which are difficult to remember and sort out, and difficult to use without confusion arising). Above all, one avoids creating the tiresome and misleading impression that one deals with mysterious concepts which transcend ordinary logic." De Finetti thinks that the conceptual scheme of the three-valued event, expressed in ordinary binary logic, could account for the quantum puzzles. In particular, he argues that this framework could serve as the basis for understanding the problem of complementarity in QM. He characterizes complementarity in terms of indeterminate three-valued events – two events are complementary if at least one of them "remains certainly indeterminate (but it is not known which...)" [46, p. 311] – and then proceeds to argue that complementary events also arise in classical phenomena though "the most celebrated example is undoubtedly that of complementarity in quantum mechanics." [46, p. 312]

We argued above that de Finetti's theory of probability could serve as a basis for interpretation of the quantum probabilities. Yet, we believe that de Finetti's discussion of QM probabilities and their relationships to classical probabilities does not do justice to the difficulties that are involved in such an endeavor. In particular, de Finetti seems to be unaware of Bell's and Kochen and Specker's theorems and the heavy constraints they impose on assignments of probabilities in the quantum realm.

16.7 Conclusions

De Finetti held that a theory of probability has to express what is inherent in the notion of probability and nothing more. Probability is a rational guide of life under uncertainty. Probabilities are coherent degrees of belief in verifiable events, and the theorems of probability are supposed to follow from the coherence conditions of degrees of belief. Unlike other subjective interpretations, probability is not supposed to be ignorance about objective probabilities. Probability reflects only subjective uncertainty, and its value is purely instrumental. We argued that in de Finetti's instrumental philosophy of probability, coherence embodies a certain kind of verificationism, and accordingly the coherence conditions of degrees of belief in events depend on their verifiability. Indeed, in the context of this philosophy it is difficult to make sense of coherent degrees of beliefs in events that are unverifiable.

We argued that de Finetti's verificationist conception of coherence has important implications. A common view has it that in the subjective interpretation, probabilities are coherent degrees of belief and in principle every event (or proposition about it) may have a probability. In de Finetti's theory, there are many degrees of belief that have no corresponding probability; for degrees of belief in unverifiable events have no coherence conditions, and accordingly no probability. The restriction of probabilities to verifiable events also entails that the coherence conditions of degrees of belief in events that are not jointly verifiable are weaker than the (familiar) coherence conditions that such events would have had, had they been jointly verifiable.

The idea that verifiability is relevant for probability was also highlighted in Pitwosky's [54, 55] discussion of George Boole's [56] "conditions of possible experience." Boole thought of probabilities as relative frequencies in a finite sample, and of the conditions of possible experience as inequalities concerning such probabilities. Pitowsky [55, p. 105] notes that *"none of Boole's conditions of possible experience can ever be violated when all the relative frequencies involved have been measured in a single sample.* The reason is that such a violation entails a logical contradiction ... But sometimes, for various reasons, we may choose or be forced to measure the relative frequencies of (logically connected) events, in several distinct samples. In this case a violation of Boole's conditions may occur."

We proposed that the restriction of probabilities to verifiable events in de Finetti's theory entails that the probability space of these events is "non-classical" (see de Finetti's big-space approach in Sect. 16.4), or that probabilities are represented by multiple, smaller probability spaces, each of which contains events that are jointly verifiable (see de Finetti's many-spaces approach in Sect. 16.4). In either case, the implication is that the inequalities that constrain the probabilities of the values of spin observables in the EPR/B experiment are different from the inequalities that would have obtained had these events been jointly verifiable; and similarly, *mutatis mutandis,* for spin-measurement outcomes. This different probability structure provides followers of de Finetti's theory with some extra flexibility. Thus, for example, their probability assignments for the values of spin observables in "hidden-variables" models for the EPR/B experiment will not be constrained by (Bell/CH – prob) (see Sect. 16.2). Accordingly, they may suppose that the probabilities of spin-measurement outcomes in the EPR/B experiment "mirror" the probabilities of the corresponding spin observables before any measurement occur (*Mirror*) and that the distribution of the values of these spin observables is independent of the measurement settings (*λ-independence*) (see Sect. 16.2), yet their probabilities of spin-measurement outcomes will not be subjected to the (Bell/CH – phys - ψ) or (Bell/CH – phys - ψ - big) (see Sects. 16.2 and 16.5). However, the heavy constrains that Kochen and Specker's and similar theorems impose substantially limit the scope of such advantages (see Sect. 16.5).

Finally, it is noteworthy that in the context of de Finetti's theory of probability it is more difficult to reconstruct Bell's argument for non-locality. First, in this context it is more difficult to relate probabilities to causality, and accordingly it is hard to motivate the violation of *Factorizability* (see Sect. 16.1.1) as a locality condition. Second, it may be impossible to formulate *λ-independence*, another main premise of Bell's theorem; for if probability is interpreted along de Finetti's theory, in some hidden-variables theories the probability of the complete pair-state in the EPR/B experiment will not exist because this state is unverifiable. Whether this is the case will depend on both the nature of the complete pair-state, which varies from one hidden-variables theory to another, and the concept of verifiability one has in mind. Yet, that it is more difficult to reconstruct Bell's argument in the context of such radical subjective theory of probability should not be surprising, as probabilities in this theory are purely subjective and instrumental and accordingly are not supposed to reflect objective facts about the world. In de Finetti's interpretation, quantum probabilities are not supposed to reflect the ontological nature of the quantum realm; they only serve as a guide for policing uncertainty and forming anticipations about events in this realm.

Acknowledgements I owe a great debt to Itamar Pitowsky. My interest in de Finetti's theory of probability and quantum probabilities was provoked by a question that he raised in 1988 about the applicability of de Finetti's theory to the interpretation of the quantum probabilities. This interest led me to write a long seminar paper on this subject, which is the foundation for both my MA thesis and the current paper. During my work on the seminar paper and the MA thesis, I learned from Itamar a great deal about the curious nature of probabilities and properties in quantum mechanics. The invitation to *The Probable and the Improbable: The Meaning and Role of Probability in Physics* workshop in Honor of Itamar prompted me to write this paper, and I am very grateful to the organizers Yemima Ben Menachem and Meir Hemmo. For helpful discussions and comments on earlier versions of this paper, I am grateful to Itamar Pitowsky, Jeremy Butterfield, Chris Belanger, Arthur Fine, Chris Fuchs, Lucien Hardy, Alan Hájek, Carl Hoefer, Duncan Maclean, Rob Spekkens, Mauricio Suarez and audiences at *The Probable and Improbable* workshop, Van Leer Institute, Jerusalem, the Centre for Time, University of Sydney, Facultad de Filosofía, Complutense University de Madrid, and the South West Ontario Philosophy of Physics (SWOPP) forum, Perimeter Institute, Waterloo, Canada.

References

1. Berkovitz, J.: The world according to de Finetti: operationality. Unpublished seminar paper for Itamar Pitowsky's course in the philosophy of probability, Hebrew University of Jerusalem (In Hebrew) (1989)
2. Berkovitz, J.: The world according to de Finetti: on interpreting quantum probabilities along de Finetti's subjective theory of probability. MA thesis, Hebrew University of Jerusalem (In Hebrew) (1991)
3. Caves, C.M., Fuchs, C.A., Schack, R.: Quantum probabilities as Bayesian probabilities. Phys. Rev. A **65**(2), 1–6 (2002); quant-ph/0106133
4. Caves, C.M., Fuchs, C.A., Schack, R.: Unknown quantum states: the quantum de Finetti representation. J. Math. Phys. **43**, 4537–4559 (2002)
5. Caves, C.M., Fuchs, C.A., Schack, R.: Subjective probability and quantum certainty. Stud. Hist. Philos. Mod. Phys. **38**(2), 255–274 (2007); quant-ph/0608190
6. Einstein, A., Podolsky, B., Rosen, N.: Can quantum-mechanical description of physical be considered complete? Phys. Rev. **47**, 777–780 (1935)
7. Bell, J.S.: On the Einstein-Podolsky-Rosen paradox. Physics **1**, 195–200 (1964); Reprinted in Bell (1987), pp. 14–21
8. Bell, J.S.: On the problem of hidden variables in quantum mechanics. Rev. Mod. Phys. **38**, 447–452 (1966); Reprinted in Bell (1987), pp. 1–13

9. Bell, J.S.: Introduction to the hidden-variable question. In: Espagnat, B. (ed.) Foundations of Quantum Mechanics. Proceedings of the International School of Physics "Enrico Fermi", pp. 171–181. Academic, New York/London (1971); Reprinted in Bell (1987), pp. 29–39. See Ref. [57]

10. Bell, J.S.: The theory of local beables. TH-2053-Cern. (1975) Reprinted in Epistemological Letters (1976), and Bell (1987), pp. 52–62. See Ref. [57]

11. Bell, J.S.: Locality in quantum mechanics: reply to critics. Epistemol. Lett. 7, 2–6 (1975); Reprinted in Bell (1987), pp. 63–66. See Ref. [57]

12. Bohm, D.: Quantum Theory. Prentice-Hall, New York (1951); Reprinted in Dover Publications, New York (1989)

13. Clauser, J.F., Horne, M.A.: Experimental consequences of objective local theories. Phys. Rev. D 10, 526–535 (1974)

14. Butterfield, J.: Bell's theorem: what it takes. Br. J. Philos. Sci. 48, 41–83 (1992)

15. Redhead, M.L.G.: Incompleteness, Nonlocality, and Realism. Clarendon, Oxford (1987)

16. Shimony, A. Bell's theorem. The Stanford Encyclopedia of Philosophy (Summer 2009 Edition), Edward N. Zalta (ed.). http://plato.stanford.edu/archives/sum2009/entries/bell-theorem (2009)

17. Kolmogorov, A.N.: Foundations of Probability. Chelsea Publishing Company, New York (1933/1950)

18. Hájek, A.: What conditional probability could not be. Synthese 137(3), 273–323 (2003)

19. Popper, K.: The logic of scientific discovery. Basic Books. Routledge, London/NewYork (1959); Reprinted in Routledge, London (1992)

20. Berkovitz, J. Action at a distance in quantum mechanics. The Stanford Encyclopedia of Philosophy (Spring 2007 Edition), Edward N. Zalta (ed.). http://plato.stanford.edu/archives/spr2007/entries/qm-action-distance (2007)

21. Reichenbach, H.: The Direction of Time. University of California Press, Berkeley (1956)

22. Fine, A.: Correlations and physical locality. In: Asquith, P., Giere, R. (eds.) PSA 1980, vol. 2, pp. 535–562. Philosophy of Science Association, East Lansing (1981)

23. Fine, A.: The Shaky Game. The University of Chicago Press, Chicago (1986)

24. Fine, A.: Do correlations need to be explained? In: Cushing, J., McMullin, E. (eds.) Philosophical Consequences of Bell's Theorem, pp. 175–194. The University of Notre Dame Press, South Bend (1989)

25. Cartwright, N.: Nature's Capacities and Their Measurements. Clarendon, Oxford (1989)

26. Chang, H., Cartwright, N.: Causality and realism in the EPR experiment. Erkenntnis 38, 169–190 (1993)

27. Fine, A.: Hidden variables, joint probability, and the Bell inequalities. Phys. Rev. Lett. 48, 291–295 (1982)

28. Lewis, D.: A subjectivist's guide to objective chance. In: Lewis, D.K. (ed.) Philosophical Papers, 1987, vol. 2. Oxford University Press, Oxford (1980)

29. Carnap, R.: On inductive logic. Philos. Sci. 12, 72–97 (1945)

30. Svetlichny, G., Redhead, M., Brown, H., Butterfield, J.: Do the Bell inequalities require the existence of joint probability distributions? Philos. Sci. 55, 387–401 (1988)

31. Bacciagaluppi, G., Miller, D., Price, H.: Time-symmetric approaches to quantum mechanics. A focus issue of Stud. Hist. Philos. Mod. Phys. 39(4), 705–916 (2008)

32. Berkovitz, J.: On predictions in retro-causal interpretations of quantum mechanics. Stud. Hist. Philos. Mod. Phys. 39(4), 709–735 (2008)

33. Miller, D.J.: Quantum mechanics as a consistency condition on initial and final boundary conditions. Stud. Hist. Philos. Mod. Phys. 39(4), 767–781 (2008)

34. Price, H.: Toy models for retrocausality. Stud. Hist. Philos. Mod. Phys. 39(4), 752–761 (2008)

35. Sutherland, R.I.: Causally symmetric Bohm model. Stud. Hist. Philos. Mod. Phys. 39(4), 782–805 (2008)

36. Bell, J.S.: On the impossible pilot wave. Found. Phys. 12, 989–999 (1982); Reprinted in Bell (1987), pp. 159–168

37. Albert, D.Z.: Quantum Mechanics and Experience. Harvard University Press, Cambridge, MA (1992)
38. De Finetti, B.: Probability, Induction and Statistics: The Art of Guessing. Wiley, London (1972)
39. De Finetti, B.: Theory of Probability: A Critical Treatment, vol. 1. Wiley, London (1974)
40. Ramsey, F.P.: Truth and probability. In: Mellor, D.H. (ed.) F.P. Ramsey: Philosophical Papers, 1991. Cambridge University Press, Cambridge (1926)
41. De Finetti, B. Philosophical lectures on probability. Collected, edited, and annotated by Alberto Mura. Translated by Hykel Hosni. Synthese Library, 340. Dordrecht: Springer (2008)
42. Hájek, A. Interpretations of probability. The Stanford Encyclopedia of Philosophy (Spring 2010 Edition), Edward N. Zalta (ed.). http://plato.stanford.edu/archives/spr2010/entries/probability-interpret (2009)
43. Eriksson, L., Hájek, A.: What are degrees of belief? Studia Logica 86(2), 185–215 (2007)
44. Bridgman, P.W.: The Logic of Modern Physics. Beaufort Books, New York (1927)
45. Berkovitz, J.: The propensity interpretation: reply to critics. A Paper Presented at the Workshop Probability in Biology and Physics, IHPST, Paris, 12–14 Feb 2009
46. De Finetti, B.: Theory of Probability: A Critical Treatment, vol. 2. Wiley, London (1974)
47. De Finetti, B.: The logic of probability. Philosophical Studies 67, 181–190 (1995). Translated by R. A. Angell from "La Logique de La Probabilité", Actualités Scientifiques et Industrielles 391, Paris: Hermann et Cle, pp. 31–9. Actes de Congrès International de Philosophie Scientifique, Sorbonne, Paris, 1935, VI. Induction et Probabilité. (1936)
48. Mura, A.: Probability and the logic of trievents. In: Galavotti, M.C. (ed.) Bruno de Finetti, Radical Probabilist, pp. 200–241. College Publications, London (2009)
49. Held, C. The Kochen-Specker Theorem. The Stanford Encyclopedia of Philosophy (Winter 2008 Edition), Edward N. Zalta (ed.). http://plato.stanford.edu/archives/win2008/entries/kochen-specker (2006)
50. Berkovitz, J., Hemmo, M.: A modal interpretation of quantum mechanics in terms of relational properties. In: Demopoulos, W., Pitowsky, I. (eds.) Physical Theory and Its Interpretation: Essays in Honor of Jeffrey Bub, Western Ontario Series in Philosophy of Science, pp. 1–28. Springer, New York (2006)
51. Von Neumann, J.: Mathematical Foundations of Quantum Mechanics. Princeton University Press, Princeton (1932/1955)
52. Bodieu, G.: Theorie Dialectique des Probabilities. Gauthiers-Villars, Paris (1964)
53. Reichenbach, H.: Philosophic Foundations of Quantum Mechanics. University of California Press, California (1944). Reprinted in Dover Publications, New York (1998)
54. Pitowsky, I.: From George Boole to John Bell: the origins of Bell's inequality. In: Kafatos, M. (ed.) Bell's Theorem, Quantum Theory and Conceptions of the Universe. Kluwer, Dordrecht (1989)
55. Pitowsky, I.: George Boole's "conditions of possible experience" and the quantum puzzle. Br. J. Philos. Sci. 45, 95–125 (1994)
56. Boole, G.: On the theory of probabilities. Philos. Trans. R. Soc. Lond. 152, 225–252 (1862)
57. Bell, J.S.: Speakable and Unspeakable in Quantum Mechanics. Cambridge University Press, Cambridge (1987)
58. Kochen, S., Specker, E.: The Problem of Hidden Variables in Quantum Mechanics. J. Math. Mech. 17, 59–87 (1967)
59. Pitowsky, I.: Betting on the outcomes of measurements: a Bayesian theory of quantum probability. Stud. Hist. Philos. Mod. Phys. 34, 395–414 (2003)
60. De Finetti, B.: Lu prévision: ses lois, ses sources subjectives. Annuls de l' Institut Henri Poincaré 7, 1–68 (1937). Translated in Kyburg, H.E. Jr., Smokler, H.E. (eds.), Studies in Subjective Probability, pp. 93–158. New York: Wiley (1964)

Chapter 17
Four and a Half Axioms for Finite-Dimensional Quantum Probability

Alexander Wilce

Abstract It is an old idea, lately out of fashion but now experiencing a revival, that quantum mechanics may best be understood, not as a physical theory with a problematic probabilistic interpretation, but as something closer to a probability calculus *per se*. However, from this angle, the rather special C^*-algebraic apparatus of quantum probability theory stands in need of further motivation. One would like to find additional principles, having clear physical and/or probabilistic content, on the basis of which this apparatus can be reconstructed. In this paper, I explore one route to such a derivation of *finite-dimensional* quantum mechanics, by means of a set of strong, but probabilistically intelligible, axioms. Stated very informally, these require that systems appear completely classical as restricted to a single measurement, that different measurements, and likewise different pure states, be equivalent (up to the action of a compact group of symmetries), and that every state be the marginal of a bipartite non-signaling state perfectly correlating two measurements. This much yields a mathematical representation of (basic, discrete) measurements as orthonormal subsets of, and states, by vectors in, an ordered real Hilbert space – in the quantum case, the space of Hermitian operators, with its usual tracial inner product. One final postulate (a simple minimization principle, still in need of a clear interpretation) forces the positive cone of this space to be homogeneous and self-dual and hence, to be the state space of a formally real Jordan algebra. From here, the route to the standard framework of finite-dimensional quantum mechanics is quite short.

A. Wilce (✉)
Department of Mathematical Sciences, Susquehanna University, Selinsgrove, PA, USA
e-mail: wilce@susqu.edu

Y. Ben-Menahem and M. Hemmo (eds.), *Probability in Physics*, The Frontiers Collection,   281
DOI 10.1007/978-3-642-21329-8_17, © Springer-Verlag Berlin Heidelberg 2012

17.1 Introduction

It is an old idea that Quantum Mechanics is best seen, not as a dynamical theory
with a problematic probabilistic interpretation, but as something like a probability
calculus *per se*. Certainly, in its C^*-algebraic formulation, QM exhibits some
striking formal similarities to classical probability theory, with the latter actually
emerging as the special case of the former in which all observables commute, that
is, are simultaneously measurable. However, this C^*-algebraic apparatus is not
something one wants to swallow whole: it requires some further motivation.
Ideally, one might hope to *derive* it from a set of well-motivated axioms having
clear physical, operational or probabilistic interpretations. This possibility animates
von Neumann's book [1], and is made both explicit and programmatic in Mackey's
work in the late 1950s [2]. There is a small literature of attempts at such
a derivation, including the seminal papers of von-Neumann and Birkhoff [3],
Zierler [4], and Piron [5], framing the quantum-logical approach to the problem,
and the work of Ludwig [6], Gunson [7], Mielnik [8], Araki [9] and others,
approaching the problem in terms of ordered linear spaces.

In recent years, this view of quantum theory has fallen far out of fashion. This
owes in part to the rhetorical success of various attacks on "merely instrumentalist"
readings of QM. The prevailing opinion, articulated with great force by John Bell
[10], has been that terms such as "measurement" have no place among the
primitives of a self-respecting physical theory. Lately, however, attitudes towards
this older, probabilistic view of quantum mechanics have begun to thaw, and there
has been a resurgence of interest in the problem of characterizing (or deriving, or
reconstructing) quantum theory in operational/probabilistic terms. Much of the
newer work in this vein is influenced by quantum information theory. Thus,
where most of the earlier work cited above focussed on the structure of
single systems, and aimed to obtain the full apparatus of infinite-dimensional
(non-relativistic) quantum mechanics, the newer work [11–18, 40] has focussed
on characterizing finite-dimensional quantum mechanics, and has a distinctive
emphasis on properties of composite systems.

Many of these last-cited works do manage to obtain, or come very close to
obtaining, the probabilistic apparatus of finite-dimensional QM from various
packages of simple, plausible postulates. The diversity of approaches taken in
these papers is noteworthy, and lends support to the idea that QM is an especially
natural example of a probabilistic theory. This paper explores, in a preliminary way,
yet another, and possibly a less arduous, route towards such a probabilistic axiomatization of finite-dimensional quantum theory, or of something reasonably
close to it. The main ideas are that (i) both classical and quantum systems are
very symmetrical; (ii) irreducible finite-dimensional systems with homogeneous,
self-dual cones are pretty close to being quantum, thanks to the Koecher-Vinberg
Theorem. Therefore, (iii) if we can somehow use symmetry assumptions to ground
homogeneity and self-duality, we'll be heading in the right direction.

In a bit more detail: the Koecher-Vinberg Theorem [19, 20] classifies homogeneous self-dual cones in finite dimensions as the positive cones of formally real Jordan algebras. It follows from the Jordan-von Neumann-Wigner classification of such algebras that, with the single exceptional example of the cone of positive 3×3 matrices over the octonions, all physical systems having an irreducible, homogeneous, self-dual cone of (un-normalized) states, are either quantum-mechanical,[1] or arise as so-called spin factors, i.e., their normalized state spaces are n-dimensional balls. Evidently, then, one path towards deriving the mathematical framework of QM from first principles goes by way of supplying an operational motivation for homogeneity and self-duality. It will then remain either to dismiss, or to make room for, spin factors and the exceptional octonionic example as physical models.[2]

Working in a standard framework in which a probabilistic model consists of a set of basic observables, together with a finite-dimensional compact, convex set of states [11, 24], I propose four strong, but probabilistically intelligible (and not unreasonable), axioms. These yield a representation of basic observables as orthonormal subsets of a finite-dimensional ordered real Hilbert space, of states, as vectors therein, and of symmetries, as unitaries acting thereon. In the quantum case, the space in question is the space of Hermitian operators with its usual tracial inner product. A single additional postulate – the "half-axiom" of my title, a simple and natural minimization principle, for which one *hopes* to find a compelling interpretation – forces the positive cone of this ordered Hilbert space to be homogeneous and self-dual with respect to the given inner product. It is worth mentioning that these axioms all deal with the structure of a *single* system; no assumption is made concerning how systems combine. This is in marked contrast to much of the recent work mentioned above, relative to which the approach taken here is rather old-fashioned.

Two disclaimers are in order before proceeding, one historical, and the other programmatic. First, the general line of attack taken here is not entirely new. The possibility of using the Koecher-Vinberg theorem is mentioned by Gunson [7] as long ago as 1967, but the suggestion seems not to have been followed up (perhaps owing to the then-prevailing focuss on infinite dimensional systems). An exception is the work of Kummer [25], which, in the context of a quite different set of axioms, also exploits the Koecher-Vingberg Theorem. Secondly, it should be stressed that the postulates discussed below are not advanced as possible "laws of thought": the aim here is not to derive QM as the *uniquely* reasonable non-classical probability theory

---

[1] Allowing here real or quaternionic cases as "quantum".

[2] In fact, there is a fairly direct route from Jordan algebras to complex Quantum Mechanics, at least in finite dimensions. A theorem of Hanche-Olsen [21] shows that the only Jordan algebras having a Jordan-algebraic tensor product with $M_2(\mathbf{C})$ – that is, with a qubit – are the Jordan parts of C^*-algebras. Since the structure of qbits can be reasonably well-motivated on directly operational grounds, the only irreducible systems in a Jordan-algebraic theory supporting a reasonable tensor product, will be full matrix algebras. Requiring that ipartite states be uniquely determined by the joint probabilities they assign to the two component systems – a condition sometimes called *local tomography* – then forces the scalar field to be $\mathbf{C}$ [9, 22, 23].

(though that would of course be very nice!), but only to find simple and transparent *characterizations* of quantum probability theory in autonomously probabilistic or information-theoretic terms, even if still as a theory with contingent elements.

Dedication. I wish to dedicate this paper to the memory of Itamar Pitowsky. In two of his last papers [26, 27], Itamar defended a view of quantum theory very much in harmony with the one taken here. I would very much have enjoyed the chance to discuss these ideas with him, and keenly regret that this will now be impossible.

17.2 Preliminaries

In order even to advance the problem of characterizing QM as a probabilistic theory, we need first to establish a general framework for probability theory that is at the same time general enough to include quantum theory as a special case, and as transparent and uncontroversial as possible. Accepting that the basic elements of *finite* classical probability theory — that is, the concept of a finite, exhaustive set E of mutually exclusive alternative outcomes, and probability weights thereon — are about as close to this state of grace as one is going to get, but jettisoning the tacit, and surely contingent, classical assumption that any pair of tests can be represented as coarse-grainings of some third test, we are left the following machinery (described at much greater length in, e.g., [11, 23, 28–30]).

Definition. A *test space* is a collection $\mathfrak{A}$ of non-empty sets E, F, ..., each considered as the outcome-set of some experiment, measurement, or *test*. Subsets of tests are termed *events*.

We allow the possibility that distinct tests may overlap, so that one outcome may belong to several tests. We write $X = X(\mathfrak{A})$ for the total *outcome space* of $\mathfrak{A}$, i.e., $X = \cup \mathfrak{A}$. Outcomes x, $y \in X$ are termed *orthogonal*, and we write $x \perp y$, if they are distinct, but belong to a single test. Note that this language is only suggestive, since at present, there is no linear structure, let alone an inner product, in view. (Note, too, that we do *not* assume that every pairwise-orthogonal set is an event.)

Definition. A *state* on a test space $\mathfrak{A}$ is a mapping $\alpha : X \to [0,1]$ such that $\sum_{x \in E} \alpha(x) = 1$ for every test $E \in \mathfrak{A}$. We take $\alpha(x)$ to represent the *probability* that x will occur in any test to which it belongs. In other words, a state is a "non-contextual" assignment of a probability to every outcome of every test.
Examples The simplest case is that in which $\mathfrak{A}$ comprises just a single test, say $\mathfrak{A} = \{E\}$. In this case, states on $\mathfrak{A}$ are simply probability weights on E, and we recover discrete classical probability theory.[3] Discrete *quantum* probability theory

---

[3] Measure-theoretic classical probability theory is also subsumed by this framework: if (S, Σ) is a measurable space, then the collection $\mathfrak{B}(S, \Sigma)$ of countable partitions of S by non-empty measurable sets in Σ is a test space, and the states on $\mathfrak{B}(S, \Sigma)$ are exactly probability measures on (S, Σ).

arises as the special case in which $\mathfrak{A}$ is the collection $\mathfrak{F}(\mathbf{H})$ of all *frames*, or unordered orthonormal bases, of a Hilbert space $\mathbf{H}$. Here the outcome-set X is the unit sphere of $\mathbf{H}$. Gleason's Theorem tells us that states on $\mathfrak{A}$ are all of the form $\alpha(x) = \langle \rho x, x \rangle = \mathrm{Tr}(\rho p_x)$ where ρ is a density operator on $\mathbf{H}$ and p_x is the projection operator $p_x(y) = \langle y, x \rangle x$ associated with a unit vector $x \in X$.

The set $\Omega(\mathfrak{A})$ of all states on a test space $\mathfrak{A}$ is obviously a convex subset of the space $\mathbb{R}^X$ of all real-valued functions on $\mathfrak{A}$'s outcome-space X. If $\mathfrak{A}$ is *locally finite*, meaning that every test $E \in \mathfrak{A}$ is a finite set, then $\Omega(\mathfrak{A})$ is closed, hence, compact in the product topology (that is, the topology of pointwise-convergence) on $\mathbb{R}^X$. In particular, then, $\Omega(\mathfrak{A})$ has a plentiful supply of extreme points, which we call *pure* states. It will be convenient, from this point on, to adopt the standing convention that *all test spaces are locally finite*.

In many contexts, it is reasonable to consider a restricted set of states. This suggests the following

Definition. A *probabilistic model* is a pair $(\mathfrak{A}, \Omega)$, where $\mathfrak{A}$ is a test space with outcome-space X and Ω is a pointwise-closed (hence, compact) convex subset of the set $\Omega(\mathfrak{A})$ of all states on $\mathfrak{A}$.

Linearization A probabilistic model $(\mathfrak{A}, \Omega)$ has a natural representation in terms of a dual pair of ordered real vector spaces, as follows. Let V, or occasionally $V(\mathfrak{A}, \Omega)$, denote the span of Ω in $\mathbb{R}^X$ (where X is $\mathfrak{A}$'s outcome-space, as usual), ordered pointwise on X; let V_+ denote the cone of non-negative functions in V. Then Ω is a *base* for V_+, in the sense that every element of V_+ is uniquely a scalar multiple of a state in Ω. The *dimension* of a model is the linear dimension of the space V. The dual space V^* of V is ordered pointwise on Ω; equivalently, $a \in V_+^*$ iff $a(\alpha) \geq 0$ for all $\alpha \in V_+$. It will be convenient, if a trifle sloppy, to identify each outcome $x \in X$ with the corresponding evaluation functional, so that if $\alpha \in \mathbf{V}$, we may write $x(\alpha)$ for $\alpha(x)$. (This amounts to the benign assumption that Ω is *outcome-separating*, i.e, that outcomes having the same probability in every state are identical.) Writing u for the *order unit* in V^* — that is, the unique functional taking value 1 identically on Ω — we have $\sum_{x \in E} x = u$ for every test $E \in \mathfrak{A}$.

To illustrate these constructions, consider again the case in which $\mathfrak{A} = \mathfrak{F}(\mathbf{H})$, the frame manual of a Hilbert space $\mathbf{H}$. Then, as a corollary to Gleason's Theorem, $V = V(\mathfrak{F}, \Omega(\mathfrak{F}))$ is the space of (quadratic forms associated with) trace-class Hermitian operators, while $V^* \simeq \mathcal{L}_h(\mathbf{H})$, the space of all bounded Hermitian operators on $\mathbf{H}$. Of course, where $\mathbf{H}$ is finite-dimensional (equivalently: where $\mathfrak{A}$ has finite rank), V and V^* are both isomorphic to the space of all Hermitian operators on $\mathbf{H}$. Note that here Ω is not outcome-separating: identifying an outcome-*qua*-unit vector $x \in X$ with the corresponding evaluation functional in V^* amounts to replacing x with the corresponding rank-one projection operator.

Composite Systems Probabilistic models can be combined in a great variety of ways. Here, we need only a few basic ideas. If $\mathfrak{A}$ and $\mathfrak{B}$ are any test spaces, $\mathfrak{A} \times \mathfrak{B}$ denotes the space of *product tests* $E \times F$ with $E \in \mathfrak{A}$ and $F \in \mathfrak{B}$. A state on $\mathfrak{A} \times \mathfrak{B}$ is *non-signaling* iff the marginal states $\omega_1(x) = \sum_{y \in F} \omega(x, y)$ and

$\omega_2(y) = \sum_{x \in E} \omega(x, y)$ are well-defined, i.e., independent of $E \in \mathfrak{A}$ and $F \in \mathfrak{B}$. In this case we have conditional states $\omega_{2|x}, \omega_{1|y}$ defined by

$$\omega_{2|x}(y) = \omega(x, y)/\omega_1(x) \text{ and } \omega_{1|y}(x) = \omega(x, y)/\omega_2(y)$$

(where these make sense, and defined to be 0 if not). Note that we have laws of total probability: for any test $E \in \mathfrak{A}$,

$$\omega_2(y) = \sum_{x \in E} \omega_{2|x}(y)\omega_1(x), \tag{17.1}$$

and similarly for $\omega_1(x)$.

Classical Representations Notwithstanding the very great generality of the probability theory just sketched, there is a sense in which any probabilistic model $(\mathfrak{A}, \Omega)$ can be given a more-or-less classical description. One way to do this, which will be useful below, is as follows. First, note that every state is the barycenter of a (Baire) probability measure on the set $S = \Omega_{ext}$ of pure states (via the Bishop-deLeauw theorem [31]); also, every outcome x defines a random variable on S by evaluation, with $x(\alpha)d\mu(\alpha) = \rho(x)$ where $\rho = \int_S \alpha d\mu(\alpha) \in \Omega$. Since $0 \le \hat{x} \le 1$, we can regard $\hat{x}$ as a "fuzzy" version of an indicator function, and the set $\{\hat{x}|x \in E\}$, as a "fuzzy" partition of S. This gives us an interpretation of $(\mathfrak{A}, \Omega)$ as a kind of impoverished probability space, in which we have a set $\mathfrak{A}$ of "fuzzy" partitions on S, insufficiently rich to discriminate amongst all possible statistical ensembles of states in S; Ω is the quotient of the simplex of all such ensembles by the relation of indistinguishability relative to $\mathfrak{A}$.[4] (Of course, this kind of representation does not play nicely with the formation of non-signaling composite systems, but this is not our concern here.)

17.3 Symmetry, Minimization, Sharpness and Self-Duality

From this point forward, I make the standing assumption that *all test spaces are locally finite*, and *all probabilistic models are finite-dimensional*. In this section, I introduce two axioms governing such models, each having a reasonably transparent operational meaning, plus one simple (but much less transparent) minimization condition. From these, I deduce that V_+^* is self-dual.

The first axiom requires that systems be highly symmetrical in that (i) all outcomes of any given test look alike; (ii) all tests look alike; (iii) all pure states look alike. This axiom is satisfied by both (discrete) classical and pure quantum systems, and, as we'll see in a moment, already leads to some surprisingly strong consequences. To make this precise, let us agree that a *symmetry* of a system $(\mathfrak{A}, \Omega)$

---

[4] This kind of classical representation has been discussed by various authors. I first encountered it in the book [32] of Holevo.

with outcome space X is a homeomorphism $g : X \rightarrow X$ such that (i) $gE \in \mathfrak{A}$ for every test $E \in \mathfrak{A}$, and (ii) $g^*(\alpha) = \alpha \circ g^{-1}$ belongs to Ω for every state $\omega \in \Omega$. An action of a group on $(\mathfrak{A}, \Omega)$ is an action by symmetries. We say that $\mathfrak{A}$ is *fully symmetric* [30] under such an action if (i) all tests have the same cardinality, and (ii) for any bijection $f : E \rightarrow F$ between tests $E, F \in \mathfrak{A}$, there exists some $g \in G$ with $f(x) = gx$ for all $x \in E$. Notice that any symmetry g of $\mathfrak{A}$ also determines an affine automorphism of $\Omega(\mathfrak{A})$ by $(g\alpha)(x) = \alpha(g^{-1}x)$. We say that g is a symmetry of the model $(\mathfrak{A}, \Omega)$ iff $g\alpha \in \Omega$ for all $\alpha \in \Omega$ and all $g \in G$. It is easy to see that g takes extreme points of Ω to extreme points of Ω. We shall say that an action of G on $(\mathfrak{A}, \Omega)$ is *continuous* iff, for every $\alpha \in \Omega$ and every outcome $x \in X$, $g \mapsto \alpha(g^{-1}x)$ is continuous as a function of G.

Axiom 1 (Symmetry). There is a compact group G acting continuously on $(\mathfrak{A}, \Omega)$, in such a way that (i) G acts fully symmetrically on $\mathfrak{A}$, and (ii) G acts transitively on Ω_{ext}.

A classical test space $\mathfrak{A} = \{E\}$ satisfies Axiom 1 trivially with $G = S(E)$, the symmetric group on E. A quantum test space $(\mathfrak{F}(\mathbf{H}), \Omega_{\mathbf{H}})$ satisfies Axiom 1 with $G = U(\mathbf{H})$, the unitary group of $\mathbf{H}$.

Call an inner product on V^* *positive*[5] iff $\langle a, b \rangle \geq 0$ for all $a, b \in V_+^*$. Note that the trace inner product on $V^* = \mathcal{L}_h(\mathbf{H})$ is positive in this sense.

Lemma 1. *Subject to Axiom 1, there exists a positive, G-invariant inner product on V^*.*
Proof. Since G acts transitively on $S := \Omega_{\text{ext}}$, we can represent the latter as G/K where K is the stabilizer of some (any) pure state α_o. This carries an invariant measure (induced by the Haar measure on G), giving the space $C(S)$ of continuous Real-valued functions on S a canonical G-invariant inner product, namely $\langle f, g \rangle = \int_{\alpha \in S} f(\alpha) g(\alpha) d\alpha$. Note that this is positive in sense defined above. As discussed in Sect. 17.1, we have a natural order-preserving embedding of V^* in $C(S)$ obtained by restricting each $a \in V^*$ to $S \subseteq V$. The restriction of the natural translation-invariant inner product on $C(S)$ to V^* gives us a positive, G-invariant inner product on the latter. $\qquad \qquad \square$

Let's agree to call the specific inner product arising from $C(G)$ the *canonical* inner product on V^*. This is related to the classical representation of V^* as a space of random variables on the space $S = \Omega_{\text{ext}}$ by

$$\langle a, b \rangle = \text{cov}(\hat{a}, \hat{b}) + \langle a, u \rangle \langle u, b \rangle.$$

Note that, relative to this inner product, we have $\| u \| = 1$, so we can alternatively write

$$\text{cov}(\hat{a}, \hat{b}) = \langle P'_u a, P_{u'} b \rangle$$

---

[5] As opposed to positive-definite, which every inner product is.

where P'_u is the orthogonal projection onto the subspace of V^* orthogonal to u.

Lemma 2 [33]. *Let,$\langle,\rangle$ be any positive, G-invariant inner product on V^*. There is an embedding $x \mapsto v_x$ of the outcome-space X into the unit sphere of V^* with $x \perp y$ implying $\langle v_x, v_y \rangle = 0$.*[6]
Proof. For each $x \in X$, set

$$q_x = x - \langle x, u \rangle u,$$

so that

$$\langle q_x, u \rangle = 0. \tag{17.2}$$

Note that $\sum_{x \in E} q_x = 0$ for all $E \in \mathfrak{A}$. Notice, too, that $L_\alpha^* q_x = q_{\alpha x}$ for all $\alpha \in G$ and all $x \in X$. Since L is unitary and G acts transitively on X, the vectors q_x have a constant norm $\| q_x \| = r$. Moreover, since G takes any orthogonal pair of outcomes to any other, $\langle q_x, q_y \rangle$ is constant for any pair $x \perp y$ in X. Call this value s_q. If $s_q = 0$, we are done: simply set $v_x = q_x / \| q_x \|$. If not, we have

$$0 = \langle q_x, 0 \rangle = \left\langle q_x, \sum_{y \in E} q_y \right\rangle = r^2 + (n-1)s_q.$$

In particular, $s_q = -\frac{r^2}{n-1} < 0$. In this case, set $v_x = q_x + cu$ where $c = r/\sqrt{n-1}$ (so that $s_q = -c^2$). Then, using (2), we have $\langle v_x, v_y \rangle = 0$. Normalizing if necessary, we can take each v_x to be a unit vector. Obviously, the mapping $x \mapsto q_x$ is injective $\langle x, u \rangle$ is constant on X; hence, so is $x \mapsto v_x$. $\square$

In order to get maximum mileage out of this, we impose a very simple, but very strong condition. To set the stage, we need the following observation.

Lemma 3. *Let S denote the constant value of $\langle x, y \rangle$ where $x \perp y$. With notation as in Lemma 3, we have, for all outcomes x and y, that*

$$\langle v_x, v_y \rangle = \langle x, y \rangle - s.$$

Proof. Letting m denote the (constant) value of $\langle x, u \rangle$, we set $q_x = x - mu$ as in the proof of Lemma 3, so that $\langle q_x, u \rangle = 0$ for all x. Recall that $v_x = q_x + cu$ where $-c^2 = s_q$, the constant value of $\langle q_x, q_y \rangle$ when $x \perp y$. Thus, we have

$$\langle v_x, v_y \rangle = \langle q_x, q_y \rangle + c^2 = \langle q_x, q_y \rangle - s_q.$$

---

[6] I remind the reader that here, $x \perp y$ means only that the outcomes $x, y \in X$ are distinct and belong to a common test; this does not (yet) imply that $\langle x, y \rangle = 0$.

Now

$$\langle q_x, q_y \rangle = \langle x - mu, y - mu \rangle = \langle x, y \rangle - m\langle x, u \rangle - m\langle u, y \rangle + m^2 = \langle x, y \rangle - m^2.$$

Considering the case where $x \perp y$, this yields

$$s_q = s - m^2.$$

Hence,

$$\langle v_x, v_y \rangle = \langle x, y \rangle - m^2 - s_q = \langle x, y \rangle - s,$$

as promised.   □

Definition. Call a G-invariant, positive inner product on V^* *minimizing* iff the constant S of Lemma 3 is in fact the minimum value of $\langle x, y \rangle$ on $X \times X$.

Note that this is certainly the case for the trace inner product on $\mathcal{L}_h(\mathbf{H})$, where $S = 0$!

Lemma 4. *For a minimizing inner product, the vectors v_x of Lemma 2 lie in the positive cone of V^*.*
Proof. Immediate from Lemma 3.   □

Provisional Axiom 2 (Minimization). There exists a minimizing G-invariant, positive inner product on V^*.

As we'll see in Sect. 17.3, *all* positive inner products on $V^* = \mathcal{L}_h(\mathbf{H})$ invariant under the unitary group of $\mathbf{H}$, are in fact minimizing. In any case, one would like to have an operational interpretation for minimization. At present, I do not have one to offer; however, it may be useful to note that, in terms of the classical representation of effects as random variables on $S = \Omega_{\mathrm{ext}}$, discussed in the previous section, we have $\langle x, y \rangle = \mathrm{cov}(\hat{x}, \hat{y}) + 1/n^2$. Thus, the canonical inner product is minimizing iff "orthogonal" (that is, distinguishable) pairs of outcomes are precisely those that minimize covariance.

There is one important class of examples in which the existence of a minimizing inner product *does* follow from the previous axioms. Call a test space $\mathfrak{A}$ *2-connected* iff every pair of outcomes $x, y \in X$ there exist tests $E, F \in \mathfrak{A}$ with $x \in E, y \in F$ and $E \cap F \neq \emptyset$. Equivalently, $\mathfrak{A}$ is 2-connected iff, for all outcomes x, y there exists an outcome z with $x \perp z \perp y$. Example: the frame manual of a Hilbert space.

Lemma 5. *If $\mathfrak{A}$ is a fully-symmetric, rank-three, 2-connected test space, then any invariant, positive inner product is minimizing.*
Proof. Let $x \not\perp y$. By 2-connectedness, we can find an outcome z with $x \perp z \perp y$. As $\mathfrak{A}$ has rank three, we have tests $E = \{x, a, z\}$ and $F = \{z, b, y\}$. Now, as all outcomes have the same norm in V^* (here, I conflate an outcome with the corresponding

evaluation functional in V^*), we see that $\langle x, t \rangle$ is maximized over outcomes t by $t = x$. Let s be the common value of $\langle e, f \rangle$ where e and f are outcomes with $e \perp f$. Noting that $x + a + z = u = z + b + y$, we have

$$\| x \|^2 + 2s = \langle x, u \rangle = s + \langle x, b \rangle + \langle x, y \rangle.$$

This yields $\langle x, b \rangle + \langle x, y \rangle = s + \| x \|^2$. Since $s < \| x \|^2$ and neither $\langle x, b \rangle$ and $\langle x, y \rangle$ can exceed $\| x \|^2$, it follows that both must exceed s. $\square$

Remark. 2-connectivity is close to requiring the sets $x^\perp, x$ ranging over outcomes, to form a projective geometry. See [33] for more on this.

Lemma 6. *Subject to Axiom 1 and Provisional Axiom 2, For every $x \in X$, $\alpha_x(y) :=$* $\langle v_x | v_y \rangle$ *defines a state on* $\mathfrak{A}$.
Proof. By Lemma 4, $\langle v_x, v_y \rangle \geq 0$ for all y. Since v_x and v_y are unit vectors, we also have $\langle v_x, v_y \rangle \leq 1$ for all y. Finally, letting $x \in E \in \mathfrak{A}$, we have, by Lemma 2, and with $v := \sum_{y \in E} v_y$, a multiple of u,[7] that

$$\langle v_x, v \rangle = \sum_{y \in E} \langle v_x, v_y \rangle = \langle v_x, v_x \rangle = 1$$

We now impose another axiom that, while decidely strong, has a clear physical meaning: it says that if we know for certain that a particular outcome will occur, then we know the system's state. In this rough sense, the measurements belonging to A are maximally informative.

Axiom 3 (Sharpness). To every outcome $x \in X$, there corresponds a unique state $\varepsilon_x \in \Omega$ with $\varepsilon_x(x) = 1$.

Note that ε_x is necessarily a pure state. Note, too, that both (discrete) classical and non-relativistic QM satisfy this postulate. For some further discussion of, and motivation for, Axiom 3, see Appendix A.

Proposition 1. *Subject to Axioms 1–3, V^*_+ is self-dual.*
Proof. Let $\langle, \rangle$ be a minimizing, G-invariant positive inner product. Positivity gives us $V^*_+ \subseteq V^{*+} \simeq V_+$. Letting v_x be defined as in Lemma 2, Lemma 7 tells us that $\varepsilon_x(y) := \langle v_x, v_y \rangle$ defines a state making x certain (since $\langle v_x, v_x \rangle = \| v_x \| = 1$). By Axiom 3, there is but one such state, which, by virtue of its uniqueness, is pure. It follows from Axiom 2 that every pure state has the form $g\varepsilon_x = \varepsilon_{gx}$ for some $g \in G$. Thus, every pure state is represented in the cone V^*_+, so that $V^{*+} \subseteq V^*_+$. $\square$

An alternative proof of self-duality, based on slightly different assumptions, is presented in Appendix C.

---

[7] Since $v_y = q_y + cu$, and $\sum_{y \in E} q_y = 0$, we have $\sum_{y \in E} v_y = ncu$, where $n = |E|$ is independent of E by virtue of $\mathfrak{A}$'s being fully symmetric.

17.4 Correlation, Filtering and Homogeneity

Having secured the self-duality of V_+^*, the next order of business is to secure its homogeneity. To this end, we introduce two further axioms. The first of these tells us that all states of a single system are consistent with that system's being part of a larger composite in a state of perfect correlation between some pair of observables. To be more precise, call a bipartite non-signaling state *correlating* iff, for some tests $E, F \in \mathfrak{A}$, and some bijection $f : E \to F$, $\omega(xy) = 0$ for all $(x, y) \in E \times F$ with $y \neq f(x)$.

Axiom 4 (Correlation). Every state is the marginal of a correlating non-signaling state.

Again, this is satisfied by both classical and quantum systems: trivially in the first case, and not-so-trivially (i.e., by the Schmidt decomposition) in the second.

Lemma 6 [34]. *Subject to Axioms 3 and 4, for every $\mu \in V_+$ there exists a test E such that $\sum_{x \in E} \mu(x)\varepsilon_x$.*
Proof. Suppose first that α is a normalized state on $\mathfrak{A}$. By Axiom 4, there exists a test space $\mathfrak{B}$ and a correlating, non-signaling state $\omega \in \Omega(\mathfrak{A} \times \mathfrak{B})$ with $\alpha = \omega_1$. Suppose ω correlates $E \in \mathfrak{A}$ with $F \in \mathfrak{B}$ along a bijection $f : E \to F$. The bipartite law of total probability [35] tells us that

$$\omega_1(x) = \sum_{y \in F} \omega_2(y)\omega_{1|y} = \sum_{x \in E} \omega_2(f(x))\omega_{1|f(x)},$$

where $\omega_{1|y} = \omega_{1|f(x)}$ is the conditional state on $\mathfrak{A}$ given outcome $y = f(x) \in F$. Since $\omega(x, y) = 0$ for $y \neq f(x)$, we have $\omega_{1|f(x)}(x) = 1$ if $y = f(x) \in F$; thus, $\omega_{2|f(x)} = \varepsilon_x$, and $\alpha = \sum_{x \in E} \alpha(x)\varepsilon_x$ as promised.

Now suppose $\mu \in V_+$. Then $\mu = r\alpha$ for some $\alpha \in \Omega$ and real constant $r \geq 0$. Expanding α as above, we have $\mu = \sum_{x \in E} r\alpha(x)\varepsilon_x = \sum_{x \in E} \mu(x)\varepsilon_x$. $\square$

The following postulate completes the set.

Axiom 5 (Filtering). For every test E and every $f : E \to (0, 1]$, there exists an order-isomorphism $\phi : V^* \to V^*$ with $\phi(x) = f(x)x$.

This says that the outcomes of a test can simultaneously and independently be attenuated by any (non-zero) factors we like by a reversible physical process. This is equivalent to saying that, for any test E, any outcome $x \in E$, and any $0 < c \leq 1$, there exists an order-automorphism ϕ such that $\phi(x) = cx$ and $\phi(y) = y$ for all $y \in E \setminus \{x\}$. This is clearly the case in both classical and quantum probability theory, and corresponds to the operationally natural idea that an outcome is always represented by a physical process, which can be subjected to a filter reducing its intensity by any specified factor.

Proposition 2. *Subject to Axioms 1–5, the cone V^*_+ is homogeneous.*[8]
Proof. Let a, b be interior points of V^*_+. By Proposition 1, V^*_+ is self-dual; hence, $\langle a|$ and $\langle b|$ are (un-normalized) states. Let us write $a(x)$ for $\langle a, x \rangle$ and similarly for b. By Lemma 6, $\langle a|$ and $\langle b|$ have decompositions $\langle a| = \sum_{x \in E} a(x) \varepsilon_x$ and $\langle b| = \sum_{y \in F} b(y) \varepsilon_y$ for some pair of tests $E, F \in \mathfrak{A}$. As $\varepsilon_x = \langle x|$, we have $\langle a| = \sum_{x \in E} a(x) \langle x|$, or, more simply, $a = \sum_{x \in E} a(x) x$, and similarly for b. Since a and b are interior points, $a(x)$ and $b(y)$ are non-zero for all x, y. Let g be a bijection matching E with F (courtesy of Axiom 4), and set $t(x) = b(gx)/a(x)$. Then, by Axiom 5, there is an order-automorphism ϕ of V^* taking x to $t(x)x$ for every $x \in E$. Hence, $\phi(a) = \sum_{x \in E} a(x) \phi(x) = \sum_{x \in E} a(x) t(x) x = \sum_{x \in E} b(gx) x$. Applying g, we have

$$g\phi(a) = \sum_{x \in E} b(gx) gx = \sum_{y \in F} b(y) y = b$$

It now follows from the Koecher-Vinberg theorem that the positive cone $V(\mathfrak{A})_+$ is the set of positive elements in a formally real Jordan algebra. It is possible that there is a more direct route to this conclusion – certainly, I have not made use of all of the available structure. For example, the full power of the assumption that $\mathfrak{A}$ is fully G-symmetric (as opposed to merely 2-symmetric) is not really exploited. Neither is it at all obvious that *every* self-dual homogeneous cone has a representation as $V(\mathfrak{A})$ with $\mathfrak{A}$ satisfying all of the foregoing axioms – again, full symmetry seems rather strong, as does Axiom 4 on correlation. It is possible that these axioms constrain the set of models much more severely.

17.5 Summary and Open Questions

Axioms 1, 3 and 4 seem natural, or at any rate, *intelligible*: one understands what they *say* about a system. Although strong (and certainly, not "laws of thought"), they do identifying a natural class of especially simple and tractable systems that we might *expect* to find well represented in nature. Axiom 5 seems natural in a slightly more restricted context, in which the measurements we make involve sending systems through filters that they may or may not pass, with probabilities that can be attenuated at our discretion. (The idea of a filter also shows up prominently in the work of Ludwig [6] and others following in the same path, e.g., [7, 8, 25].) In a broader sense, Axioms 1, 3 and 5 capture, in part, the idea that a system should look *completely* classical, as restricted to a single measurement. In particular, a reversible process allowable in classical probability theory should be implementable by a reversible "physical" process acting on $V(\mathfrak{A})$.

---

[8] A different route to homogeneity, via slightly different axioms, is discussed in Appendix A.

Provisional Axiom 2 is obviously more problematic, but on the evidence, seems likely to be satisfied by a wide range of systems. In order better to understand the scope and significance of this postulate, one would like to endow the canonical invariant, positive inner product on V^* — or, more broadly, *any* such inner product — with some operational, perhaps information-theoretic meaning. It would doubtless be very instructive to find a model satisfying the remaining axioms, for which this product is *not* minimizing. (Even better, of course, would be to prove no such example exists.)

Another interesting issue is that of how one can construct (by hand, as it were) tensor products compatible with the foregoing axioms. As mentioned in the introduction, the axioms considered here deal with the structure of individual models, imposing no condition at all on how these combine (the correlation condition of Axiom 4, as I've formulated it, is plainly a constraint on the structure of the model). Nevertheless, it is interesting to ask under what conditions systems satisfying axioms 1–5, or any subset of these, have non-signalling tensor products (containing all product states!) that also satisfy these axioms. Where this desideratum is met, we would seem to come within hailing distance of Hardy's axioms [16]. See [24] for some further discussion of the problems involved in constructing a class of test spaces closed under such a tensor product.

Acknowledgement I wish to thank Howard Barnum for reading and commenting on an earlier draft of this paper, and, more especially, for introducing me to the papers of Koecher and Vinberg, on which the present exercise depends. Thanks also to C. M. Edwards for pointing out the paper [21] of Hanche-Olsen.

Appendix A: Entropy and Sharpness

The following considerations may offer some independent motivation for Axiom 1. There are two natural ways to extend the definition of entropy to states on a test space. If α is a state on a locally finite, finite-dimensional test space $\mathfrak{A}$, then Minkowsky's theorem tells us that α has a finite decomposition as a mixture $\alpha = \sum_i t_i \alpha_i$ of pure states $\alpha_1, ..., \alpha_n$. Define the *mixing entropy* of α, $S(\alpha)$, to be the infimum of $H(t_1, ..., t_n) = -\sum_i t_i log(t_i)$ over all such convex decompositions of α. Alternatively, one can consider the *local entropy* $H_E(\alpha) = H(\alpha|_E) = -\sum_{x \in E} \alpha(x) log(\alpha(x))$. Define the *measurement entropy* of α, $H(\alpha)$, to be the infimum value of the local measurement entropies H_E over all tests E.

Suppose now that the group G figuring in Axiom 2 is compact. One can then endow $\mathfrak{A}$ with the structure of a compact topological test space in the sense of [30]. Assuming that all states in Ω are continuous as functions $X \to \mathbf{R}$, it follows [36, Lemma 6] then the infimum defining H is actually achieved, i.e., $H(\alpha) = H_E(\alpha)$ for some test $E \in \mathfrak{A}$. An easy consequence is that $H(\alpha) = 0$ iff $\alpha(x) = 1$ for some $x \in X(\mathfrak{A})$. One can also show [36] that $S(\alpha) = 0$ iff α is a limit of pure states. Consequently, if the set of pure states is closed, we have $S(\alpha) = 0$ iff α is pure.

In both classical and quantum cases, $S = H$. One might consider taking this as a general postulate:

Postulate A: $H(\alpha) = S(\alpha)$ for every state $\alpha \in \Omega$.

An immediate consequence is that, subject to the topological assumptions discussed above, a pure state (with mixing entropy $S(\alpha) = 0$) must have local measurement entropy $H_E(\alpha) = 0$ for some test E, whence, there must be some outcome $x \in E$ with $\alpha(x) = 1$. Conversely, for every $x \in X$, if $\alpha(x) = 1$, then $H_E(\alpha) = 0$ for any E containing outcome x, whence, $H(\alpha) = 0$. But then $S(\alpha) = 0$ as well, and α is therefore pure. If $\mathfrak{A}$ is *unital*, meaning that every outcome has probability 1 in at least one state, then it follows that $\mathfrak{A}$ is actually sharp. Moreover, we see that every pure state has the form ε_x for some x In this case, the second half of Axiom 2 follows automatically from the first. Further discussion of Postulate A can be found in the paper [36], where theories satisfying it are termed *monoentropic*.

Appendix B: An Alternative Route to Homogeneity

We say that the space V is *weakly self-dual* iff there exists an *order-isomorphism* – that is, a positive, invertible linear map with positive inverse – $\phi : V^* \to V$. Note that such a map corresponds to a positive bilinear form $\omega : V^* \times V^* \to \mathbf{R}$ via $\omega(x, y) = \phi(x)(y)$, hence, to a non-signaling bipartite state on $\mathfrak{A}$. We call a bipartite state ω an *isomorphism state* iff the positive linear map $\hat{\omega} : V^* \to V$ given by $\hat{\omega}(x)(y) = \omega(x, y)$ is invertible. One can show [37] that any such state is pure. Note that as u belongs to the interior of V_+^*, if ω is an isomorphism state, we must have $\omega_1 = \hat{\omega}(u)$ in the interior of V_+. This suggests the following alternative to Axioms 5:

Postulate B: Every interior state is the marginal of an isomorphism state

Lemma [37]. *Subject to Postulate B alone, V is weakly self-dual and homogeneous.*
Proof. For there to exist an isomorphism state, V must be weakly self-dual. For homogeneity, let α and β belong to the interior of V_+. Then Postulate B implies that there exist isomorphism states ω and μ with $\alpha = \hat{\omega}(u)$ and $\beta = \hat{\mu}(u)$. Thus, $\beta = (\mu \circ \omega^{-1})(\alpha)$. As $\mu \circ \omega^{-1}$ is an order-automorphism of V, it follows that the cone is homogeneous. ☐

Postulate B is similar in flavor to Axiom 4, but seems somewhat awkward in its reference only to states in the interior of V_+. It would be desirable to find a single, natural principle implying both of these axioms. Further work in this direction can be found in [37]

Appendix C: An Alternative Route to Self-Duality

An alternative proof of Proposition 1 (the self-duality of V_+) appeals to the fact [38, Lemma 1.0] that a finite-dimensional ordered space A is self-dual w.r.t a given inner product iff every vector $a \in A$ has a *unique* Jordan decomposition $a = a_+ - a_-$ with $\langle a_+, a_- \rangle = 0$. We'll need the following

Lemma. *Suppose A carries a positive inner product, with respect to which every element of A has an orthogonal Jordan decomposition. Then A_+ is self-dual.*
Proof. It suffices to show that the orthogonal Jordan decomposition is unique. Suppose $a_+ - a_- = b_+ - b_-$ are two orthogonal Jordan decompositions of an element $a \in A$, and that the inner product is positive. We $a_+ - b_+ = a_- - b_- =: x \in A$, so that

$$0 \leq \| x \|^2 = \langle a_+ - b_+, a_- - b_- \rangle = -(\langle b_+, a_- \rangle + \langle a_+, b_- \rangle).$$

But since the inner product is positive, this last quantity is non-positive: evidently, we must have

$$\langle a_+, b_- \rangle = \langle a_-, b_+ \rangle - 0,$$

whence, $x = 0$, whence, $a_+ = b_+$ and $a_- = b_-$: the decomposition is unique, as advertised. □

Let us say that a model $(\mathfrak{A}, \Omega)$ is *spectral* iff it satisfies the conclusion of Lemma 6 — that is, if every state $\mu \in \Omega$ can be expanded as $\sum_{x \in E} \mu(x) \varepsilon_x$ where $E \in \mathfrak{A}$ and, for each $x \in E$, ε_x is a state with $\varepsilon_x(x) = 1$.

Theorem A. *Suppose $V(\mathfrak{A}, \Omega)$ is spectral, that $\mathfrak{A}$ is 2-symmetric, and that Provisional Postulate 2 holds. Then $V(\mathfrak{A})$ is self-dual.*
Proof. If $f : E \to \mathbf{R}$, where $E \in \mathfrak{A}$, let $a_f = \sum_{x \in E} f(x) x$. Note that this gives us a positive linear mapping $R^E \to V(\mathfrak{A})^*$. That $\mathfrak{A}$ is spectral implies that every *positive* element of V^* has a representation as a_f for some $f \geq 0$ on some $E \in \mathfrak{A}$. Notice that $u = a_1$ for the constant function $1 : E \to \mathbf{R}$ on *any* test $E \in \mathfrak{A}$.

Now let $v_x = q_x + cu$, where $q_x = x - \langle x, u \rangle u = (1 - \langle x, u \rangle) x$, as in Lemma 2, so that $v_x \perp v_y$ for $x \neq y$ in E. If $f \in \mathbf{R}^E$, let $v_f = \sum_{x \in E} f(x) v_x$. Note that

$$v_f = \sum_{x \in E} f(x) v_x = \sum_{x \in E} f(x)(1 - \langle x, u \rangle + nc) x.$$

Setting $g \equiv 1 - \langle x, u \rangle + nc$ (noting that this is constant!), we have $v_f = \sum_{x \in E} f(x) g x = a_{fg}$. In particular, $a_g = v = ncu$, so that $g \neq 0$. Thus, we have $a_f = a_{fg/gg} = v_{f/g}$. Thus, if $g \neq 0$, every $a = a_f$ in V^* has an *orthogonal* resolution with respect to an orthonormal set $\{v_x | x \in E\}$ for some $E \in \mathfrak{A}$. Finally, since (by our provisional Postulate 2) we have $v_x \geq 0$ for every $x \in E$, every vector with an orthogonal resolution relative to the set $\{v_x | x \in E\}$ has an orthogonal Jordan decomposition.   □

Appendix D: Invariant Positive Inner Products on $\mathcal{L}_h(\mathbf{H})$

Let $\mathbf{H}$ be a complex Hilbert space of dimension n, with frame manual $\mathfrak{F}$ and unit sphere X. We seek to classify the unitarily invariant inner products on $\mathcal{L}_h(\mathbf{H})$ that are positive on the positive cone of the latter, and to show that all of these are automatically minimizing.

As remarked above, Gleason's Theorem provides an isomorphism between the space $V(\mathfrak{F})$ of signed weights on $\mathfrak{F}$, and the space $\mathcal{L}_h(\mathbf{H})$ of Hermitian operators on $\mathbf{H}$: for every $\alpha \in V(\mathfrak{F})$, there is a unique $W_\alpha \in \mathcal{L}_h(\mathbf{H})$ with $\alpha(x) = \langle W_\alpha x, x \rangle$ for all $x \in X$. We also have a dual isomorphism $V^*(\mathfrak{F}) \simeq \mathcal{L}_h(\mathbf{H})$, sending each $a \in V^*(\mathfrak{F})$ to an Hermitian operator A_a with $\mathrm{Tr}(A_a W_\alpha) = a(\alpha)$ for all $\alpha \in V(\mathfrak{F})$. Note that in this representation, the order unit is represented by the identity operator 1 on $\mathbf{H}$. If U is a unitary operator on $\mathbf{H}$, understood as acting on X, then the natural action on $V(\mathfrak{F})$ is given by $U(\alpha)(x) = \alpha(U^{-1}x)$ for all $\alpha \in V(\mathfrak{F})$ and all $X \in X$. Thus, we have $\langle W_{U\alpha}x, x \rangle = \langle W_\alpha U^{-1}x, U^{-1}x \rangle$, whence, $W_{U\alpha} = UW_\alpha U^*$ for all states α. In other words, the natural representation of $U(\mathbf{H})$ on $V(\mathfrak{F}(\mathbf{H})) \simeq \mathcal{L}_h(\mathbf{H})$ is exactly its usual adjoint action. It follows that the dual action of $U(\mathbf{H})$ on $V^*(\mathfrak{F})$ is again the adjoint action $A \mapsto U^*AU$. Noting that 1 and $1^\perp$, the space of trace-0 Hermitian operators, are both invariant under this action, it follows that the two are orthogonal with respect to any unitarily invariant inner product on $V^*(\mathfrak{F})$. Also, since the adjoint representation of $U(\mathbf{H})$ on $1^\perp$ is irreducible [39, p. 20], it follows from Schur's Lemma that up to normalization, there is only one unitarily invariant inner product on the latter – in other words, any invariant inner product on $1^\perp$ has the form $\langle a, b \rangle = \frac{\lambda}{n}\mathrm{Tr}(ab)$ for some $\lambda > 0$, with $\lambda > 1$ corresponding to the normalized trace inner product. Hence, an invariant inner product on $\mathbf{V} = \langle 1 \rangle \oplus 1^\perp$ is entirely determined by the normalization of 1 and the choice of λ. Taking $\|1\| = 1$, we have that, for any $a = s1 + a_o$ and $b = t1 + b_o$, where $a_o, b_o \in 1^\perp$ and $s, t \in \mathbf{R}$, we have

$$\langle s1 + a_o, t1 + b_o \rangle = st + \frac{\lambda}{n}\mathrm{Tr}(a_o b_o).$$

We require that $\langle a, b \rangle \geq 0$ for all positive $a, b \in V^*$. The spectral theorem tells us that this is equivalent to requiring that $\langle p_x, p_y \rangle \geq 0$ for all rank-one projections P_x, P_y $(x, y, \in X)$. Writing $P_x = \frac{1}{n}1 + (P_x - \frac{1}{n}1)$, and similarly for P_y, we have

$$\langle P_x, P_y \rangle = \frac{1}{n^2} + \lambda \mathrm{Tr}\left(\left(P_x - \frac{1}{n}\mathbb{1} \right) \left(P_y - \frac{1}{n}\mathbb{1} \right) \right)$$

$$= \frac{1}{n^2} + \frac{\lambda}{n} \mathrm{Tr}\left(P_x P_y - \frac{P_x + P_y}{n} + \frac{1}{n^2}\mathbb{1} \right)$$

$$= \frac{1}{n^2} + \frac{\lambda}{n} \left(\mathrm{Tr}(P_x P_y) - \frac{2}{n} + \frac{1}{n} \right)$$

$$= \frac{1}{n^2} + \frac{\lambda}{n} \left(|\langle x, y \rangle_o|^2 - \frac{1}{n} \right)$$

$$= \frac{1-\lambda}{n^2} + \frac{\lambda}{n} |\langle x, y \rangle_o|^2,$$

where $\langle , \rangle$ is the inner product on $\mathbf{H}$. This will be non-negative for all choices of unit vectors x and y (in particular, for x and y orthogonal) iff $0 < \lambda \leq 1$ – in which case, the minimum value of $\langle P_x, P_y \rangle$ occurs exactly when $x \perp y$, so such an inner product is automatically minimizing.

References

1. von Neumann, J.: Mathematical Foundations of Quantum Mechnanics. Springer, Berlin (1932); English translation Princeton University Press (1952)
2. Mackey, G.: Mathematical Foundations of Quantum Mechanics. Addison Wesley, Reading (1963)
3. Birkhoff and von Neumann: The logic of quantum mechanics. Ann. Math. **37**, 823–843 (1936)
4. Zierler, N.: Axioms for non-relativistic quantum mechanics. Pacific J. Math. **11**, 1151–1169 (1961)
5. Piron, C.: Axiomatique quantique. Helvetica Phys. Acta **37**, 439–468 (1964)
6. Ludwig, G.: An Axiomatic Basis for Quantum Mechanics, vol. I, II. Springer, Berlin (1985). 1987
7. Gunson, J.: On the algebraic structure of quantum mechanics. Commun. Math. Phys. **6**, 262–285 (1967)
8. Mielnik, B.: Theory of filters. Commun. Math. Phys. **15**, 1–46 (1969)
9. Araki, H.: On a characterization of the state space of quantum mechanics. Commun. Math. Phys. **75**, 1–24 (1980)
10. Bell, J.: Against "measurement". Physics World, pp. 33–40 (1990)
11. Barnum, H., Barrett, J., Leifer, M., Wilce, A.: Cloning and broadcasting in generic probabilistic theories, arXiv: 0611295; also a general no-broadcasting theorem. Phys. Rev. Lett. **99**, 240501–240505 (2007)
12. Barnum, H., Barrett, J., Leifer, M., Wilce, A.: Teleportation in general probabilistic theories, arXiv:0805.3553. (2008)
13. Dakic, B., Brukner, C.: Quantum theory and beyond: is entanglement special? arXiv:0911.0695. (2009)
14. D'Ariano, G.M.: Probabilistic theories: what is special about quantum mechanics? In: Bokulich, A., Jaeger G (eds.), Philosophy of Quantum Information and Entanglement, Cambridge University Press, Cambridge, 2010 (arXiv:0807.438, 2008)
15. Goyal, P.: An information-geometric reconstruction of quantum theory. Phys. Rev. A. **78** (2008) 052120

16. Hardy, L.: Quantum theory from five reasonable axioms, arXiv:quant-ph/00101012. (2001)
17. Mananes, L., Muller, M.P.: A derivation of quantum theory from physical requirements. New J. Phys. **13** (2011), 063001 (arXiv:1004.1483v1, 2010)
18. Rau, J.: On quantum vs. classical probability. Ann. Phys. **324**, 2622–2637 (2009) (arXiv:0710.2119v1, 2007)
19. Koecher, M.: Die geoodätischen von positivitaätsbereichen. Mathematische Annalen **135**, 192–202 (1958)
20. Vinberg, E.B.: Homogeneous cones. Doklady Academii Nauk SSR **141**, 270–273 (1960); English translation, Soviet Mathematics–Doklady **2**, 1416–1619 (1961)
21. Hanche-Olsen, H.: Jordan Algebras with Tensor Products are C* Algebras. Springer Lecture Notes in Mathematics 1132, pp. 223–229. Springer Verlag, Berlin (1985)
22. Barrett, J.: Information processing in generalized probabilistic theories. Phys. Rev. A **75** (2007) 032304
23. Kläy, M., Randall, C.H., Foulis, D.J.: Tensor products and probability weights. Int. J. Theor. Phys. **26**, 199–219 (1987)
24. Wilce, A.: Symmetry and composition in probabilistic theories. Electron. Notes Theor. Comput. Sci. **270**, 191–207 (2011) (arXiv:0910.1527)
25. Kummer, H.: A constructive approach to the foundations of quantum mechanics. Found. Phys. **17**, 1572–9516 (1987); The foundations of quantum theory and noncommutative spectral theory I, II. Found. Phys. **21**, 1021–1069 (1991), 1183–1236
26. Bub, J., Pitowsky, I.: Two dogmas about quantum mechanics. In: Saunders, S., Barrett, J., Kent A., Wallace, C (eds.) Many Worlds? Everett, Quantum Theory and Reality, Oxford Universty press, Oxford, pp. 433–460 (2010)
27. Pitowsky, I.: Quantum mechanics as a theory of probability. In: Demopolous, W., Pitowsky, I. (eds.) Physical Theory and its Interpretation. Springer, Dordrecht (2006) (arXiv:quant-ph/0510095, 2005)
28. Foulis, D., Randall, C.H.: What are quantum logics, and what ought they to be? In: Betrametti, E., van Fraassen, B. (eds.) Current Issues in Quantum Logic. Plenum, New York (1981)
29. Wilce, A.: Quantum logic and probability, The Stanford Encyclopedia of Philosophy (Winter, 2002 Edition), E. Zalta (ed.), URL = <http://www.science.uva.nl/~seop/entries/qtquantlog/>
30. Wilce, A.: Test spaces. In: Gabbay, D., Engesser, K., Lehman, D. (eds.) Handbook of Quantum Logic, vol. II. Elsevier, Amsterdam (2009)
31. Alfsen, E.M.: Compact Convex Sets and Boundary Integrals. Springer, Berlin (1970)
32. Holevo, A.S.: Probabilistic and Statistical Aspects of Quantum Theory. North-Holland, Amsterdam (1982)
33. Tran, Q., Wilce, A.: Covariance in quantum logic. Int. J. Theor. Phys. **47**, 15–25 (2008)
34. Wilce, A.: Formalism and interpretation in quantum theory. Found. Phys. **40**, 434–462 (2010)
35. Abramsky, S., Coecke, B.: A categorical semantics of quantum protocols. Proceedings of the 19th Annual IEEE Symposium on Logic in Computer Science, pp. 415–425. (2004) (arXiv: quant-ph/0402130)
36. Barnum, H., Barrett, J., Clark, L., Leifer, M., Spekkens, R., Stepanik, N., Wilce, A., Wilke, R.: Entropy and information causality in general probabilistic theories. New J. Phys. **12**, 1367–2630 (2010) (arXiv:0909.5075. 2009)
37. Barnum, H., Gaebbler, C.P., Wilce, A.: Ensemble steering, weak self-duality, and the structure of probabilistic theories, arXiv:0912.5532. (2009)
38. Bellisard, J., Iochum, B.: Homogeneous self-dual cones, versus Jordan algebras. The theory revisited. Annales de l'Institut Fourier, Grenoble **28**, 27–67 (1978)
39. Vingerg, E.B.: Linear Representations of Groups. Birkhauser, Basel (1989)
40. Chiribella, G., D'Ariano G.M., Perinotti, P.: Informational derivation of quantum theory. Phys. Rev. A **84**, 012311 (2011) 012311-012350

Chapter 18
Probability in the Many-Worlds Interpretation of Quantum Mechanics

Lev Vaidman

Abstract It is argued that, although in the Many-Worlds Interpretation of quantum mechanics there is no "probability" for an outcome of a quantum experiment in the usual sense, we can understand why we have an illusion of probability. The explanation involves: (a) A "sleeping pill" gedanken experiment which makes correspondence between an illegitimate question: "What is the probability of an outcome of a quantum measurement?" with a legitimate question: "What is the probability that 'I' am in the world corresponding to that outcome?"; (b) A gedanken experiment which splits the world into several worlds which are identical according to some symmetry condition; and (c) Relativistic causality, which together with (b) explain the Born rule of standard quantum mechanics. The Quantum Sleeping Beauty controversy and "caring measure" replacing probability measure are discussed.

18.1 Introduction

Itamar and I shared a strong passion for understanding quantum mechanics. We did not always view it in the same way but I think we understood each other well. In fact I am greatly indebted to Itamar. Being a physicist working on foundations of quantum mechanics I always thought that philosophical arguments are crucial for understanding quantum mechanics. However, my first philosophical work [1] was rejected over and over by philosophical journals and philosophers. While Hillary Putnam, Abner Shimony, Michael Redhead and others did not see the point I was making, it was Itamar who first appreciated my contribution and opened for me the way to the philosophy of science [2].

L. Vaidman (✉)
School of Physics and Astronomy, Tel Aviv University, Tel Aviv, Israel
e-mail: vaidman@post.tau.ac.il

Y. Ben-Menahem and M. Hemmo (eds.), *Probability in Physics*, The Frontiers Collection,   299
DOI 10.1007/978-3-642-21329-8_18, © Springer-Verlag Berlin Heidelberg 2012

There are three conceptually different scenarios of what happens in the process of quantum measurement. The first option is that there is a genuinely random (chance) event which makes one outcome happen without any possibility to know which one prior to the measurement. This is the case of collapse: the von Neumann type II evolution of the quantum wave, or the stochastic event in a theory of dynamical collapse [3, 4]. The second option is that the quantum wave description of the system is deterministic, there is no collapse, but it is incomplete. There are hidden variables specifying the outcome prior to the measurement, which, however, we cannot know in principle. The most successful proposal of this kind is causal interpretation [5]. The third option is that the evolution is deterministic, there is no collapse of the quantum wave and the quantum wave is the complete description of the system. Then, all outcomes take place and this is the many-worlds interpretation (MWI) [6].

The concept of probability is directly applicable in the first scenario. There is genuine chance and genuine uncertainty. If, say, A is a possible outcome, then we can talk about the probability that A will happen. Indeed, A might or might not happen. At the end of the process we will definitely know if A took place.

In the second scenario there is no random chance. Prior to experiment Nature knows the outcome, it is encoded in some (hidden) variable. There are no several options, only one. However, since the theory postulates that "hidden variables" cannot be known to the experimentalist, he has an ignorance-type probability: he does not know the value of the hidden variable which specifies the outcome of the experiment. His concept of probability is: the probability that A will happen is the probability that the hidden variables now are such that A will take place.

The situation is the most difficult in the third scenario. There is no randomness, there is no chance: A happens with certainty, but other non compatible outcomes happen with certainty too, so a standard concept of probability addressing the dilemma A or not A is not applicable here. We have no uncertainty, everything is known. We have a complete description prior to the measurement and the process of measurement is some known deterministic evolution; so we know the complete description now and forever. These leads us to the conclusion that we do not have probability here in the usual sense. But this can be expected since the picture of multiple worlds is rather unusual. I will argue that we have here an illusion of probability, an illusion behaving very much like the usual probability.

18.2 The "Tale of a Single-World Universe"

Before starting the analysis of probability in the MWI I have to clarify exactly what is the MWI since it has numerous, sometimes contradictory, presentations in the literature. The MWI, as I understand it [7], is the claim that All is the Universal Wave Function evolving according to the laws of standard quantum mechanics without collapse, together with the explanation of the correspondence between the Wave Function and our experience.

In order to explain our experience I find it useful to introduce the "Tale of a single-world Universe". Let us assume that we are the only civilization and that we live under a very strong dictatorship which has laws against quantum measurements. It is forbidden to perform quantum experiments in which there is a nonzero probability for more than one outcome. Manufacture of Geiger counters is banned, quantum random number generators [8] are forbidden, and a special police prevents world splitting devices of the kind that can be found in Tel-Aviv university [9]. There are even laws that under the threat of death enforce disposal of neon light bulbs after 6 months of operation, to avoid operating an old bulb, which, when flicking, splits our world.

In this tale Nature does not arrange quantum experiments accidentally: no macroscopically different superpositions of a macroscopic object ever develop. In such a Universe there is no difference between the MWI and the textbook interpretation: in both, the wave function evolves according to the Schrödinger equation since collapse takes place in the measurement-type situations, but in our tale these situations never take place. The wave functions of all macroscopic objects remain well localized all the time.

The connection of the Wave Function to our experience in such a Universe is through a three dimensional picture which is generated by the Wave Function. Indeed, the three dimensional map of the density of wave functions of all particles will form a familiar picture of macroscopic objects around us as well as our bodies moving in time in a classical manner.

I am aware that there are claims that the Wave Function cannot describe the reality because it is defined in configuration space [10]. In classical mechanics a similar complaint is easily rejected because we can consider each particle separately in three dimensional space, instead of one point in the configuration space of all N particles. In quantum mechanics it is more difficult, since we cannot neglect entanglement. Although in our tale macroscopic bodies are never entangled, electrons are surely entangled with nuclei in atoms and atoms entangled in molecules. Still, the picture is in three dimensions. Even if an electron in my finger is entangled with nuclei and other atoms in the molecule of my skin, its density in three dimensions is well localized. This picture in three dimensional space is what corresponds to our observations. What makes the representation of the Wave Function in space special, relative to some abstract Hilbert space representation, is that the interactions in Nature are local. Since our observations are also kind of interactions, they are local too.

18.3   Illusion of Probability

Once we understand the link between the Wave Function and experience in a single-world Universe, we can proceed to analyze a Universe in which quantum measurements are not forbidden. The quantum measurements will lead to a super-position of branches of the Wave Function, each one of them corresponding to what we experience as a "world". Until the next measurement, the link with our observations works in each branch as in a single-world Universe. The locality and strength of interactions in Nature ensure that parallel branches do not interfere (decoherence). Given the information we have at present, we can follow our branch to the past before the last quantum measurement, but we can follow it to the future only until the next measurement. In our branch we remember past events of performing quantum measurements and obtaining particular results. It seems to us that the outcomes came out randomly, although we know that there were no random evolution in Nature. The branch was split deterministically to two or more branches. We now experience only one of them and it seems to us that there was a random outcome of the quantum measurement.

If we imagine a hypothetical theory in which the wave function collapses every time a macroscopic object evolves into a superposition of macroscopically different states such that all macroscopic objects (whatever "macroscopic" means) remain well localized, then the memory and experience of the observers in the single world described by such a theory will be identical to the experience of observers in one branch of the many-world universe. In the Universe with collapse there is a genuine probability concept of random chancy events. In the MWI universe there is deterministic evolution, and no objective "chancy" probability. But the experiences in a particular branch (we can follow the branch in the MWI backwards in time) are identical to the experiences of genuine probability of the observers living in the physical universe with collapses of the quantum wave. This explains the illusion of probability in the MWI.

So, one approach to introducing "probability" to the MWI is to point out that the observer in a branch of the Universe in the framework of the MWI and the observer in the single-world Universe with collapse postulate are described by the same mathematical object and thus have the same experience. Since in the theory with collapse, the probability concept is clear, we can associate the same concept in the branch of the MWI with the observer who is planning to perform a quantum experiment.

Although I do not think that the probability can be *derived* in the framework of the MWI as Deutsch advocates [11], I do think that one can argue more why the illusion of probability in the MWI works so well. To illustrate this let us consider two gedanken experiments. The experiments will include steps which seem technologically unimaginable, yet they do not require changes in any physical law.

18.4 Gedanken Experiment I: Complete Symmetry

Three identical space stations A, B and C were built and put on the same orbit around Earth in a symmetrical way, see Fig. 18.1. Bob wants to travel to space and he arranges an automatic device which will send him to one of these stations after he goes to sleep. The device consists of a spin-1 particle in the state

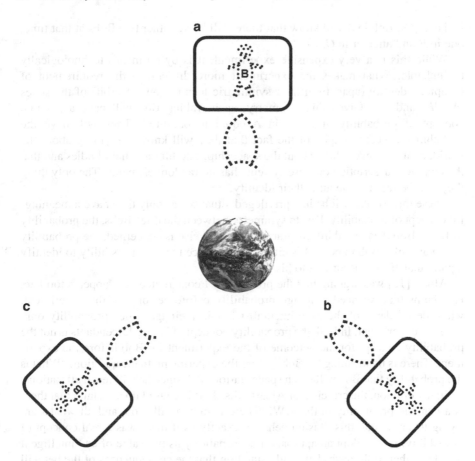

Fig. 18.1 Bob's descendants in symmetric state in three space stations

$$\frac{1}{\sqrt{3}}(|1\rangle + |0\rangle + |-1\rangle),\qquad(18.1)$$

measuring device of the spin component and the spaceship which will move him, while he is asleep, to one of the stations according to the outcome of the spin measurement. Bob, who accepts the MWI of quantum mechanics is certain that, at a later time, there will be three Bobs. The quantum state will be

$$\frac{1}{\sqrt{3}}(|A\rangle|1\rangle + |B\rangle|0\rangle + |C\rangle|-1\rangle),\qquad(18.2)$$

where $|A\rangle$ signifies the quantum state of Bob in A as well as the state of spaceship which brought him to A and everything else which interacted with Bob and his spaceship and became correlated to his wave function in A, and similarly for

$|B\rangle|$ and $|C\rangle$. Bob in A will know that there will be yet other two Bobs at that time, one in B and another in C.

While this is a very expensive experiment, it is by no means technologically unthinkable. What makes the experiment more difficult is the requirement of complete identity (apart from their symmetric location on the orbit) of the states $|A\rangle$, $|B\rangle$ and $|C\rangle$. Given this symmetry, each waking Bob will have a genuine concept of probability of being in A equal to one-third. They will have the probability concept in spite of the fact that they will know everything about the world, or at least everything about the spaceship, satellites and their bodies and that the complete description of these systems has no random elements. The only thing they will be ignorant about is their identity.

These three Bobs will be in a privileged situation, as only they have a meaningful concept of probability. Due to symmetry between the three Bobs, the probability of being Bob A is one-third. Insofar as everyone else is concerned, the probability for a particular Bob to be in A is either 1 or 0, since the only possibility to identify a particular Bob is according to his location.

Albert [12] was arguing that the probability I constructed here appears too late. He claims that we need to assign probability before performing the experiment, while descendants of the experimentalist obtain their ignorance probability only after the experiment. Indeed, the probability concept of Bob's descendants is not the probability concept for the outcome of the experiment for Bob before the experiment. There is no meaning for Bob, before the experiment, to the question: "What is the probability that Bob will reach space station A?", since he will reach all stations.

In my opinion, the criticism of Albert falls short because I do not claim that there is a genuine probability in the MWI. There is only an illusion, and all what I am trying to say is that this illusion behaves exactly as if there was a real concept of probability. If we adopt an approach for probability as the value of an "intelligent bet" [13], then Bob makes bets understanding that the consequences of the bet will be relevant for his descendants (Bobs after the experiment). They will get the reward of the bet (and they will have initially less money if Bob spent money on the bet). They will have an ignorance concept of probability, so they will be pleased to find out that the bet was placed. The Bob before the experiment cared about Bobs after the experiment, due to symmetry of the situation, in an equal way. This, together with the fact that all Bobs like to bet, provide the rational for his betting.

18.5 Gedanken Experiment II: Derivation of the Born Rule

The sleeping pill trick [1] provides the way to talk about probability in the MWI, which, in my view is the main difficulty to be resolved. In recent years, however, even more attention was given to the issue of the Born rule in the framework of the MWI, i.e. not just to justifying the probability concept when all outcomes of the experiment are realized, but also assigning the correct values of probability for different outcomes. In the above idealized symmetric setup we do get the correct probability, one-third, but we need to work much more for a general case.

Let us note that even in Experiment I there is no complete symmetry: I gave names to Bob's (identical) stations, one of them is "A" and others are not. In a completely symmetrical situation there cannot be different names. So, the symmetry is not complete. We only assume that all relevant aspects of the three stations are completely identical, but we accept a possibility, and in fact a necessity, that there are other properties of the stations, like pictures on their surface which are different. In a scientific theory we have an idea as to what is relevant and what is not. A hypothesis that a different text drawn on otherwise identical space stations will change the outcome of the experiment described above does not seem to be scientific. So, we have to make our setup symmetric only in relevant details.

Let us modify the above setup trying to keep relevant aspects symmetric, in a way which will lead us to the Born rule. We still have our three identical space stations, but now, the observer, Bob, is moved only to space station A. We also send Charlie and John to stations B and C and perform similar operations there to keep at least partially the symmetry between the stations, see Fig. 18.2. While Bob is asleep (as well as Charlie and John), a device of the type described above causes a particle to be in a superposition in all three stations. Then, in all three stations automatic devices perform measurements of the presence of the particle there. According to the outcome of such measurement in station A, Bob is moved to a room "yes" if the particle is found in A and to a room "no" if the particle is not found there. Similar operations are performed in stations B and C. Now, upon awakening, each Bob will have a genuine ignorance probability concept regarding the question: "In which room am I?" Each Bob will have a reason to declare probability one-third for being in the room "yes", because there are three worlds and only in one of them this Bob is in the room "yes".

The statement that "there are three worlds" needs clarification. In one world Bob cannot view himself in a superposition, so, it was clear in Experiment I that there are three worlds: in the first Bob is in A, in the second he is in B, and in the third he is in C. In Experiment II, in the first world, A, Bob is in room "yes"; in the worlds B and C he is in the room "no". If we follow Everett's original "'Relative state' formulation of quantum mechanics" [6], we might say that for Bob there are only two worlds: in one of them he is in the "yes" room and in another he is in the "no" room. In my approach [7] macroscopic objects and especially people cannot be in a superposition of macroscopically different states in one world. So, the measurements of Charlie and John in stations B and C ensure that there are three worlds: A, B, and C, with Bob, Charlie, and John, in their "yes" rooms correspondingly, while the others are in "no" rooms. The symmetry argument here is suggestive but not rigorous. We need three worlds which are symmetric in all relevant aspects. It is not obvious that the fact that in two of the worlds (B and C) there is one and the same Bob, while in the world A there is another Bob, is not relevant for our analysis.

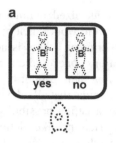

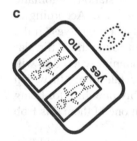

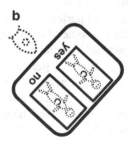

Fig. 18.2 Bob's descendant in room "yes", while Charlie's and John's descendants are in rooms "no" in superposition with two other similar options

Accepting the suggestive symmetry argument that gives Bob probability one third to find himself in room "yes", provides the Born rule for this case: $p_{yes} = \frac{1}{3} = \langle \mathbf{P}_A \rangle$. Now we can add to the MWI the locality and causality postulates.

The MWI yields: *There is nothing but the wave function.*

Locality provides: *Outcomes of local experiments depend only on local values of the wave function.*

Causality of relativistic quantum theory yields: *Any action in a space-like separated region cannot influence an outcome of local experiment.*

From this it follows that Bob should assign probability $p_{yes} = \frac{1}{3}$ for all states which can be obtained from (2) through actions at regions which are space-like separated from the measurement in A. For example, if Charlie and John in stations B and C do not perform measurements of the presence of the particle, the symmetry is broken: there will be two worlds instead of three, but the probability to find the particle in A remains one-third. This gedanken experiment shows that the probability of finding the particle in A for a quantum state which allows "symmetrization", i.e., there exists symmetric state with N parts with the same density matrix at location A, is $\frac{1}{N}$.

A more general question is the probability of an outcome of any quantum measurement in a particular location, not just the measurement of projection operator of a particle on this location. A celebrated example is a Stern Gerlach measurement of a spin component. To cover this case we can consider unitary evolution which creates a spin state via absorption of a photon. Then, the spin component measurement is equivalent to a photon projection measurement [14]. The concept of "symmetrization" becomes: existence of a symmetric situation with identical systems in symmetrically located N locations with the same density matrix at location A.

Generalizing the argument for an arbitrary state *a la* Deutsch [11] or by using Gleason theorem [15] we can derive the Born rule. However, I do not see how to make this derivation rigorous. If we could make a similar argument for Experiment I, in which symmetry is robust, this could provide a rigorous derivation of the probability in the MWI. However, the locality argument cannot be applied there. Bob is not localized in this experiment. When he is asked what is the probability that he is in A, the situation in B and C matters. For example, if Bob knows that nobody will ask this question in B and C, he should give the answer 1 instead of $\frac{1}{3}$. (Note that when Charlie and John refrain from making measurements in B and C it changes nothing in Experiment II.)

So, although the Born rule fits the MWI very well, I do not see how to *derive* it without some (plausible) assumptions of what is relevant for the probability of an outcome of a quantum measurement. But then, similar, if not simpler, symmetry arguments yield the Born rule also in the framework of collapse interpretations, so I do not think that the MWI has an advantage relative to this question. I adopt the MWI because it removes randomness and nonlocality from physics.

18.6 Quantum Sleeping Beauty

It is harder to approach the probability issue in the framework of the MWI than in other interpretations of quantum mechanics and I found only one situation in which the MWI helps to analyze a probabilistic question. This is the story of "Sleeping Beauty" [16].

> Some researchers put Beauty to sleep. During the 2 days of her sleep they will briefly wake her up either once or twice, depending on the toss of a fair coin (Heads: once; Tails: twice). After each waking, they will put Beauty back to sleep with a drug that makes her forget that waking. Every wakening the Beauty is asked: What is your credence for the outcome Heads?

This problem raised a great controversy: is the answer one-third or one half? Although I believe that one can argue convincingly that the answer is one-third without help of quantum mechanics [17], the MWI provides an even more convincing argument [18]. I implement the fair coin toss via quantum experiment with probability half, which is an ultimate fair coin. Then one can unambiguously describe the situation as the unitary evolving quantum wave of Beauty and quantum measuring device. This makes the problem easier to analyze than in the case of a "chancy" fair coin. For simplicity I will add to the story that before wakening, the Beauty is moved to the room "Heads" or "Tails" according to the result of the quantum measurement. The rooms are identical inside, so, when she is asked the question she is in one of three different locations in space and time, but she will have the same memory state and identical environment. For the case of the quantum measurement, the question: "Is it Heads or Tails?" is senseless, since both options are realized. The actual question is: "Is it Heads or Tails in the world the Beauty is asked the question?" (Compare with Groisman's [19] approach to resolve the Sleeping Beauty controversy for the classical coin.) Beauty knows that in the Universe there are three events in which she is asked this question. The measures of existence of worlds in all these events are equal. Since only one of these events corresponds to Heads, she assigns probability one-third for Heads.

The Quantum Sleeping Beauty also generated a considerable controversy. To my surprise the answer one-third is not in the consensus. Peter Lewis [20] claimed that quantum coin tossing leads to the Beauty's answer of one half. He insisted, especially, that it has to be one half in the framework of the MWI [21]. Very recently Bradley [22] also claimed (but did not show) that the MWI approach leads to the answer one half, adding that in his view this is good news for the MWI. Papineau and Dura-Vila [23] criticized Lewis, but argued that accepting my approach to probability in the MWI strengthens Lewis's claim for one half.

Most of Lewis's arguments rest on assigning pre-branching uncertainty in the MWI advocated by Saunders [24] and Wallace [25] which I strongly deny. Lewis briefly mentions that one half is obtained in my approach too, arguing that there is an analogy with a process with two consecutive coin tosses. (I just learned that Peterson [26] argued against this analogy.) I could not see such an analogy: the only second coin toss in the Sleeping Beauty story I can imagine is her guess about which wakening, out of three, is now. I cannot understand the rational for Lewis's coin toss between the two Tail wakening. The answer one half seems to follow from an error similar to the Bertrand Box paradox [27].

18.7   Caring Measure Instead of Probability

In the MWI there is no genuine probability. Instead of probability measure, I introduced "measure of existence" of a world. Measure of existence, apart from its relation to probability, describes the ability of a particular world to interfere with other worlds [2]. I defined a "behavior principle" [1, 7] according to which an experimenter performing quantum experiments cares about his descendants according to their measures of existence. This principle answers the (naive) criticism that a believer in the MWI would agree to play "Russian Quantum Roulette" [28]. "Measure of existence" is a philosophically problematic term, so "caring measure" is frequently used instead of "measure of existence" [29].

Albert [12] recently criticized this approach, providing deliberately ridiculous alternative measure (proportional to the measure of existence of the branch times fatness of the observer in this branch) and arguing that this caring measure is as good as the other one. Albert's criticism might apply if caring measure is considered as a standing alone proposal. In my approach the foundation of the caring measure is the post-measurement genuine probability of the descendants of the experimentalist. In a typical quantum experiment, splitting of worlds (creation of superposition of macroscopically different wave packets of macroscopic systems) happens before the time the experimentalist (i.e. all of his versions) will become aware of the outcome. So, there will be a stage with genuine probability concept. At that moment all the descendants will be happy if actions according to the behavior principle have been performed.

I am encouraged by recent support coming from Tappenden [30], who approved attaching post-measurement uncertainty to (the illusion) of pre-measurement probability naming it the Born-Vaidman Rule. Tappenden uses it for the analysis of confirmation of the MWI; see also Greaves and Myrvold [31]. I might agree with these arguments, but for me, the strongest confirmation of the MWI lays in the non-probabilistic consequences of quantum theory, such as spectrum of a hydrogen atom. Experiments confirming this type of predictions are so successful that only extreme deviation from the Born-Vaidman rule might question the MWI.

18.8 Conclusions

I have argued that there are two main issues related to probability in the MWI. First is how to talk about the probability of an outcome in a measurement when all outcomes are actualized, and the second is what is the status of the Born rule.

If we disregard the first problem, then the second is simple: The MWI tells us that All is the Wave Function. Locality tells us that the result of any experiment in one location can depend only on the property of the Wave Function in this location. The expectation value of the projection of the Wave Function on this location is the only property which cannot be changed by actions elsewhere, so causality tells us that the probability is the function of this projection. Then, from a symmetry argument and/or Gleason theorem the Born rule is derived.

The resolution of the first problem is the statement that there is no genuine probability concept in the MWI, and the "sleeping pill" argument explains well why we have an illusion of probability which is essentially indistinguishable from real probability. The combination of two arguments would solve the whole problem, but unfortunately, it requires additional, although very plausible, assumptions regarding what might influence the probability of an outcome. Note, however, that with such an assumption the Born rule can be derived in the framework of other interpretations of quantum theory as well.

I understand that Itamar and I viewed the problem of the Born rule in a similar way, i.e. that some assumptions are necessary in order to prove the Born rule using the Gleason theorem [32]. On the other hand, Itamar and Meir Hemmo were sceptical about resolution of the first issue. I believe that the reason why I do not see the difficulties they encountered is that I consider a direct connection between the Wave Function and our experience without insisting on giving values to various observables, the topic which was at the center of Itamar's research. His view on the MWI was according to the lines of the many minds interpretation [33] in which the situation is very different, since, *a priori*, pre-measurement uncertainty seems to be possible and it is a non-trivial fact that actually it is not. Since I never tried to introduce the pre-measurement uncertainty in the first place, I had no reason to be discouraged by this result.

This work has been supported in part by the Israel Science Foundation Grant No. 1125/10.

References

1. Vaidman, L.: About Schizophrenic experiences of the neutron or why we should believe in many-worlds interpretation of quantum theory. University of South Carolina Preprint, Columbia, SC.. http://philsci-archive.pitt.edu/8564/ (1990)
2. Vaidman, L.: On schizophrenic experiences of the neutron or why we should believe in the many-worlds interpretation of quantum theory. Int. Stud. Philos. Sci. **12**, 245–261 (1998)
3. Ghirardi, G.C., Rimini, A., Weber, T.: Unified dynamics for microscopic and macroscopic systems. Phys. Rev. D **34**, 470–491 (1986)
4. Pearle, P.: Combining stochastic dynamical state-vector reduction with spontaneous localization. Phys. Rev. A **39**, 2277–2289 (1989)
5. Bohm, D.: A suggested interpretation of the quantum theory in terms of 'hidden' variables, I and II. Phys. Rev. **85**, 166–193 (1952)
6. Everett III, H.: 'Relative state' formulation of quantum mechanics. Rev. Mod. Phys. **29**, 454–462 (1957)
7. Vaidman, L.: Many-worlds interpretation of quantum mechanics. In: Zalta, E N. (ed.) Stanford Encyclopedia of Philosophy http://plato.stanford.edu/entries/qm-manyworlds/ (2002)
8. Quantis, true random number generator exploiting quantum physics. http://www.idquantique. com/true-random-number-generator/products-overview.html/
9. Tel-Aviv world splitter. http://qol.tau.ac.il/tws.html
10. Maudlin, T.: Can the world be only wavefunction? In: Saunders, S., Barrett, J., Kent, A., Wallace, D. (eds.) Many Worlds? Everett, Quantum Theory, and Reality, pp. 121–143. Oxford University Press, Oxford (2010)
11. Deutsch, D.: Quantum theory of probability and decisions. Proc. R. Soc. Lond. A **455**, 3129–3137 (1999)
12. Albert, D.: Probability in the Everett picture. In: Saunders, S., Barrett, J., Kent, A., Wallace, D. (eds.) Many Worlds? Everett, Quantum Theory, and Reality, pp. 355–368. Oxford University Press, Oxford (2010)
13. Pitowsky, I.: Betting on the outcomes of measurements: a Bayesian theory of quantum probability. Stud. Hist. Philos. Mod. Phys. **34**, 395–414 (2003)

14. Aharonov, Y., Vaidman, L.: Nonlocal aspects of a quantum wave. Phys. Rev. A **61**, 052108 (2000)
15. Pitowsky, I., Finetti, D.: Extensions, quantum probability and Gleason's theorem. In: Hartman, S., Beisbart, C. (eds.) Probability is Physics. Oxford University Press, Oxford (2011)
16. Elga, A.: Self-locating belief and the Sleeping Beauty problem. Analysis **60**, 143–147 (2000)
17. Vaidman, L., Saunders, S.: On sleeping beauty controversy. Oxford University Preprint, Oxford. http://philsci-archive.pitt.edu/324/ (2001)
18. Vaidman, L.: Probability and the many worlds interpretation of quantum theory. In: Khrennikov, A. (ed.) Quantum Theory: Reconsideration of Foundations, pp. 407–422. Vaxjo University Press, Sweden (2001)
19. Groisman, B.: The end of Sleeping Beauty's nightmare. Br. J. Philos. Sci. **59**, 409–416 (2008)
20. Lewis, P.: Quantum Sleeping Beauty. Analysis **67**, 59–65 (2007)
21. Lewis, P.: Reply to Papineau and Dura-Vila. Analysis **69**, 86–89 (2009)
22. Bradley, D.J.: Confirmation in a branching world: the Everett interpretation and Sleeping Beauty. Br. J. Philos. Sci. **62**, 323–342 (2011)
23. Papineau, D., Dura-Vila, V.: A thirder and an Everettian: a reply to Lewis's 'Quantum Sleeping Beauty'. Analysis **69**, 78–86 (2009)
24. Saunders, S.: Time, quantum mechanics, and probability. Synthese **114**, 373–404 (1998)
25. Wallace, D.: Epistemology quantized: circumstances in which we should come to believe in the everett interpretation. Br. J. Philos. Sci. **57**, 655–689 (2006)
26. Peterson, D.: Qeauty and the books: a response to Lewis's quantum Sleeping Beauty problem. Synthese **181**, 367–374 (2011)
27. Bertrand, J.: Calcul des probabilités. Gauthier-Villars et fils, Paris (1889)
28. Lewis, P.: What is it like to be Schrödinger's cat? Analysis **60**, 22–29 (2000)
29. Greaves, H.: Understanding Deutsch's probability in a deterministic multiverse. Stud. Hist. Philos. Mod. Phys. **35**, 423–456 (2004)
30. Tappenden, P.: Evidence and uncertainty in Everett's multiverse. Br. J. Philos. Sci. **62**, 99–123 (2011)
31. Greaves, H., Myrvold, W.: Everett and evidence. In: Saunders, S., Barrett, J., Kent, A., Wallace, D. (eds.) Many Worlds? Everett, Quantum Theory, and Reality, pp. 264–307. Oxford University Press, Oxford (2010)
32. Hemmo, M., Pitowsky, I.: Quantum probability and many worlds. Stud. Hist. Philos. Mod. Phys. **38**, 333–350 (2007)
33. Hemmo, M., Pitowsky, I.: Probability and nonlocality in many minds interpretations of quantum mechanics. Br. J. Philos. Sci. **54**, 225–243 (2003)

Index

Titles in this Series

Quantum Mechanics and Gravity
By Mendel Sachs

Quantum-Classical Correspondence
Dynamical Quantization and the Classical Limit
By Josef Bolitschek

Knowledge and the World: Challenges Beyond the Science Wars
Ed. by M. Carrier, J. Roggenhofer, G. Küppers and P. Blanchard

Quantum-Classical Analogies
By Daniela Dragoman and Mircea Dragoman

Life - As a Matter of Fat
The Emerging Science of Lipidomics
By Ole G. Mouritsen

Quo Vadis Quantum Mechanics?
Ed. by Avshalom C. Elitzur, Shahar Dolev and Nancy Kolenda

Information and Its Role in Nature
By Juan G. Roederer

Extreme Events in Nature and Society
Ed. by Sergio Albeverio, Volker Jentsch and Holger Kantz

The Thermodynamic Machinery of Life
By Michal Kurzynski

Weak Links
The Universal Key to the Stability of Networks and Complex Systems
By Csermely Peter

The Emerging Physics of Consciousness
Ed. by Jack A. Tuszynski

Y. Ben-Menahem and M. Hemmo (eds.), *Probability in Physics*, The Frontiers Collection,   319
DOI 10.1007/978-3-642-21329-8, © Springer-Verlag Berlin Heidelberg 2012